Systems Engineering: Availability and Reliability

Systems Engineering: Availability and Reliability

Editors

Katarzyna Antosz
Jose Machado
Dariusz Mazurkiewicz
Dario Antonelli
Filomena Soares

MDPI • Basel • Beijing • Wuhan • Barcelona • Belgrade • Manchester • Tokyo • Cluj • Tianjin

Editors

Katarzyna Antosz
Rzeszów University of
Technology
Poland

Jose Machado
University of Minho
Portugal

Dariusz Mazurkiewicz
Lublin University of
Technology
Poland

Dario Antonelli
Politecnico di Torino
Italy

Filomena Soares
University of Minho
Portugal

Editorial Office
MDPI
St. Alban-Anlage 66
4052 Basel, Switzerland

This is a reprint of articles from the Special Issue published online in the open access journal *Applied Sciences* (ISSN 2076-3417) (available at: https://www.mdpi.com/journal/applsci/special_issues/Systems_Engineering_Availability_and_Reliability).

For citation purposes, cite each article independently as indicated on the article page online and as indicated below:

LastName, A.A.; LastName, B.B.; LastName, C.C. Article Title. *Journal Name* **Year**, *Volume Number*, Page Range.

ISBN 978-3-0365-3623-1 (Hbk)
ISBN 978-3-0365-3624-8 (PDF)

Contents

About the Editors

Katarzyna Antosz

Katarzyna Antosz is a professor at Rzeszow University of Technology, Poland. She obtained her Habilitation degree in technical sciences in the discipline of mechanical engineering in 2019. She is a member of Lean Learning Academy Polska, the Polish Production Management Association, as well as a member of the international "Association of Engineering, Project, and Production Management (EPPM)" and IEEE. Her professional interests, research topics and international collaborations mainly focus on production engineering, systems reliability, predictive maintenance, lean maintenance and intelligent decision support systems, Industry 4.0, artificial intelligence and data mining. She is the author of more than 100 scientific papers published as monographs, conference proceedings and papers. She is a contractor of EU projects and projects implemented in industry.

Jose Machado

Jose Machado concluded his Habilitation degree in February 2019 at the University of Minho, Portugal. He received his PhD degree in Mechanical Engineering, Automation, from both the University of Minho, Portugal and from Ecole Normale Superieure de Cachan, France, in 2006. He is Deputy Director of MEtRICs Research Centre and Assistant Professor in the Mechanical Engineering department at the University of Minho. He has authored, or co-authored, more than 220 refereed journal and conference proceeding papers. He coordinates and has coordinated—and participated as collaborator—in several Research and Technology Transfer Projects on Mechatronics and Automation domains. His main interests are related to Industry 4.0, more specifically on the design and development of Cyber–Physical Systems, design and analysis of dependable controllers for obtaining dependable mechatronic systems, and mechatronic systems design, with special focus on medical or biomedical applications, wellbeing and/or rehabilitation.

Dariusz Mazurkiewicz

Dariusz Mazurkiewicz is a professor of mechanical engineering at the Lublin University of Technology in Lublin, Poland. Previously he has been a visiting scholar or research fellow of the Cambridge University Engineering Department (Cambridge, UK), Kobe University (Kobe, Japan) and the System Research Institute (Polish Academy of Sciences, Warsaw, Poland). His research interests include production engineering, especially with respect to maintenance, and technical infrastructure reliability. His research skills and experience are in maintenance and reliability, predictive maintenance, IIoT, numerical modelling, transportation systems, mining engineering technology, production engineering, data mining, artificial neural networks and fuzzy logic, innovation and regional innovation policy. He is an expert of the European Commission, Research Executive Agency and a scientific Board member for six top ranking international journals.

Dario Antonelli

Dario Antonelli has been an Associate Professor at the Politecnico di Torino (Italy) in the subject of 'Technology and Production Systems' since 2001. He has been team leader for several EU research projects, developing supporting systems for collaborative industry networks and for human–robot collaboration. His present research activity is aimed at creating a collaborative work environment in which humans and robots coexist and interact. His other research interests include modelling large-scale distributed manufacturing systems and simulating production systems and manufacturing processes. He is also interested in new learning methodologies, in particular participative learning and in the applications of Virtual and Augmented Reality for industrial training.

Filomena Soares

Filomena Soares is an Associate Professor with Habilitation at the Industrial Electronics Department of the University of Minho and researcher at R&D Algoritmi Centre (Portugal). Her main scientific interests are in the areas of System Monitoring and Control, with applications in biomedical processes. Her research interests mainly include motor and cognitive rehabilitation, using serious games and robots to foster the communication with impaired children/adults. She is interested in new teaching/learning methodologies, in particular blended-learning and virtual and remote laboratories.

applied sciences

Editorial

Systems Engineering: Availability and Reliability

Katarzyna Antosz [1,*], Jose Machado [2], Dariusz Mazurkiewicz [3], Dario Antonelli [4] and Filomena Soares [5]

[1] Faculty of Mechanical Engineering and Aeronautics, Rzeszow University of Technology, Aleja Powstańców Warszawy 8, 35-959 Rzeszow, Poland
[2] MEtRICs Research Center, Department of Mechanical Engineering, School of Engineering, University of Minho, 4800-058 Guimarães, Portugal; jmachado@dem.uminho.pt
[3] Mechanical Engineering Faculty, Lublin University of Technology, Nadbystrzycka 36, 20-618 Lublin, Poland; d.mazurkiewicz@pollub.pl
[4] Department of Management and Production Engineering, Polytechnic University of Turin, Corso Duca degli Abruzzi 24, 10138 Torino, Italy; dario.antonelli@polito.it
[5] Centro Algoritmi, Campus of Azurém, University of Minho, 4800-058 Guimaraes, Portugal; fsoares@dei.uminho.pt
* Correspondence: katarzyna.antosz@prz.edu.pl

Citation: Antosz, K.; Machado, J.; Mazurkiewicz, D.; Antonelli, D.; Soares, F. Systems Engineering: Availability and Reliability. *Appl. Sci.* **2022**, *12*, 2504. https://doi.org/10.3390/app12052504

Received: 9 February 2022
Accepted: 22 February 2022
Published: 28 February 2022

Publisher's Note: MDPI stays neutral with regard to jurisdictional claims in published maps and institutional affiliations.

1. Introduction

Current trends in Industry 4.0 are largely related to issues of reliability and availability. As a result of these trends and engineering systems' complexity, research and development needs now refer to new solutions in the integration of intelligent machines or systems, with emphasis on changes in production processes aimed at increasing production efficiency or equipment reliability. The emergence of innovative technologies and new business models based on innovation, cooperation networks, and the enhancement of endogenous resources is assumed to be a strong contribution to the development of competitive economies all around the world. Innovation and engineering, focused on sustainability, reliability, and availability of resources, have a key role in this context. The scope of this Special Issue is closely associated with that of the ICIE'2020 conference. This conference and the journal's Special Issue aims to present the current innovations and engineering achievements of top world scientists and industrial practitioners in the thematic areas related to:

- Reliability and risk assessment;
- Innovations in maintenance strategies;
- Production process scheduling, management and maintenance;
- Systems analysis, simulation, design and modeling.

The rapid development of industry, especially observed in recent decades, has significantly complicated the functioning of production systems and thus intensified the maintenance process. Nowadays, the reliability of hardware resources and employee safety, as well as low environmental risk, is required while implementing production processes with maximum efficiency, effectiveness and flexibility. At the same time, the awareness of the role of maintenance services in preventing breakdowns and early detection of initial problems with machines and systems is growing, which, in turn, increases the life of production systems, affecting the achievement of a high level of productivity.

Detecting and diagnosing faults in the early stages of damage is necessary to prevent the incorrect operation of the machine park and its breakdowns during operation. Hence, the growing role of availability and reliability research. Regular monitoring of the actual technical condition of the equipment and the operational efficiency of technological systems allows ensuring the maximum interval between repairs, thus extending the time of effectively using the existing hardware resources. It also minimizes the number and cost of unplanned downtime caused by machine failures, increasing the availability of both individual machines and entire production lines. It also allows making the right decision—repair or replacement—using cost–benefit analysis methods using appropriate optimization tasks, for example, when determining the maintenance strategy of the machine as a whole or separately for each category of its structural elements. The implementation of such tasks, also based on the methodology of optimal maintenance, requires appropriate knowledge about the course of operation of a given machine or its element in the past, currently and in a certain forecast of future use and its impact on potential failures and overall service life. Measurement systems monitoring the operation of machines and their parameters as well as diagnostic systems are thus extremely helpful. They all record significant amounts of data that require appropriate processing into knowledge and inference. An extremely dynamic digital transformation is taking place here.

2. Reliability and Risk Assessment

With the rapid development of high integrations in large complex systems, such as aircraft, satellite and railway systems, due to the increasingly complex coupling relationship between components within the system, local disturbances or faults may cause global effects on the system by fault propagation. Therefore, there are new challenges in safety analysis and risk assessment for complex systems. Some of them with their proposed solutions are described below.

As presented in [1], to this day, research articles highlighting analytical methods on small data sets are practically limited. From the literature review presented [1], it is revealed that most of the studies conducted thus far using scarce failure data are less. The methods used for reliability analyses for small number data are mostly the Bayesian approach, FMECA and Monte Carlo method. Using these methods, the small failure data of any machine or system can be grouped and easily used for further reliability analysis. Maintenance philosophies involve performing maintenance after given time intervals, typically after a fixed running hour for an engine. In spite of the scheduled maintenance, failure of the engine is inevitable, thereby decreasing the availability of dumpers and reducing the production cost. Reliability analysis of engine subsystems is essential for formulating the maintenance strategies that will reduce the downtime of the engine and enhance its availability. The main obstacle was the deficiency of adequate data for the appropriate statistical analyses. A data set containing a small sample size of failure data limits the possibility of precise decision-making. The study [1] gives specific guidelines for using CBH and meta-analysis testing, which emphasizes the failure data to predict reliability and MTBF. The researchers perform the reliability assessment using the grouped TBF data, during which suitable maintenance strategies could be formed. It provides a roadmap of reliability analysis for any machinery having fewer failure data.

For the development of the technical solution—a reliable robotic system to facilitate precision and safety—the ISO 14971 standard for risk management in healthcare devices was followed in [2]. This paper presents the risk management strategy for the development of the technical solution. Moreover, by evaluating the associated risks of the procedure, the medical and technical requirements for the proposed technical solution are defined. It also presents an analytical hierarchy process that is introduced subsequently in a four-phase quality function deployment, which, in turn, leads to the HeRo conceptual design, showing the advantages of the design with respect to risk reduction. The residual risks introduced by the robotic system were also evaluated, showing a reduction in hazard occurrence and severity, which validates the HeRo concept as a possible technical solution.

An approach based on the hidden Markov model is proposed [3] for risk performance reasoning. The unobservable state process in the approach aims to model the underlying risk performance, while the observation process was formed from the time series of risk factors. Within the framework, the log-likelihood probability was used as the measure of similarity between historical and current data of risk reasoning factors. Based on scalar quantization regulation and risk performance quantization regulation, the risk performance reasoning approach with different step sizes was conducted on the operational case, the performance of which was evaluated in terms of effectiveness and accuracy. The results obtained show significant improvement in the reasoning capacity and satisfactory performance for numerical risk reasoning and categorical performance reasoning. The proposed model is able to provide a reference for risk performance monitoring and threat pre-warning during the bauxite shipping process. The risk performance analyzed in [3] is essential in relation to the bauxite shipping process, whose associated risk remains to be officially recognized. The output of this paper can support quantitative risk decision-making, as opposed to previously used empirical decision-making, thereby laying a foundation for risk pre-warnings and process safety.

Aiming at analyzing and evaluating the inherent risks of the complex system with coupling correlation characteristics objectively, Jiang et al. [4] propose a novel risk assessment and analysis method for correlations in the complex system based on multi-dimensional theory. Firstly, the formal description and coupling degree analysis method of the hierarchical structure of complex systems is established. Moreover, considering the three safety risk factors of fault propagation probability, potential severity, and fault propagation time, a multi-dimensional safety risk theory is proposed in order to evaluate the risk of each element within the system affecting the overall system. Furthermore, critical safety elements are identified based on Pareto rules, As Low as Reasonably Practicable and safety risk entropy to support the preventive measures. Finally, an application of an avionics system is provided to demonstrate the effectiveness of the proposed method. Compared with the current methods and technologies, the method proposed [4] mainly reflects the advantages of two aspects. On the one hand, the hierarchical model is modeled in a matrix manner, and the association relationship of each element in the complex system is quickly and accurately analyzed, which reduces the skill requirements of analysts. On the other hand, it provides a feasible and multi-faceted analysis method for the risk assessment of systems in view of fault propagation, which is the core judgment criterion for identifying critical risk factors and is of great significance for ensuring system safety.

3. Innovations in Maintenance Strategies

The key issue in the use of technical facilities is the problem of making optimal decisions. Along with the development of computer aiding techniques and the progressive digitization of production, numerous tools have appeared to facilitate the work of technical services, allowing for the collection and processing of data, information and knowledge about objects and operational processes, thus enabling and facilitating effective decisions regarding the technical systems in use. This applies to both the technical, organizational and economic areas.

Optimally made decisions are one of the key conditions for the proper functioning of maintenance services, which means the necessity to determine the value of selected features. Correct exploitation of technical infrastructure facilities requires linking decision-making processes with the performance of maintenance or repair work. There are many mathematical models that constitute the basis for the quantitative assessment of the method and scope of the operation of technical objects. However, there are still several research challenges requiring innovations in maintenance strategies.

For example, as mentioned in [5], the supply chain of petroleum products faces major challenges, i.e., demand growth and the complexity of fluid transportation. The petroleum supply chain contains multiple stations extending from oil wells in exploration and production areas to the final destination. Each station has its own difficulties and challenges in the contribution towards the success of the safe and continuous supply. Undoubtedly, meeting a regional demand for petroleum products requires an uninterrupted, safe operation. In addition, the unloading of petroleum products is a complex and potentially dangerous operation since the unloading system contains complex interdependency components. Any failures in one of its components lead to a cut in the petroleum supply chain. Therefore, it is important to assess and evaluate the reliability of the unloading system in order to improve its availability. In this context, Mohammed et al. [5] present the operation philosophy of the truck unloading system, failure modes of the components within the system, and a bottom-up approach to analyze the reliability of the system. In addition, it provides reliability data, such as failure rates and the mean time between failures of the system components. Furthermore, the reliability of the whole system was calculated and is presented for different time periods. The critical components, which are major contributors towards the system's reliability, were identified. To enhance the system's reliability, a reliability-based preventive maintenance strategy for the critical components was implemented. In addition, the preventive maintenance scheduling was identified based on the reliability plots of the unloading system. The best schedule for preventive maintenance of the system was determined based on the reliability function to be every 45 days for maintaining the system reliability above 0.9.

In order to maximize inventory benefits or minimize costs, the reliability and cost of inventory control models need to be identified and analyzed. These importance measures are one important approach to recognize and evaluate system weaknesses. However, the importance measures have fewer applications in inventory systems' reliability. Considering the cost, Chen et al. [6] discuss the reliability change of performance parameters with the importance measures in inventory systems. The calculation methods of differential importance and Birnbaum importance are studied in the inventory control model with shortages. By comparing the importance values of various parameters in the model, the optimization analysis of the inventory model can be used to identify the key parameters so as to effectively reduce the total inventory cost. The importance order and the identification of key parameters are helpful to increase the operational efficiency of the inventory control and provide effective methods for improving inventory management. Lastly, a case study with a shortage and limited inventory capacity is used to demonstrate the proposed model. One of the most important contributions of this paper [6] is that based on the research of inventory systems, it was found that there was almost no literature on the reliability of an inventory system. Combining the concept of reliability with the inventory system, an inventory system reliability model was proposed. It could enrich the research in the field of inventory system reliability.

The growing competitiveness of the market, coupled with the increase in automation driven by the advent of Industry 4.0, highlights the importance of maintenance within organizations. At the same time, the amount of data capable of being extracted from industrial systems has increased exponentially due to the proliferation of sensors, transmission devices and data storage via the Internet of Things. These data, when processed and analyzed, can provide valuable information and knowledge about the equipment, allowing a move towards predictive maintenance being considered one of the most innovative maintenance strategies. Maintenance is fundamental to a company's competitiveness since actions taken at this level have a direct impact on aspects such as cost and quality of products. Hence, equipment failures need to be identified and resolved. Artificial Intelligence tools, in particular Machine Learning, exhibit enormous potential in the analysis of large amounts of data, now readily available, thus aiming to improve the availability of systems, reduce maintenance costs and increase operational performance and support in decision making. This is why Cardoso and Ferreira [7] apply Machine Learning to a set of data made available online and the specifics of this implementation are analyzed, as well as the definition of methodologies, in order to provide information and tools to the maintenance area. Although the results obtained compare well with those presented so far in the literature, the biggest disadvantage in using the presented methodology lies in the definition of the features. If the selection of features is not the most correct, the results obtained can lead to wrong predictions. For future work, the application of feature learning concepts will be considered instead of feature engineering, which appears to be promising to improve the results obtained [7]. This section may be divided into subheadings. It should provide a concise and precise description of the experimental results, their interpretation, as well as the experimental conclusions that can be drawn.

4. Production Systems Scheduling, Management and Maintenance

Advancements in technology, such as information and communication technologies, have changed the traditional manufacturing systems practices. This is especially true for different manufacturing systems due to their ability to cater to their needs such as Big Data, interoperability, timely delivery, etc.

Production and maintenance tasks apply for access to the same resources. Maintenance-related machine downtime reduces productivity, but the costs incurred due to unplanned machine failures often outweigh the costs associated with predictive maintenance. Costs incurred due to unplanned machine failure include corrective maintenance, reworks, delays in deliveries, breaks in the work of employees and machines. Therefore, scheduling production and maintenance tasks should be considered jointly.

The problem of generating a predictive schedule with given constraints is considered by Paprocka et al. [8], with an objective to develop a scheduling method that reflects the operation of the production system and the nature of disturbances. The original value research results presented is the development of the method of a basic schedule generation with the application of Ant Colony Optimization. A predictive schedule is built by planning the technical inspection of the machine at the time of the predicted failure-free time. The numerical simulations are performed for job/flow shop systems. In the future, the presented method for generating predictive schedules will be compared with the genetic algorithm, as well as immune and clonal selection algorithms. The presented algorithm may, however, contribute to the development of a method that reflects the operation of the production system and the nature of disturbances and improves the system operation.

In production systems maintenance, fault propagation behavior analysis is the basis of fault diagnosis and health maintenance. As presented by Mu et al. [9], traditional fault propagation studies are mostly based on a priori knowledge of a causality model combined with rule-based reasoning, disregarding the limitations of experience and the dynamic characteristics of the system that cause deviations in the identification of critical fault sources. Thus, the authors [9] propose a dynamic analysis method for fault propagation behavior of machining centers that combines fault propagation mechanisms with model structure characteristics. They use the design structure matrix to establish the fault propagation hierarchy structure model. Considering the correlation of fault time, the fault probability function of a component is obtained, and the fault influence degree of nodes are calculated. By introducing the Copula and Coupling degree functions, the fault influence degree of the edges between the same level and different levels are calculated, respectively. As a result, a fault propagation intensity model was constructed by integrating the edge betweenness and then used as an index to analyze real-time fault propagation behavior. Finally, a certain type of machining center is taken as an example for a specific application. This study can provide a reference for the fault maintenance and reliability growth of a machining center. According to the fault propagation intensity of the components, the critical fault propagation paths and nodes of a machining center can be identified, provide a reference for the fault maintenance and encourage reliability growth of machining centers. This paper [9] demonstrates the effectiveness and practicability of the proposed method through the application of the specific case.

Ramakurthi et al. [10] point out that rising energy prices, increasing maintenance costs and strict environmental regimes have augmented the already existing pressure on the contemporary manufacturing environment. Although the decentralization of the supply chain has led to rapid advancements in manufacturing systems, finding an efficient supplier simultaneously from the pool of available ones as per customer requirements and enhancing the process planning and scheduling functions are the predominant approaches that still need to be addressed. Therefore, they have decided [10] to address this important issue by considering a set of gear manufacturing industries located across India as a case study. An integrated classifier-assisted evolutionary multi-objective evolutionary approach is proposed for solving the objectives of makespan, energy consumption and increased service utilization rate, interoperability and reliability. To execute the approach initially, text-mining-based supervised machine-learning models were adopted for the classification of suppliers into task-specific suppliers. Following this, with the identified suppliers as input, the problem was formulated as a multi-objective Mixed-Integer Linear Programming model, and finally, a Hybrid Multi-Objective Moth Flame Optimization algorithm was proposed to optimize process planning and scheduling functions. Numerical experiments have been carried out with the formulated problem as well.

5. Systems Analysis, Simulation, Design and Modeling

The activities performed by the maintenance department are usually a combination of technical, administrative and management activities carried out during the life cycle of a given technical object. The effective implementation of these activities requires appropriate data management—data collection, proper analysis and the use of appropriately effective models to support decision making. We use the data for many different purposes, e.g., to determine the moment of generating service orders, monitor the quality of performed activities, optimize and plan activities or develop plans for the delivery dates of materials and spare parts, i.e., in the field of broadly understood logistics for the operation of technical facilities and systems.

This is why proper data management and knowledge management are extremely important issues for the implementation and effective maintenance strategies when it is possible not only to determine the current wear of devices or their structural elements but also to predict whether and when a failure or catastrophic wear will occur. It is also possible to effectively detect the cause of a failure or identify performance or product quality problems. To achieve this, adequate analysis, design and modeling methods are required.

This is why Aliev and Antonelli [11] have focused on developing a framework using I4.0 enabling technologies to improve reliability and safety in human–robot collaboration applications. The proposed framework allows a robot's condition to be continuously monitored during human–robot collaboration. The monitoring deploys IoT connectivity, a data acquisition system, physical cyber-systems and ML tools to perform analytics. The paper is divided as follows: the relevant equipment parameters are first identified, a description of the data acquisition framework is then given, an application to an assembly case study in which all the necessary data are collected is presented, and finally, the analysis results of the considered case study are presented and discussed. The case study was performed on benchmark tasks for collaborative assembly processes. An automatic machine learning tool was used to perform online monitoring and predict outages of the industrial robots during a human–robot collaboration process. Such an online monitoring system allows more reliable human–robot collaboration applications to be created, unplanned downtime during task execution to be eliminated, and the trust of humans during interaction with a robot and the lifetime of the robot to be maximized. The proposed framework demonstrates data management techniques on an industrial robot that is considered as a physical=cyber system.

Achievements of accurate robotic arm-based bike frame quality checks with the use of a 3D mathematical model are discussed by Lin et al. [12]. Unlike the traditional way to find coefficients of a space sphere, the proposed model requires only three check point coordinates to achieve the sphere axis coordinate and its radius. In the practical work, the contact sensor combined with the robotic arm is used to realize the compliance items measurement in shaft length, internal diameter, verticality, parallelism, etc. The proposed model is validated based on both mathematic verification and an actual bike frame check. The stylus probe used in the proposed model presents a simple and accurate performance. However, successful measurement depends on the activity range of the robotic arm that certain features of bike frames should be reached by the stylus probe. In future work, the optical sensors may provide an alternative solution, although more complex signal processing algorithms should be addressed.

The quality and reliability of consumables, including gear oils, resulting in the failure-free operation of the transmission components in heavy trucks are discussed by Gil et al. [13].

As oil viscosity is essential for all lubricated tribopairs for wear and friction reduction in all vehicles with a gearbox, it may be influenced by the contamination that wear products can impart on the oil. Oil contamination can also affect lubrication efficiency in the boundary friction conditions in gearboxes where slips occur (including bevel and hypoid gearboxes). Therefore, the present research [13] focused on this issue. An obvious hypothesis was adopted, where it was theorized that exploiting the contaminants that are present in gear oil may affect how the lubricating properties of gear oils deteriorate. Laboratory tests were performed on contaminants that are commonly found in gear oil. The study was designed to identify a number of different solid particles that are present in oil. The quantitative contamination of the gear oils that contained solid particles and the curves representing the friction coefficients of fresh oils with a history of exploitation were compared. Exploitation was shown to have a significant impact on the contamination of gear oils. It was revealed that the contamination and the mileage had no effect on the tested oils. The research [13] showed that the presence of contaminants is not catastrophic and that in order to fully examine the oils and to determine the critical moment, the oils with a much greater operational mileage should be tested in order to establish the relationship between the number of particles and their tribological properties.

Various factors are considered in system design, such as efficiency, costs, safety, and environmental effect. Availability is also one of the important issues in system design. The availability indicates how much a system approaches ideal operation without production loss caused by equipment failures or undesired external events. Availability estimation is frequently performed in the oil and gas, chemical and power plant industries to find the optimum design option, to predict the production level and to evaluate maintenance and operating policies. Precise availability estimation is important because it directly influences the owner's decision. This is why Seo et al. [14] have investigated the availability gap between the early and late design stages by estimating it with the design stages to find a practical manner of availability estimation in the early design stage. The sensitivity analysis was conducted to analyze the key factors in the results. The most crucial factor was the redundant equipment. Although this study [14] did not consider the whole system, this gives an important guide to progress the next step for the accurate availability estimation in the early design stage.

Steel tapes with certain special markings generated on their surface are often used for metrological and technological purposes, for example, to measure displacement. The smoothness of the movement of a moving precision tape and the stability of the tape area where symbols are generated are very important factors that affect the quality parameters of a generated tape. One of the most important characteristics of the raster generation device is the smoothness of the stretching of the tape, which is important for the overall operation of the system. The smoothness of the stretching of the tape affects errors in the position of the raster element being generated and the control of activation of the laser beam. Determining not only the amplitude but also the frequency of the belt stretching oscillations is important. For this purpose, [15] analyzes the tape movement system consisting of electromechanical tape pulling and its constant stretching mechanisms as well as a tape deflection mechanism, which operates in sliding friction. This system was mounted on a massive granite base placed on a foundation using passive vibration insulation supports. A research and data processing method together with the results of experimental research of a mock-up system were developed and presented to examine the raster generation method and the generation device. This method may be used to produce a precision metrological scale on stainless steel tape. The generation process takes place in the dynamic mode because both the steel tape and the laser raster generation head are constantly moving during the process. The main aim of the research [15] was to develop a system for measuring the displacement of the tape in the raster generation device, to examine the model of that system and to evaluate the possible impact of external and internal factors on raster generation in the dynamic mode.

The use of numerical methods for simulation, design or modeling, such as finite element analysis (FEA), has proven to be an advantageous tool to predict the mechanical behavior of many materials. Up to now, the application of FEA related to elastomers and other cork composites has been utilized to access static, dynamic loading and impact behavior. One of the first requirements for the application of isolation pads is to evaluate their capacity to support static loadings. Regarding elastomers, one of the crucial steps during FEA is the definition of material properties. Typically for large strains, elastomer's properties are defined through the application of non-linear models. However, if a linear stress–strain relationship at small strains is observed, Hooke's Law can be adopted for that strain range. As expressed by Lopes et al. [16], like other types of elastomers, different geometries of the same cork–rubber material present different mechanical behavior when subject to compression between bonded plates. To validate the application of Hooke's Law on cork–rubber materials, under compression at small strains, a set of experimental and numerical analyses has been conducted [16]. Using finite element analysis, a methodology is described to relate frictionless and frictional compression between a cork–rubber sample and loading plates. Based on that, the performance of square cross-section blocks with other dimensions can be evaluated. The results obtained by this approach [16] showed a good agreement with experimental compression tests and with outputs from other models available in the literature relating Young and apparent compression moduli. However, future research should address the effect of higher shape factors, other cross-section shapes (rectangular and other polygons), and the friction coefficient between sample and loading surfaces. Moreover, applying this knowledge and relating it to the dynamic compression behavior of isolation pads could be a topic of interest.

6. Conclusions

The availability and reliability of engineering systems is an important issue for modern companies, especially from the Industry 4.0 challenges and requirements. In addition, in recent years, the concept of sustainable development has also been gaining importance, as sustainable, intelligent production should be taken into account. This means developing products manufactured through production processes that have a minimal negative impact on the environment, save energy and natural resources and are safe for workers and the community and economically viable. This means that the goal of sustainable production is to achieve a balance between the environmental, social and economic dimensions. Creating a sustainable production environment also requires the elimination of breakdowns and energy waste and, as a concept, is strongly associated with sustainable maintenance. From a practical point of view, this requires changes in the activities performed in the maintenance area. Overall, this means the ability to monitor, control and process data combined to create intelligent, learning, self-diagnosing and self-adapting technology machines. This kind of technological intelligence in maintenance can reduce the need for operators to act, improving safety and reducing unnecessary costs; however, they can be still considered an important research challenge in systems engineering.

Funding: This research received no external funding.

Institutional Review Board Statement: Not applicable.

Informed Consent Statement: Not applicable.

Data Availability Statement: Not applicable.

Conflicts of Interest: The authors declare no conflict of interest.

References

1. Dinkar, B.; Mukhopadhyay, A.; Chattopadhyaya, S.; Sharma, S.; Alam, F.; Machado, J. Statistical Reliability Assessment for Small Sample of Failure Data of Dumper Diesel Engines Based on Power Law Process and Maximum Likelihood Estimation. *Appl. Sci.* **2021**, *11*, 5387. [CrossRef]
2. Pisla, D.; Calin, V.; Birlescu, I.; Hajjar, N.; Gherman, B.; Radu, C.; Plitea, N. Risk Management for the Reliability of Robotic Assisted Treatment of Non-resectable Liver Tumors. *Appl. Sci.* **2020**, *10*, 52. [CrossRef]
3. Wu, J.; Jin, Y.; Hu, S.; Fei, J.; Zhang, Y. Approach to Risk Performance Reasoning with Hidden Markov Model for Bauxite Shipping Process Safety by Handy Carriers. *Appl. Sci.* **2020**, *10*, 1269. [CrossRef]
4. Jiang, Z.; Zhao, T.; Wang, S.; Ren, F. A Novel Risk Assessment and Analysis Method for Correlation in a Complex System Based on Multi-Dimensional Theory. *Appl. Sci.* **2020**, *10*, 3007. [CrossRef]
5. Mohammed, A.; Ghaithan, A.; Al-Saleh, M.; Al-Ofi, K. Reliability-Based Preventive Maintenance Strategy of Truck Unloading Systems. *Appl. Sci.* **2020**, *10*, 6957. [CrossRef]
6. Chen, L.; Kou, M.; Wang, S. On the Use of Importance Measures in the Reliability of Inventory Systems, Considering the Cost. *Appl. Sci.* **2020**, *10*, 7942. [CrossRef]
7. Cardoso, D.; Ferreira, L. Application of Predictive Maintenance Concepts Using Artificial Intelligence Tools. *Appl. Sci.* **2021**, *11*, 18. [CrossRef]
8. Paprocka, I.; Krenczyk, D.; Burduk, A. The Method of Production Scheduling with Uncertainties Using the Ants Colony Optimisation. *Appl. Sci.* **2021**, *11*, 171. [CrossRef]
9. Mu, L.; Zhang, Y.; Liu, J.; Zhai, F.; Song, J. Dynamic Analysis Method for Fault Propagation Behaviour of Machining Centres. *Appl. Sci.* **2021**, *11*, 6525. [CrossRef]
10. Ramakurthi, V.; Manupati, V.; Machado, J.; Varela, L. A Hybrid Multi-Objective Evolutionary Algorithm-Based Semantic Foundation for Sustainable Distributed Manufacturing Systems. *Appl. Sci.* **2021**, *11*, 6314. [CrossRef]
11. Aliev, K.; Antonelli, D. Proposal of a Monitoring System for Collaborative Robots to Predict Outages and to Assess Reliability Factors Exploiting Machine Learning. *Appl. Sci.* **2021**, *11*, 1621. [CrossRef]
12. Lin, H.; Yu, B.; Wang, J.; Lai, J.; Wu, J. Achievement of Accurate Robotic Arm-based Bike Frame Quality Check Using 3D Geometry Mathematical Model. *Appl. Sci.* **2019**, *9*, 5355. [CrossRef]
13. Gil, L.; Przystupa, K.; Pieniak, D.; Kozlowski, E.; Antosz, K.; Gauda, K.; Izdebski, P. Influence of Contamination of Gear Oils in Relation to Time of Operation on Their Lubricity. *Appl. Sci.* **2021**, *11*, 11835. [CrossRef]
14. Seo, Y.; Jung, J.; Han, S.; Kang, K. Availability Estimation of Air Compression and Nitrogen Generation Systems in LNG-FPSO Depending on Design Stages. *Appl. Sci.* **2020**, *10*, 8657. [CrossRef]
15. Fursenko, A.; Kilikevicius, A.; Kilikeviciene-, K.; Borodinas, S.; Kasparaitis, A.; Matijosius, J. Investigation of Roller-Tape Contact Pair Used in Precision Mechatronic System. *Appl. Sci.* **2020**, *10*, 4041. [CrossRef]
16. Lopes, H.; Silva, S.; Machado, J. Analysis of the Effect of Shape Factor on Cork-Rubber Composites under Small Strain Compression. *Appl. Sci.* **2020**, *10*, 7177. [CrossRef]

applied sciences

Article

Statistical Reliability Assessment for Small Sample of Failure Data of Dumper Diesel Engines Based on Power Law Process and Maximum Likelihood Estimation

Brajeshkumar Kishorilal Dinkar [1], Alok Kumar Mukhopadhyay [1], Somnath Chattopadhyaya [2], Shubham Sharma [3,*], Firoz Alam [4] and José Machado [5,*]

[1] Department of Mining Machinery Engineering, Indian Institute of Technology (Indian School of Mines), Dhanbad 826004, India; bridinkar@iitism.ac.in (B.K.D.); akm_emm@yahoo.co.in (A.K.M.)
[2] Department of Mechanical Engineering, Indian Institute of Technology (Indian School of Mines), Dhanbad 826004, India; somnathchattopadhyaya@iitism.ac.in
[3] Department of Mechanical Engineering, IK Gujral Punjab Technical University, Main Campus-Kapurthala, Kapurthala 144603, India
[4] School of Aerospace, Mechanical and Manufacturing Engineering, RMIT University, P.O. Box 2476, Melbourne, VIC 3001, Australia; firoz.alam@rmit.edu.au
[5] MEtRICs Research Center, Campus of Azurém, University of Minho, 4800-058 Guimarães, Portugal
* Correspondence: rs.shubhamsharma@ptu.ac.in (S.S.); jmachado@dem.uminho.pt (J.M.)

Citation: Dinkar, B.K.; Mukhopadhyay, A.K.; Chattopadhyaya, S.; Sharma, S.; Alam, F.; Machado, J. Statistical Reliability Assessment for Small Sample of Failure Data of Dumper Diesel Engines Based on Power Law Process and Maximum Likelihood Estimation. *Appl. Sci.* **2021**, *11*, 5387. https://doi.org/10.3390/app11125387

Academic Editor: Jordi Cusido

Received: 19 May 2021
Accepted: 7 June 2021
Published: 10 June 2021

Publisher's Note: MDPI stays neutral with regard to jurisdictional claims in published maps and institutional affiliations.

Abstract: Dumpers or dump trucks are used all over the world to move overburden from many opencast mines. Diesel engines are the main driving force behind the trucks. The frequency of damage due to the failure of diesel engines is enormous. Therefore, efforts are necessary to analyze failure to reduce the downtime periods. A detailed analysis of engine failure at the subsystem level needs to be done. Reliability analysis and maintenance planning remain the norm in this regard. The obstacle faced while analysing the reliability of dumpers was the availability of a large number of data failures. In this paper, this issue is addressed by using Common Beta Hypothesis test and Meta-analysis test. The engine is divided into five subsystems. The result shows that all five subsystems pass the CBH test and Meta-analysis test. Accordingly, the failure data is grouped. The trend test of grouped failure data shows that the Failure data of two subsystems follows the independent and identically distributed characteristics while the remaining three do not follow it. The reliability is estimated for all five subsystems. Finally, fuel supply subsystems show the highest reliability while the lowest value is seen for self-starting subsystems.

Keywords: time between failure (TBF); common beta hypothesis (CBH) test; meta-analysis; level of heterogeneity; reliability; mean time between failure (MTBF)

1. Introduction

The main drive units used in dump trucks are diesel engines. Dump trucks or dumpers are used to transport heavy materials around the world. The frequency of breakdown causing the failure of diesel engines is adequate. A vital concern in the engine system's performance under certain operational conditions is to guarantee the satisfactory uninterrupted operation of the equipment [1–7]. However, failure of components is unescapable and takes place due to the ongoing wear and tear process in working parts of the system. This deterioration can result in unexpected failures of the system which will incur a significant increase in repair cost than in scheduled maintenance or repair. To control the impact of cost, it is necessary to evaluate the reliability of the equipment and its components. Such a study will be useful for making maintenance decisions and incorporate adaptive changes in maintenance policies. The main hurdle is the availability of a large amount of failure data [6–9]. The general pattern is that the small sample is not representative of the data and there is every possibility that any statistical treatment is misleading when a small number of failures are used. In the present study, a roadmap is provided using which reliability analysis could be possible for a small amount of failure data for any machinery.

To this day, research articles highlighting analytical methods on small data sets are practically limited. D.H. Olwell et.al [10] completed limited data with advanced information using the Weibull probability distribution. The paper conducted a firing analysis of 2000 motors used in missiles in field conditions using the Maximum Likelihood Estimation (MLE) method and the Bayesian method [10–12]. R M Mayer et.al [13] pooled the data from multiple data sets to get a large amount of failure data for statistical interpretation. The paper emphasizes that grouping of failure data is valid only when the data is collected with sufficient reliability. G. Wang et al. [14] used Failure mode effects and critical analysis (FMECA) for analyzing small sample of failure data of diesel engines. L. Qin et al. [15] analyzed the reliability of bearings based on performance attenuation data. E. J. Ahn et al. [16], described the methods used in systematic reviews and meta-analyses in medical sciences. W. Dai et al. [17] made an effective method for reliability assessment using signal features of the machining process. W. Si et al. [18] suggested reliability model for repairable systems having incomplete failure time data. X. Xintao et al. [19] proposed a model of improved maximum entropy probability distribution for estimation of reliability of bearings. F.V. Garcia et al. [20] discussed in their paper the methods to improve failure data used for high-speed marine diesel engine using Failure Modes, Effects, and Criticality Analysis. L. Zhang et al. [21] used Bayesian method for reliability evaluation of very few failure data. The researchers performed the reliability analysis on wet friction plate used in hydraulic control. S. Darmanto et al. [22] analyzed the reliability of diesel engines as a driver for the fire water pump. The researchers have determined reliability and the rate of failure of the diesel engine [22–28].

Recently, Y. He [25] suggested using a combined forecasting model to increase the amount of fault data samples. This increase in data is utilized for reliability analysis of Sanitation vehicles having Small Sample Data.

From the above literature review, it is revealed that most of the studies so far conducted using scarce failure data are less. The methods used for reliability analyses for small number data are mostly the Bayesian approach, FMECA and Monte Carlo method. Studies on reliability analysis with a very small sample amount of failure data on engine subsystems have been carried out. The present study uses CBH which has not been addressed so far for statistical treatment of small failure data. Additionally, the Meta-analysis test used in this paper has been used only in the medical field and not in the industrial field. Using the methods mentioned above the small failure data of any machine or system (in this case diesel engine) can be grouped and easily used for further reliability analysis.

Maintenance philosophies involve performing maintenance after given time intervals, typically after a fixed running hour for an engine. In spite of the scheduled maintenance, failure of the engine is inevitable, thereby decreasing the availability of dumpers and reducing the production cost. Reliability analysis of engine subsystems is essential to formulate the maintenance strategies which will reduce the downtime of the engine and enhance its availability. The main obstacle was the deficiency of adequate data for the appropriate statistical analyses. A data set containing a small sample size of failure data limits the possibility of precise decision-making. The current study gives specific guidelines for using CBH and meta-analysis testing, which emphasizes the failure data to predict reliability and MTBF. The researchers perform the reliability assessment using the grouped TBF data using which suitable maintenance strategies could be formed. It provides a roadmap of reliability analysis for any machinery having less failure data.

2. Research Methodology

An engine is made up of components, each of which is vital to the operation of an entire engine. There are certain major failures which can be prevented by replacing certain parts of the engine in the work site itself. The High oil consumption which is commonly caused by the hose pipe burst or hose pipe leakage can be prevented by replacing the hose pipe timely. The presence of metallic pieces in a lube oil filter can heavily damage the condition of an engine. Hence, lube oil should be replaced along with the bearing oil filters regularly. If the lubrication oil is not changed timely its viscosity will increase leading to overheating. Overheating will cause expansion of the piston liners which will ultimately leads to engine seizure. Hence, timely replacement of lube oil can prevent engine seizure. The problem of overheating is the most common problem occurring in the engine. It may also occur due to insufficient working of the cooling fan and radiator. Proper and timely maintenance of the radiator will prevent the overheating problem. The reliability analysis is desired to prevent any catastrophic failure which may be fatal. Chart 1 shows all the steps used in this paper for reliability analysis. The following methodology is used to perform reliability analysis of an appreciably small amount of failure data.

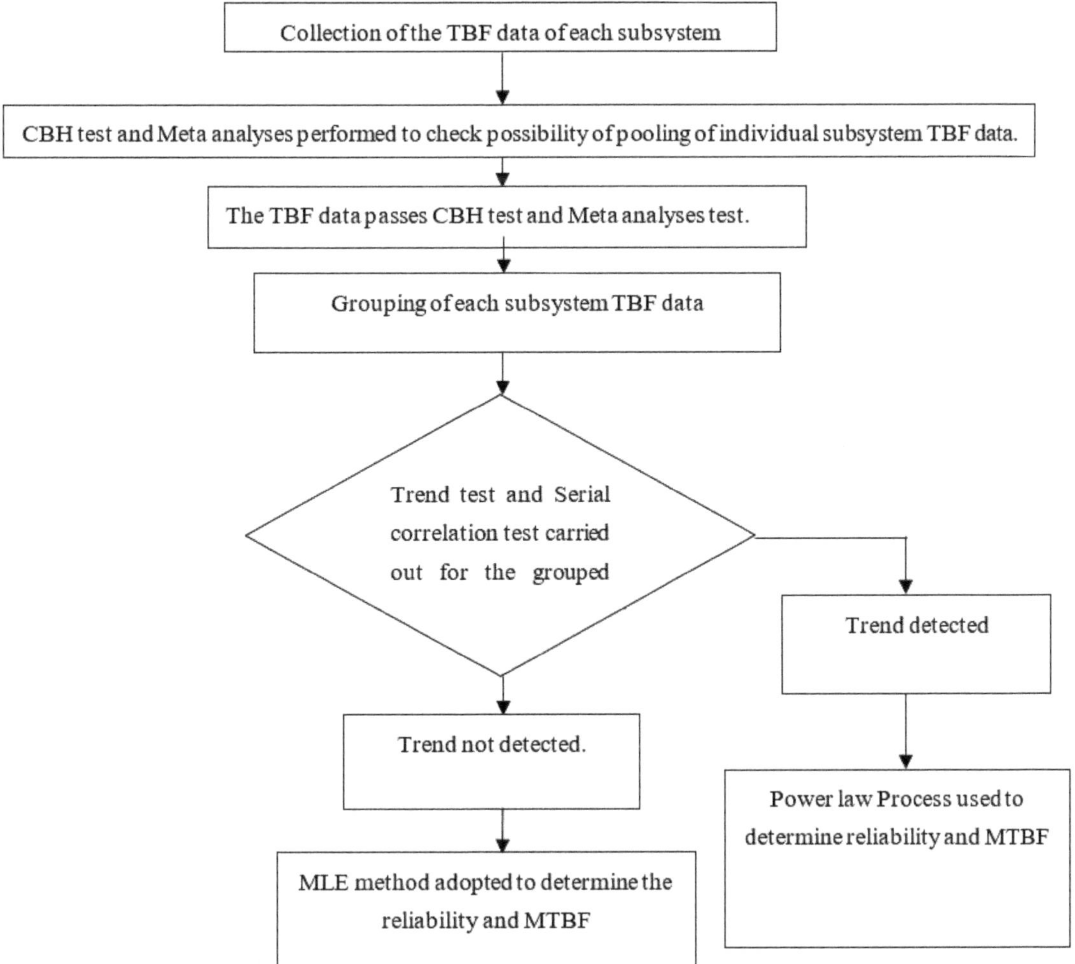

Chart 1. Steps followed for estimating reliability analysis.

The TBF data from the three engines is collected from the management log book of the surface mine. All three engines are of the same type. For statistical analysis, the engine is divided into main subsystems such as air supply, lubrication, self-starting, fuel supply and cooling subsystems. The number of TBF data collected from the project was found to be low. TBF data are pooled to increase the number of failure data. Grouping of failure data magnifies the sample size of each subsystem of three engines. Before aggregating TBF data, a CBH test and a meta-analysis test are applied on the TBF data of all five subsystems to examine the difference between the failure data of individual engines. In the CBH test, the consistency of an inter-arrival failure rate of each subsystem is evaluated [6]. If the failure rate is consistent between the three engines of each subsystem, then the failure data can be grouped. To combine the findings (in this case failure data) for independent studies (in this case three engines) meta-analysis test is used. In this analysis level of heterogeneity is checked among the three engines failure data of each subsystem. Heterogeneity in meta-analysis refers to the variation in the three engines failure data of each subsystem. Next *iid* characteristics of the TBF data of all five subsystems are tested. The relationship between cumulative time and cumulative number failures is considered for trend tests using group TBF data from all five subsystems. For the serial correlation test, the graph between the failure of i th number and (i-1) th number is considered. Based on the results of the trend test, Reliability and MTBF is determined for all five subsystems using either the MLE method or Power Law Process (PLP) model. The Power Law Process (PLP) model is basically a popular infinite NHPP model utilized to determine the reliability of repairable systems on the basis of the analysis of the observed failure data [29,30].

3. Experimentation

3.1. Collection of Field Data

The engines under study are turbocharged compression ignition (C.I.) engines with 12 cylinders, V-type and a maximum power rating of 900 H.P, rotating at 2100 rpm. In CI engines, air is compressed in the combustion chamber such that the injected liquid fuel can easily catch fire and burn progressively for power generation. Figure A1 shows a view of the dumper engine under study (see Appendix A). TBF data of each subsystem is collected for a period of three years from the mechanical register book of the mechanical open pit mine. The failure data in Table 1 were found to be less than 7 for each subsystem.

Table 1. TBF data (in hours) of engine subsystems of three engines.

Sl.No.	The Sub System	Failed Component	TBF for Engine 1	Failed Component	TBF for Engine 2	Failed Component	TBF for Engine 3
1	Air suppl	Compressor Dryers Oil Remover	2655 633 4112	Dryers Electric motor Pressure Gauge Motor	422 600 2036 1479	Condensate Trap Moisture Separator Turbo charger Electric motor	77 3585 1673 646
2	Self-starting	Battery Filter Starter gear Solenoid Motor Gear pump	1246 44 856 2328 3913 759	Solenoid Motor Filter Gear pump Battery Starter gear	423 185 761 1197 1116 3450	Starter gear Battery Gear pump Filter Motor Solenoid	1920 797 550 191 917 1595
3	Fuel supply	Pulsation damper Magnetic screen Injector Throttle Cam	423 240 525 934 3856	Shut down valve Fuel tank Magnetic screen Injector	96 346 914 2036	Cam Pulsation damper Shut down valve Fuel tank Magnetic screen	112 1449 290 225 2828
4	Lubrication	Turbocharger bearings Valves Tappets and push-rods	2566 2278 426	Oil pump parts Tappets and push-rods Valves Camshaft and bearings Timing Gears	1584 757 200 991 916	Oil pump parts Turbocharger bearings Timing Gears Cylinder walls Piston rings Valves	855 1115 1503 1367 990 2926

Table 1. *Cont.*

Sl.No.	The Sub System	Failed Component	TBF for Engine 1	Failed Component	TBF for Engine 2	Failed Component	TBF for Engine 3
5	Cooling	Radiator Cooling Fans Pressure Cap and Reserve Tank Water Pump	3827 2356 577	Thermostat Bypass System Freeze Plugs Head Gaskets and Intake Manifold Gaskets Heater Core	1823 1177 680 1424 3236	Fins Radiator Hoses Water Pump Freeze Plugs Radiator Cooling	170 108 219 934 329 2149

The occurrence of failures has been calculated and shown in Table 2. The pie chart has been drawn to depict the frequency of failure. Figures 1–3 show pie charts for all five subsystems of three engines.

Table 2. Data of occurrence of failures for three engines.

Sub-Systems	TBF (hours) of Engine-1 in %	TBF (hours) of Engine-2 in %	TBF (hours) of Engine-3 in %
Air Supply	21%	16%	20%
Self-starting	27%	26%	20%
Fuel supply	17%	12%	17%
Lubrication	15%	16%	30%
Cooling	20%	30%	13%

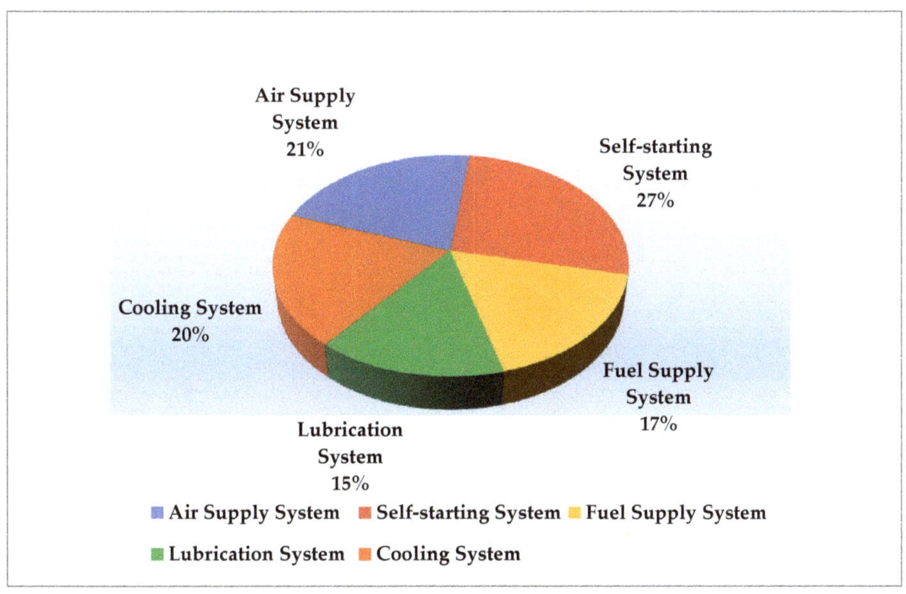

Figure 1. TBF Hours in percentage in comparison to total time for Engine 1.

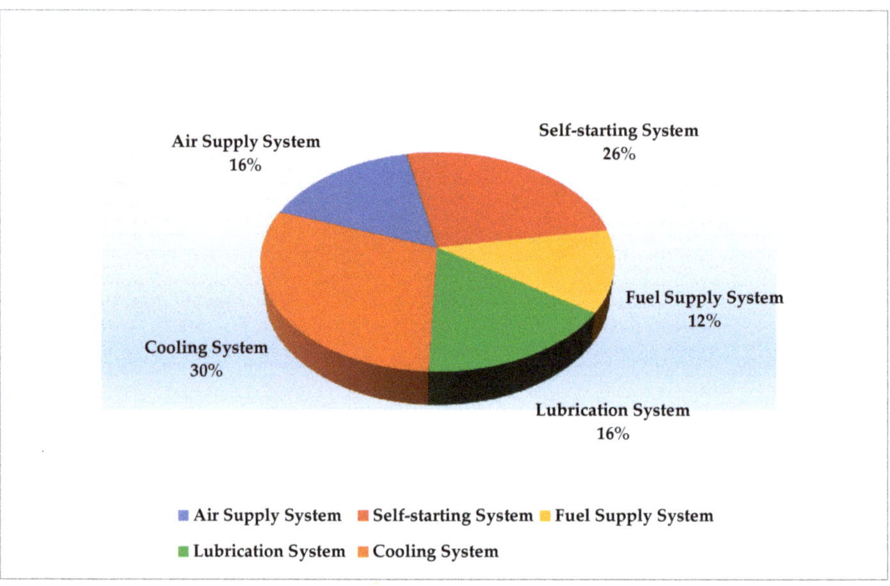

Figure 2. TBF Hours in percentage in comparison to total time for Engine 2.

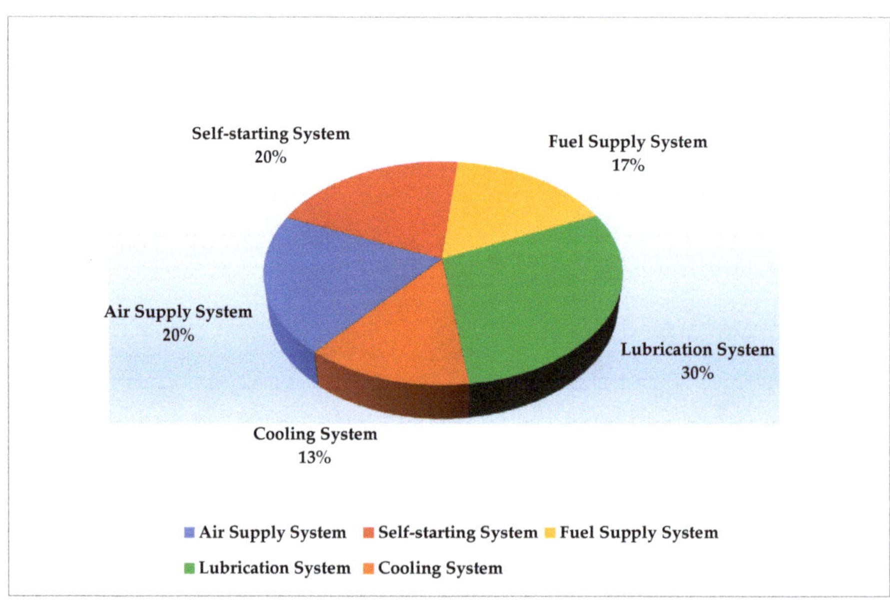

Figure 3. TBF Hours in percentage in comparison to total time for Engine 3.

3.2. Grouping of Data

3.2.1. Common Beta Hypothesis Test

The TBF data have been collected for dumper engine subsystems of three engines of three years duration are used to have pictorial representation in the form of the pie chart [27–29]. To increase the number of TBF data, the TBF data of three same types of engines for each subsystem are grouped together. The grouping of data is validated by using Common Beta Hypothesis (CBH) test [2].

In the CBH test, all three engines are under test. An intensity function of each subsystem is given by Equation (1).

$$\mu_q(t) = \lambda_q \beta_q \, t^{\beta_q - 1} \tag{1}$$

where q is the number of engines, i.e., 1, 2, and 3. The intensity functions of each engine is compared by comparing the β_q of each system. Let $\widetilde{\beta_q}$ denote the conditional maximum likelihood estimate of β_q, which is given by [31–33]:

$$\beta_q = \frac{\sum_{q=1}^{k} M_q}{\sum_{q=1}^{k} \sum_{t=1}^{M_q} \ln\left[\frac{T_q}{X_{tq}}\right]} \tag{2}$$

β_q is the shape parameter of each subsystem
$K = 1, 2$ and 3 is the number of engines.
M_q = number of subsystem failures of each engine.
T_q = total working hours of each engine.
X_{iq} is the i th time-to-failure on the q th engine system
The shape parameter $\beta*$ average value is given by

$$\beta* = \frac{M}{\sum_{q=1}^{K} \frac{M_q}{\beta_q}} \tag{3}$$

where,

$$M = \sum_{q=1}^{K} M_q \tag{4}$$

For calculation of yield statistics D,

$$L = \sum_{q=1}^{K} M_q \ln\left(\widetilde{\beta_q}\right) - M \ln(\beta^*) \tag{5}$$

$$a = 1 + \frac{1}{6(K-1)} \left[\sum_{q=1}^{K} \frac{1}{M_q} - \frac{1}{M}\right] \tag{6}$$

Calculate the statistic D, such that:

$$D = \frac{2L}{a} \tag{7}$$

The statistic D is distributed as a chi-squared random variable with a degree of freedom $(3 - 1) = 2$. It is estimated using Equation (7). The chi-squared tables are referred to to find the critical points.

3.2.2. CBH Test of Engine Subsystems

The data used to calculate the chi-squared value D for the CBH test are given in Table 3. "Start" refers to the time the engine was first put into service, which is 0. The cumulative time between failure hours of all subsystems of individual engines is calculated (from the values given in Table 1). For a given engine, the maximum cumulative time of its subsystem failures (between all five subsystems) is considered to be the life of the engine during data collection. This is shown in Table 3 below the "End" event. "Failures" mention cumulative TBF of individual subsystem taken from Table 1.

Table 3. Common Beta Hypothesis test failure data in hours.

S.No.	The Subsystems	Event	Failures for Engine 1	Failures for Engine 2	Failures for Engine 3
1	Air supply	Start	0	0	0
		End	9146	8340	8756
		Failures	• 2655 • 3288 • 7400	• 422 • 1022 • 3058 • 4537	• 77 • 3662 • 5335 • 5981
2	Self-starting	Start	0	0	0
		End	9146	8340	8756
		Failures	• 1246 • 1290 • 2146 • 3741 • 4474 • 8387	• 423 • 608 • 1369 • 2566 • 3682 • 7132	• 1920 • 2717 • 3267 • 3458 • 4375 • 5970
3	Fuel supply	Start	0	0	0
		End	9146	8340	8756
		Failures	• 423 • 663 • 1188 • 2122 • 5978	• 96 • 412 • 1326 • 3362	• 112 • 1561 • 1851 • 2076 • 4904
4	Lubrication	Start	0	0	0
		End	9146	8340	8756
		Failures	• 2566 • 4844 • 5270	• 1584 • 2341 • 2579 • 3570 • 4486	• 855 • 1970 • 3473 • 4840 • 5830
5	Cooling	Start	0	0	0
		End	9146	8340	8756
		Failures	• 3827 • 6183 • 6760	• 1823 • 3000 • 3680 • 5104	• 170 • 278 • 497 • 1431 • 1760 • 3909

3.2.3. Meta-Analysis Test Steps

To check the level of heterogeneity meta-analysis test is used. It is a statistical technique for combining findings from independent studies. In the present study, variability of the failure data among the three engines for each subsystem is tested using Meta-analysis. Variability means differences in statistical results obtained between the individual failed data and pooled failure data for a particular subsystem [31–33].

In Table 4, the column "downtime hours" describes the total downtime hours of a particular engine for the problem related to a specific subsystem mentioned at the top of the table. The total run of the engine column indicates the total time in hours the engine has worked.

Table 4. CBH test values of all five subsystems.

The Subsystems	Values of D				Results
	Values Numerically Calculated	Software Calculated	Lower Critical Value	Upper Critical Value	
Air supply	0.88	0.88	0.10	5.99	Passed CBH test
Self-starting	0.78	0.79	0.10	5.99	Passed CBH test
Fuel supply	0.26	0.22	0.10	5.99	Passed CBH test
Lubrication	0.21	0.21	0.10	5.99	Passed CBH test
Cooling	4.45	4.45	0.10	5.99	Passed CBH test

The outcome or Effect size (E.S) column is calculated [12] as

$$\frac{Downtime\ hours}{total\ run\ of\ engine} \tag{8}$$

Standard Error (SE) for each engine is calculated using the formula,

$$SE = \frac{\sqrt{Downtime\ hours}}{total\ run\ of\ engine} \tag{9}$$

$$Rate\ of\ outcome = Outcome \times 100 \tag{10}$$

The failure data for each subsystem has been weighed (w) against its variance, and it is calculated using

$$w = \frac{1}{SE^2} \tag{11}$$

weighted effected size for each engine is a computed by-product of effect size and study weight, i.e.,

$$(w \times es) \tag{12}$$

Other important variables, $w \times es^2$ is calculated for each engine required for calculating Q statistics. Q test measures the diversity of studies and t acts as a test. It is calculated as the weight of the squared differences between the individual effects of the collected failure data and the effects collected across the failure data using Equation (13).

The formula is

$$Q = \sum \left(w \times es^2 \right) - \sum \frac{(W \times ES)^2}{\sum W} \tag{13}$$

Finally, the level of heterogeneity, i.e., i^2 is calculated using Equation (14). The i^2 is a percentage of the total variability between the failure data.

The formula is

$$i^2 = \frac{(Q - df)}{Q} \times 100 \tag{14}$$

where "df" is degrees of freedom which is equal to n − 1, and where n is the number of engines under study (in this case, it is 3 − 1 = 2).

4. Results and Discussions

After going through the recent studies on reliability analysis on small failure data, it is evident that the CBH test and Meta-analysis test has not been seen as a possible solution for small failure data. Although Meta-analysis has been considered for medical studies, it has not been considered for machines. This paper uses CBH that has not been considered so far as statistical treatment for a small amount of failure data. Additionally, the Meta-analysis used in this paper has been used only in the medical field and not in the industrial sector.

4.1. CBH Test

The values of D of the five subsystems were calculated using the CBH test. The three engines together are shown in Table 4. Mathematically calculated values, as well as software values, are shown in the table. The test values for all five subsystems falls between the lower (0.10) and upper value (5.99). Hence, the TBF data for each subsystem of three engines pass CBH suggesting pooling of TBF data for further analysis.

4.2. Meta-Analysis Test

It can be observed from Table 5 that the level of heterogeneity was a negative value for self-starting, fuel supply, lubrication and cooling subsystems. Negative level of heterogeneity values can be treated as equal to zero [11]. The level of heterogeneity value for the air supply subsystem is 2.23% which is very low [12]. The zero value for four subsystems and the low level of heterogeneity value for one subsystem indicated that there is no variability among the failure data of three engines for all the five subsystems. It suggests that all the samples came from the same underlying distribution thereby supporting the result of the CBH test, which allows the pooling of failure data of three engines for each subsystem.

Table 5. Level of Heterogeneity (i^2) values of all five subsystems.

Air Supply Subsystem									
Engines	Downtime Hours	Total Run of Engine	Outcome	S.E.	Rate	W	W × ES	W × ES^2	Level of Heterogeneity (i^2)
1	18.5	30,641	0.000604	0.00014	0.06	50,749,777.35	30,641	18.5	
2	26	27,857	0.000933	0.000183	0.09	29,846,632.65	27,857	26	2.23%
3	21	29,520	0.000711	0.000155	0.07	41,496,685.71	29,520	21	

Self–Starting subsystem									
Engines	Downtime Hours	Total run of engine	Outcome	S.E.	Rate	W	W × ES	W × ES^2	Level of Heterogeneity (i^2)
1	46.5	30,641	0.001518	0.000223	0.15	20,190,772	30,641	46.5	−170.23%
2	47.5	27,857	0.001705	0.000247	0.17	16,337,104	27,857	47.5	which is taken
3	42	29,520	0.001423	0.00022	0.14	20,748,343	29,520	42	as 0

Fuel supply subsystem									
Engines	Downtime Hours	Total run of engine	Outcome	S.E.	Rate	W	W × ES	W×ES^2	Level of Heterogeneity (i^2)
1	33	30,641	0.001077	0.000187	0.10	28,450,633	30,641	33	−452.77%
2	25.5	27,857	0.000915	0.000206	0.09	23,515,529	2,1525.86	19.70	which is taken
3	28.5	29,520	0.000965	0.000195	0.09	26,406,982	25,494.55	24.61	as 0

Lubrication subsystem									
Engines	Downtime Hours	Total run of engine	Outcome	S.E.	Rate	W	W × ES	W × ES^2	Level of Heterogeneity (i^2)
1	34	30,641	0.00111	0.00019	0.11	27,613,849	30,641	34	
2	38.5	27,857	0.001382	0.000223	0.13	20,156,168	27,857	38.5	−52.67% which is taken as 0
3	41.3	29,520	0.001399	0.000218	0.13	21,100,010	29,520	41.3	

Cooling subsystem									
Engines	Downtime Hours	Total run of engine	Outcome	S.E.	Rate	W	W × ES	W × ES^2	Level of Heterogeneity (i^2)
1	25	30,641	0.000816	0.000163	0.08	37,554,835	30,641	25	
2	28.1	27,857	0.001009	0.00019	0.10	27,616,101	27,857	28.1	−58.85% which is taken as 0
3	32	29,520	0.001084	0.000192	0.10	27,232,200	29,520	32	

The failure data of all five subsystems of the engine showed consistency by confirming CBH test. Additionally, meta-analysis test supports the result of CBH test which allows the pooling of failure data of three engines. The pooled data for each subsystem are shown in Table 6. This pooled data can be further successfully used for reliability analysis.

Table 6. Grouped TBF data for engine subsystems.

Sl. No.	Subsystem	TBF (hours)
1	Air supply	2655, 633, 4112, 422, 600, 2036, 1479, 77, 3585, 1673, 646.
2	Self-starting	1246, 44, 856, 2328, 3913, 759, 423, 185, 761, 1197, 1116, 3450, 1920, 797, 550, 191, 917, 1595
3	Fuel supply	423, 240, 525, 934, 3856, 96, 316, 914, 2036, 112, 1449, 290, 225, 2828
4	Lubrication	2566, 2278, 426, 1584, 757, 238, 991, 916, 855, 1115, 1503, 1367, 990, 2926.
5	Cooling	3827, 2356, 577, 1823, 1177, 680, 1424, 3236, 170, 108, 219, 934, 329, 2149.

4.3. Trend Test and Serial Correlation Test

The graph is plotted for all five subsystems of the engine between cumulative time between successive failures and the cumulative number of failures using Grapher software. The linearity of the graph will validate that collected data has no trend and they are independent and identically distributed. Next, with TBF data, a plot between $(i-1)$ th TBF and i th TBF is drawn for all five subsystems. The scattered plot will reveal whether that the data have no trend and no serial correlation exists [7]. The grouped TBF data of Table 6 is considered for plotting the graph. The trend test plots of five subsystems. They are shown in Figure 4a–e.

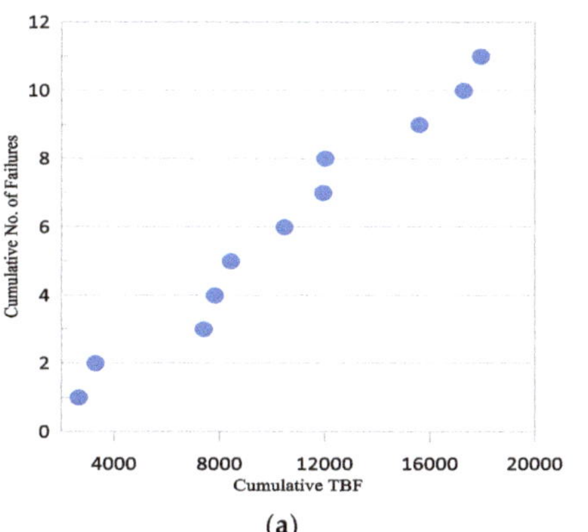

(a)

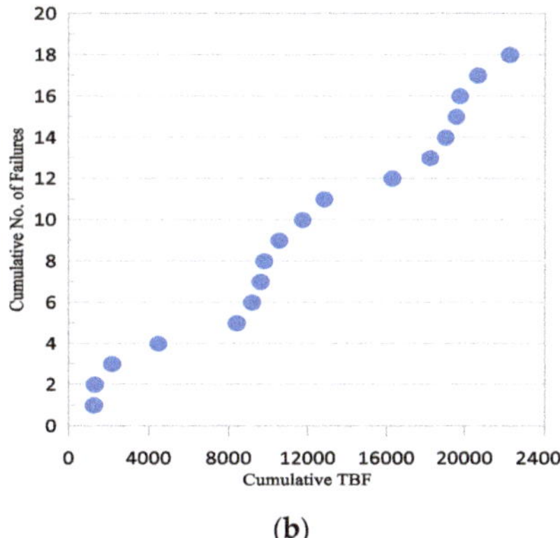

(b)

Figure 4. *Cont.*

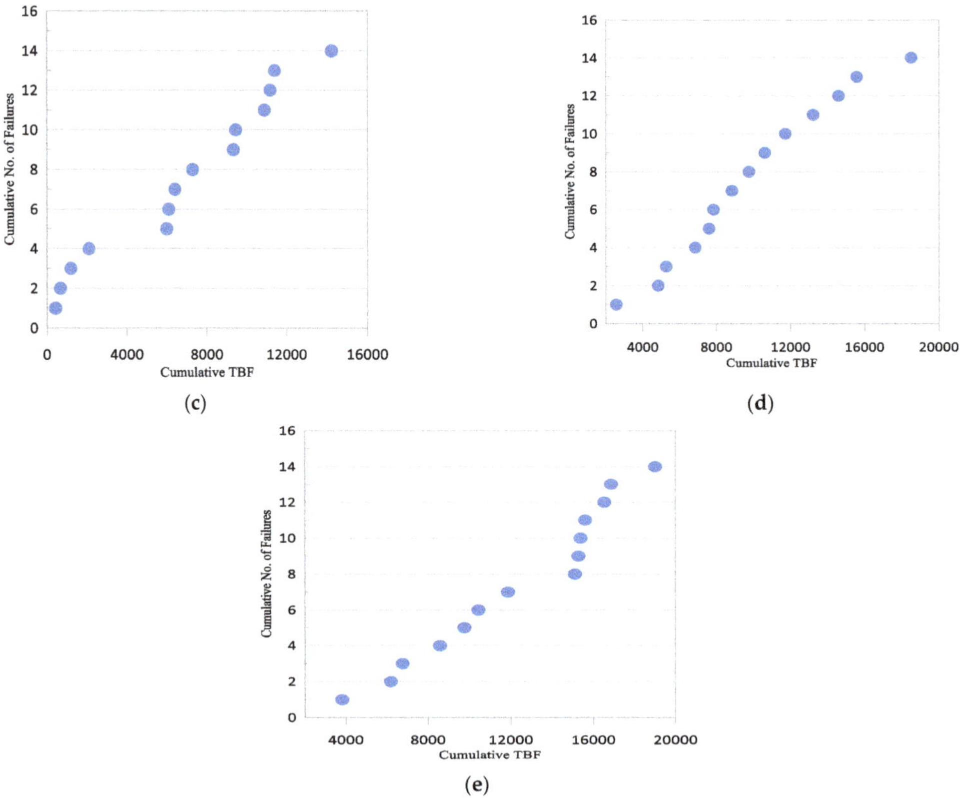

Figure 4. (**a**). Trend test of TBF data for air supply subsystem; (**b**). Trend test of TBF data for self-stating subsystem; (**c**). Trend test of TBF data for fuel supply subsystem; (**d**). Trend test of TBF data for lubrication subsystem and (**e**). Trend test of TBF data for cooling subsystem.

The plots in Figure 5a–e show the serial correlation tests of all five subsystems. From the plots above, no trend is observed in air supply and lubrication subsystems as the plotted points are in a straight line. The trend is seen for self-starting, fuel supply and cooling subsystems. No serial correlation is found for all five subsystems due to the scattered nature of the graphs (Figure 5a–e). Hence, self-starting, fuel supply and cooling subsystems do not follow *iid* characteristics whereas the air supply and lubrication subsystem follows it.

4.4. Reliability Analysis

The grouped TBF data of self-starting, fuel supply and cooling subsystems are identified as not independently and identified distributed. The TBF data of the air supply and the lubrication subsystems were distributed independently and evenly. The MLE method is used to estimate the reliability and MTBF. PLP model is used for reliability estimation of subsystems having non-IID data. The reliability is estimated at an arbitrary value after 1000 h (for comparison) and also Mean Time Between Failure (MTBF) is calculated. Table A1 shows the values of reliability and MTBF for all five subsystems (See Appendix A).

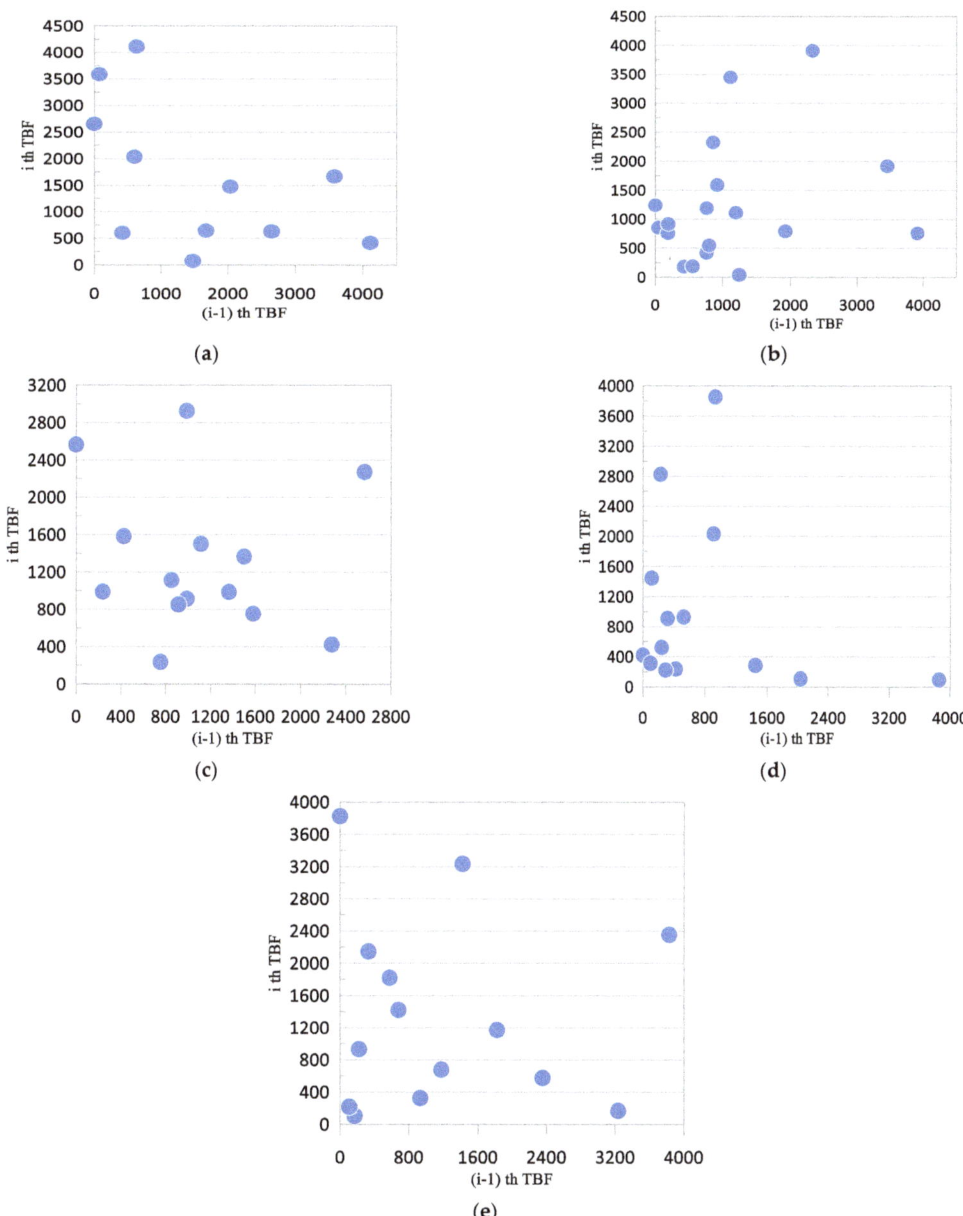

Figure 5. (**a**). Serial correlation test of TBF data for air supply subsystem; (**b**). Serial correlation test of TBF data for self-starting subsystem; (**c**). Serial correlation test of TBF data for fuel supply subsystem; (**d**). Serial correlation test of TBF data for lubrication subsystem and (**e**). Serial correlation test of TBF data for cooling subsystem.

Figures 6 and 7 show the value of reliability and MTBF for all five subsystems. The value of reliability is highest for fuel supply subsystems and lowest for self-starting subsystems. The lowest MTBF value is of Self-starting subsystem which is 1186.47 h and the highest is of the air supply subsystem which is 1525.50.

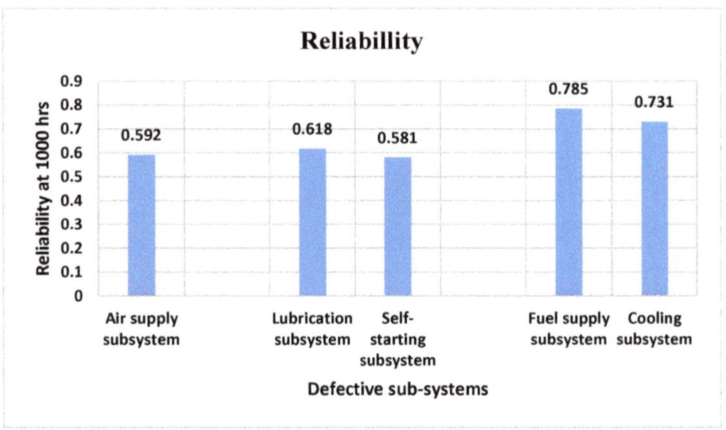

Figure 6. Reliability for five subsystems.

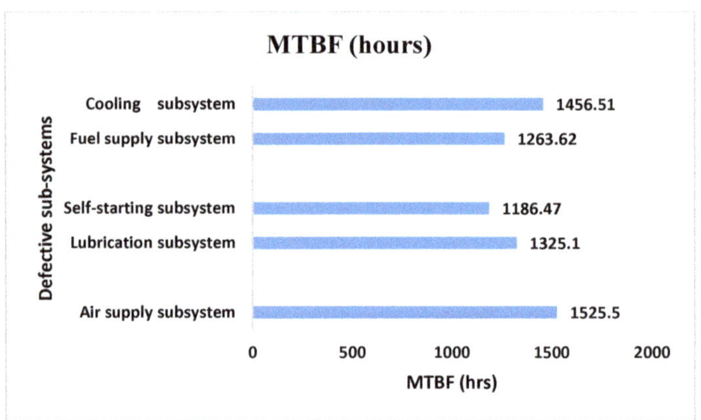

Figure 7. MTBF for five subsystems.

5. Conclusions

The problem associated with reliability analysis using a very small number of failure data has been solved in this paper. This research work provides a guide which can be used for reliability analysis of any repairable system and its subsystems when a very small sample size of failure data is available. Using CBH, the consistency of failure data of the system can be checked. Further to support the CBH test results, using meta-analysis the level of heterogeneity can be found out for systems and subsystems. After passing the above two tests, the very small failure data can be pooled. The pooled TBF data can be effectively further tested for trend analysis.

By using the MLE method and PLP model reliability analysis can be carried out. The values of reliability and MTBF are estimated. The value of MTBF can be utilized in scheduling the maintenance of the engine. Additionally, the subsystem with the lowest reliability, i.e., self-starting subsystems should be taken extra care of during maintenance.

The test values for all five subsystems falls between the lower (0.10) and upper value (5.99). Hence, the TBF data for each subsystem of three engines pass CBH suggesting pooling of TBF data for further analysis.

The zero value for four subsystems and the low level of heterogeneity value for one subsystem indicated that there is no variability among the failure data of three engines for all the five subsystems. It suggests that all the samples came from the same underlying distribution, thereby supporting the result of CBH test which allows the pooling of failure data of three engines for each subsystem.

The failure data of all five subsystems of the engine showed consistency by confirming the CBH test. Additionally, meta-analysis test supports the result of the CBH test which allows the pooling of failure data of three engines.

The trend is seen for self-starting, fuel supply and cooling subsystems. No serial correlation is found for all five subsystems and thus, self-starting, fuel supply and cooling subsystems do not follow *iid* characteristics whereas the air supply and lubrication subsystem follows it.

The value of reliability is highest for fuel supply subsystems and lowest for self-starting subsystems. The lowest MTBF value is of the Self-starting subsystem which is 1186.47 h, and the highest is of the air supply subsystem which is 1525.50 h.

Due to reliability analysis and a reliability-based maintenance schedule, the downtime and catastrophic failure of dumpers can be reduced.

Author Contributions: Conceptualization, B.K.D., A.K.M., S.C., S.S., F.A., J.M. methodology, B.K.D., A.K.M., S.C., S.S.; formal analysis, B.K.D., A.K.M., S.C., S.S., J.M.; investigation, B.K.D., A.K.M., S.C., S.S., F.A., J.M.; resources, B.K.D., A.K.M., S.C., S.S.; writing—original draft preparation, B.K.D., A.K.M., S.C., S.S.; writing—review and editing, B.K.D., A.K.M., S.C., S.S., F.A., J.M.; supervision, B.K.D., A.K.M., S.C., S.S.; funding acquisition, S.S., J.M. All authors have read and agreed to the published version of the manuscript.

Funding: The authors are grateful to FCT—Fundação para a Ciência e Tecnologia who financially supported this work through the RD Units Project Scope: UIDP/04077/2020 and UIDB/04077/2020.

Institutional Review Board Statement: Not applicable.

Informed Consent Statement: Not applicable.

Data Availability Statement: The data presented in this study are available on request from the corresponding author.

Conflicts of Interest: The authors declare no conflict of interest.

Appendix A

Figure A1. Dumper engine under study.

Table A1. Reliability and MTBF for all five subsystems.

S.No.	Defective Sub-System	Reliability at 1000 h	MTBF (hours)
1	Air supply subsystem	0.592	1525.50
2	Lubrication subsystem	0.618	1325.10
3	Self-starting subsystem	0.581	1186.47
4	Fuel supply subsystem	0.785	1263.62
5	Cooling subsystem	0.731	1456.51

References

1. Crow, L.H. Reliability analysis for complex repairable systems. In *Reliability and Biometry*; Proschan, F., Serfling, R.J., Eds.; SIAM: Philadelphia, PA, USA, 1974; pp. 379–410.
2. Standard, M. Military-Handbook-189. In *Reliability Growth Management*; 1981.
3. Littlewood, B.; Ascher, H.; Feingold, H. Repairable Systems Reliability: Modelling, Inference, Misconceptions and their Causes. *J. R. Stat. Soc. Ser. A* **1985**, *148*, 165. [CrossRef]
4. Ascher, H.; Feingold, H. *Repairable System Reliability*; Marcel Dekker: New York, NY, USA, 1984.
5. Jones, B.; Bethea, R.M.; Duran, B.S.; Boullion, T.L. Statistical Methods for Engineers and Scientists. *J. R. Stat. Soc. Ser. A* **1985**, *148*, 390. [CrossRef]
6. Crow, L.H. Evaluating the Reliability of Repairable Systems. In Proceedings of the Annual Reliability and Maintainability Symposium, Los Angeles, CA, USA, 23–25 January 1990; pp. 275–279.
7. Kumar, U.; Klefsjö, B. Reliability analysis of hydraulic systems of LHD machines using the power law process model. *Reliab. Eng. Syst. Saf.* **1992**, *35*, 217–224. [CrossRef]
8. Ebeling, C.E. *An Introduction to Reliability and Maintainability Engineering*; Mcgra, T., Ed.; Tata McGraw-Hill Publishing Co., Ltd.: New York, NY, USA, 2000.
9. Higgins, J.; Thompson, S.G.; Deeks, J.; Altman, D. Measuring inconsistency in meta-analyses. *BMJ* **2003**, *327*, 557–560. [CrossRef] [PubMed]
10. Olwell, D.H.; Sorell, A. Warranty Calculations for Missiles with only Current-Status Data, Using Bayesian Methods. In Proceedings of the Annual Reliability and Maintainability Symposium, Philadelphia, PA, USA, 22–25 January 2001; pp. 144–148.
11. Crombie, I.; Davies, H.T. *What Is Meta-Analysis?* 2nd ed.; Hayward Medical Publishing: Evidence Based Medicine: Little Bookham, UK, 2009; pp. 1–8.
12. Neyeloff, J.L.; Fuchs, S.C.; Moreira, L.B. Meta-analyses and Forest plots using a microsoft excel spreadsheet: Step-by-step guide focusing on descriptive data analysis. *BMC Res. Notes* **2012**, *5*, 52. [CrossRef] [PubMed]
13. Mayer, R.M.; Larean, J.P.; Crowford, S.L.; Anderson, M.T. *Review of Literature on Model Assisted Probability Detection*; PNNL-23714.Richland; Pacific Northwest National Laboratory: Richland, WA, USA, 2014.
14. Wang, G.; Zhou, Y.; Zhang, Q.; Wang, S. The Small Sample Failure Distribution Model of Diesel Engine Component Parts Using the FMECA Approach. *Int. J. Modeling Optim.* **2017**, *7*, 19–23.
15. Qin, L.; Shen, X.; Chen, X.; Gao, P. Reliability Assessment of Bearings Based on Performance Degradation Values Under Small Samples. *J. Mech. Eng.* **2017**, *63*, 248–254. [CrossRef]
16. Ahn, E.J.; Kang, H. Introduction to Systematic Review and Meta-Analysis. *Korean J. Anaesthesiol.* **2018**, *71*, 103–112. [CrossRef] [PubMed]
17. Dai, W.; Chi, Y.; Lu, Z.; Wang, M.; Zhao, Y. Research on Reliability Assessment of Mechanical Equipment Based on the Performance-Feature Model. *Appl. Sci.* **2018**, *8*, 1619. [CrossRef]
18. Si, W.; Yang, Q.; Monplaisir, L.; Chen, Y. Reliability Analysis of Repairable Systems with Incomplete Failure Time Data. *IEEE Trans. Reliab.* **2018**, *67*, 1043–1059. [CrossRef]
19. Xintao, X.; Zhen, C.; Lijun, Z.; Xiaowei, Y. Estimation on Reliability Models of Bearing Failure Data. *Math. Probl. Eng.* **2018**, *2018*, 1–21. [CrossRef]
20. Garcia, F.V.; Rubio, J.A.P.; Grau, J.H.; Hernandez, D.A. Improvements of a Failure Database for Marine Diesel Engines Using the RCM and Simulations. *Energies* **2019**, *13*, 104. [CrossRef]
21. Zhang, L.; Jin, G.; You, Y. Reliability Assessment for Very Few Failure Data and Weibull Distribution. *Math. Probl. Eng.* **2019**, *2019*, 1–9. [CrossRef]
22. Darmanto, S.; Atmanto, I.S.; Permana, B.H. Reliability Analysis of Diesel Engine at LNG Plant Using Counting Process. *J. Vocat. Stud. Appl. Res.* **2019**, *1*, 1–4.
23. Liu, B.; Xie, L. An Improved Structural Reliability Analysis Method Based on Local Approximation and Parallelization. *Mathematics* **2020**, *8*, 209. [CrossRef]
24. Laumann, K. Criteria for qualitative methods in human reliability analysis. *Reliab. Eng. Syst. Saf.* **2020**, *194*, 106198. [CrossRef]
25. He, Y. Research on Reliability Evaluation Method of Sanitation Vehicles under Small Sample Data. In Proceedings of the 2020 6th International Conference on Advances in Energy Resources and Environment Engineering, Chongqing, China, 20–22 November 2020; In IOP Conference Series: Earth and Environmental Science. IOP Publishing: England and Wales, UK, 2021; Volume 647, p. 12171.

26. Agrawal, A.K.; Murthy, V.M.S.R.; Chattopadhyaya, S. Investigations into reliability, maintainability and availability of tunnel boring machine operating in mixed ground condition using Markov chains. *Eng. Fail. Anal.* **2019**, *105*, 477–489. [CrossRef]
27. Agrawal, A.; Chattopadhyaya, S.; Murthy, V.; Legutko, S.; Krolczyk, G. A Novel Method of Laser Coating Process on Worn-Out Cutter Rings of Tunnel Boring Machine for Eco-Friendly Reuse. *Symmetry* **2020**, *12*, 471. [CrossRef]
28. Bisht, V.; Kumar, S.; Agrawal, A.K.; Siddiqui, M.A.H.; Chattopadhyaya, S. Prediction of breakdown hours of load haul dumper by long short term memory network. In *IOP Conference Series: Earth and Environmental Science*; IOP Publishing: England and Wales, UK, 2021; Volume 1104, p. 12006.
29. Vashishta, S.; Kumar, S.; Agrawal, A.K.; Siddiqui, M.A.H.; Chattopadhyaya, S. Reliability and Maintainability Analysis of LHD Loader at Saoner Mines, Nagpur, India. In *IOP Conference Series: Earth and Environmental Science*; IOP Publishing: England and Wales, UK, 2019; Volume 691, p. 12013.
30. Kumar, G.; Agrawal, A.K.; Chattopadhyaya, S.; Siddiqui, M.A.H. Reliability and maintainability analysis of universal drill machine at Saoner mines, Nagpur, India. In *IOP Conference Series: Earth and Environmental Science*; IOP Publishing: England and Wales, UK, 2021; Volume 691, p. 12014.
31. Turk, L.I.A. Testing the Performance of the Power Law Process Model Considering the Use of Regression Estimation Approach. *Int. J. Softw. Eng. Appl.* **2014**, *5*, 35–46.
32. Greenland, S.; O'Rourke, K. Meta-Analysis. In *Modern Epidemiology*, 3rd ed.; Rothman, K.J., Greenland, S., Lash, T.L., Eds.; Wolters Kluwer Health/Lippincott Williams and Wilkins: Philadelphia, PA, USA, 2012; p. 652.
33. Rossi, R.J. *Mathematical Statistics: An Introduction to Likelihood Based Inference*; John Wiley & Sons: Malden, MA, USA, 2018; p. 227, ISBN 978-1-118-77104-4.

Article

Risk Management for the Reliability of Robotic Assisted Treatment of Non-resectable Liver Tumors

Doina Pisla [1], Vaida Calin [1,*], Iosif Birlescu [1,*], Nadim Al Hajjar [2], Bogdan Gherman [1], Corina Radu [2] and Nicolae Plitea [1]

[1]   CESTER, Technical University of Cluj-Napoca, 400114 Cluj-Napoca, Romania;
    doina.pisla@mep.utcluj.ro (D.P.); bogdan.gherman@mep.utcluj.ro (B.G.); nicolae.plitea@mep.utcluj.ro (N.P.)
[2]   Iuliu Hatieganu University of Medicine and Pharmacy of Cluj-Napoca, 400012 Cluj-Napoca, Romania;
    nadim.alhajjar@umfcluj.ro (N.A.H.); corina.radu@umfcluj.ro (C.R.)
*   Correspondence: calin.vaida@mep.utcluj.ro (V.C.); iosif.birlescu@mep.utcluj.ro (I.B.);
    Tel.: +40-264-401-765 (V.C. & I.B.)

Received: 26 November 2019; Accepted: 17 December 2019; Published: 19 December 2019

Abstract: Hepatic cancers represent an important worldwide health issue where surgery alone in most cases is not a feasible therapeutic solution since most tumors are non-resectable. Despite targeted therapies showing positive results in other areas of cancer treatment, in the case of liver tumors, no low-risk delivery methods have been identified. Based on a risk assessment approach, this paper proposes a technical solution in the form of a robotic system capable of achieving a reliable delivery method for targeted treatment, focusing on the patient safety and therapeutic efficiency. The design of the robotic system starts from the definition of the design constraints with respect to the medical protocol. An analytical hierarchy process is used to prioritize the data correlated with the technical characteristics of a new robotic system, aiming to minimize risks associated with the medical procedure. In a four-phase quality function deployment, the technical solution is evaluated with respect to the quality characteristics, functions, subsystems, and components aiming to achieve a safe and reliable system with high therapeutic efficiency. The results lead to the concept of HeRo, a parallel robotic system for the reliable targeted treatment of non-resectable liver tumors.

Keywords: risk management; safety assurance; medical parallel robot; robotic assisted cancer treatment

---

1. Introduction

One of the most lethal forms of cancer in the world is hepatocellular carcinoma (HCC), which represents the most common primary malignant liver tumor. HCC represents more than 5% of all cancer localizations on a world scale, being the fifth most common malignant localization in males and the ninth in females [1]. The number of HCC deaths per year is almost equal to the incidence with 0.93/1 lethality index. The best curative options, involving the complete removal of the malignant cells from the body, are surgical resection of the tumor and liver transplant [2]. However, only 20% of patients with HCC can be subjected to one of these procedures due to various reasons related to the disease (localization, size, vascularization, number of tumors, spread) or patient (general state, other associated diseases, i.e., cirrhosis) [1]. When resection and liver transplant are not viable options, there are multiple locoregional treatments (curative or palliative ones) which have been investigated in medical centers all over the world, including TACE (trans-arterial chemoembolization), HDR (high-dosage radiation) brachytherapy, RFA (radiofrequency ablation), and delivery of chemotherapeutic drugs injected directly inside the tumor [3]. According to a team of clinicians from the "Iuliu Hatieganu" University of Medicine and Pharmacy in Cluj-Napoca, the palliative locoregional treatments may increase the survival time of the patient by "down-staging" the disease (and improving the patient condition), allowing them to become liver transplant candidates. With all the recent advancements in nuclear medicine and medicinal chemistry, these targeted approaches have a high potential to increase the patient survivability and quality of life, but they are hindered by a common constraint: the placement accuracy of the therapy delivery tool is critical, with a maximum acceptable positioning error of 2 mm which in most cases cannot be achieved manually [4]. According to [5] the combined positioning error, assuming 1° angular positioning error and 1 mm Cartesian positioning error from the zero compensation position (values that may be exceeded when the needle is inserted manually, leading to worse outcomes), of a needle inserted in the patient body (with zero needle deflection) is approximately 2.5 mm for 50 mm depth and approximately 5 mm at 200 mm depth.

An efficient solution that can overcome the human limitations in such techniques is represented by the use of robotic systems which, based on careful preplanning of the procedure, enable tumor targeting with increased accuracy [6]. The most promising results have been achieved in the treatment of tumors located in the prostate [7], lung [8], or breast [9], but all studies reflected that the procedure difficulty increases as the tumors are located deeper in the body (as the access path is longer and needle orientation errors, especially, multiply) and is more complicated for organs with complex vascular structure. In fact, the therapeutic feasibility of HDR brachytherapy for HCC treatment is not yet proven, mostly due to the lack of accurate technical solutions capable of delivering the therapeutic agent in a reliable manner which facilitates patient safety and decreases the overall risk of the medical procedure.

When considering the development of a technical solution designed to facilitate targeted treatment of HCC, besides accuracy, other characteristics should be considered which are strongly related to the patient safety. The technical solution (e.g., a robotic system) must comply with various technical and medical requirements (e.g., the robot must be able to properly manipulate the medical instruments with minimum risk of causing harm; some critical components must be sterilized and, therefore, modularity is required, etc.). The accuracy and patient safety, considering the robotic assisted medical procedure, should also be enhanced by using a real-time imaging technique. Since the proposed procedure is minimally invasive, a pneumoperitoneum is created beforehand, and most likely, the liver position will change (relative to the position defined preoperatively based on the volumetric data). For the proposed technical solution, another robot will be used to guide an intra-operatory ultrasound probe (I-US) to visualize the tumor and the needle insertion (a robotic solution is desired to have an optimum positioning control between the I-US imaging plane and the needle trajectory). The end result (in the form of a modular robotic system) must be validated by showing that the benefits (of the technical solution) outweigh the risk, i.e., by using the robotic system for the medical procedure, the risk of patient harm must be minimized, whereas the therapeutic efficiency must be maximized. Therefore, to design a robotic system that complies with the safety and therapeutic efficiency needs, the authors identified the risks associated with the medical procedure and designed the robotic system by using engineering tools such as analytical hierarchy process (AHP) and quality function deployment (QFD).

Due to the promising outcomes in some areas of medial robotics (especially in percutaneous procedures [4,6–9]) some medical experts (from the "Iuliu Hatieganu" University of Medicine and Pharmacy in Cluj-Napoca) believe that future advances in surgery and oncology may come also from the development of technical solutions that help the clinicians in performing the therapies. On the one hand, present advances in nuclear medicine and medicinal chemistry may provide increasingly better therapeutic agents, but on the other hand, technical solutions (e.g., robotic systems) may provide better ways to deliver the therapeutic agents. In fact, there is a tendency of growth in the robotics market towards non-industrial robots, which will attract more than 160 billion USD by 2021 [10]. One major area of non-industrial robotics is healthcare robotics according to The European Commission [11] through the Eurobotics AIBSL forum, which identified three major areas of interest where medical robotics would play an important role and included them in the strategic development agenda for the next five years. The first area is clinical robotics, defined as robotic systems that interact directly with the patient supporting the "care" and "cure" processes. An important category of clinical robots is represented by the surgical ones. Depending on the specifics of the application, the requirements for surgical robots are expressed in terms of safety involving risk analysis and essential performances.

The risk and effectiveness should be rigorously specified because a robotic device is a machine that can hurt the operators and the patient, being in close contact with the latter.

This paper is structured as follows: Section 2 presents the risk management strategy for the development of the technical solution. Moreover, by evaluating the associated risks of the procedure, the medical and technical requirements for the proposed technical solution are defined. Section 3 presents an analytical hierarchy process which is introduced subsequently in a four-phase quality function deployment which, in turn, leads to the HeRo conceptual design, showing also the advantages of the design with respect to risk reduction. Section 4 presents the discussion of the obtained results, and finally, Section 5 presents the conclusions and further work.

2. Materials and Methods

For the development of the technical solution (a reliable robotic system to facilitate precision and safety) which may enable the use of HDR brachytherapy in HCC treatment, the ISO 14,971 [12] standard for risk management in healthcare devices was followed. Figure 1 shows a flow chart which describes the risk management for the early stages of device (or technical solution) development (before prototyping).

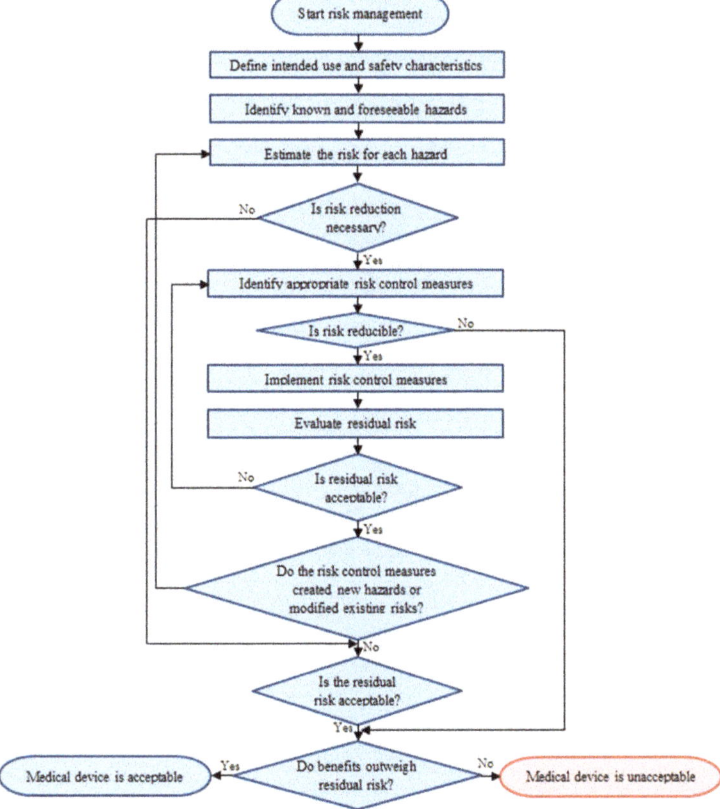

Figure 1. Risk management flowchart.

The general strategy detailed in Figure 1 is to establish the limitations of the medical procedure (as if it was performed manually) by analyzing the risks and to proceed to determine whether the risks can be diminished or not by developing a technical solution design to facilitate the medical procedure. The process *starts* with the risk assessment, which has three main steps. The first step refers to defining the intended use and the safety characteristics of the medical procedure, which in turn helps to identify all the known and foreseeable hazards (which is the second step in the risk assessment). At this point, it is important to distinguish between risks associated with the HDR brachytherapy procedure (for the HCC therapy) and other health-related risks (e.g., diabetes imposes risks for the surgical procedure but should not be considered in this analysis). The third step is the estimation of the risk of each hazard, which serves at the ground base for the risk reduction in the risk management process. Since HDR brachytherapy is not a feasible therapeutic method for HCC treatment (despite its positive results in treating cancers in other body areas), a strong assumption is made which states that by reducing the current risks associated with the procedure, HDR brachytherapy may become a viable therapeutic tool for HCC; consequently, risk reduction is necessary. Assuming that the risk associated with the medical procedure may be reduced by using a technical solution, the risk reduction process (focused on implementing risk control measures and evaluating thereafter the residual risk) is related to determining the medical and technical characteristics of the emerging technical therapeutic solution. Therefore, a medical protocol for the technical solution (the medical characteristics) and the design constraints (the technical characteristics) must be established. In the later stages of the risk management, the technical solution is evaluated in order to determine if new hazards were introduced or if there exist any modified hazards (hazards which were not eliminated but changed in various aspects). At the end, if the overall residual risks are managed (in acceptable ranges), if the benefits outweigh the residual risks, the risk management process ends with an acceptable medical device as a result.

2.1. Definition of the Medical Task

The first step in the risk management process (see Figure 1) is to define the intended use and safety characteristics, i.e., the medical task. The proposed therapeutic procedure aims to achieve an efficient and reliable treatment (which minimizes the risk) of non-resectable HCC tumors by taking into account all the existing medical and technical constraints, focusing on three targeted treatment options: HDR brachytherapy (using, e.g., a 1.6 mm gauge needle), intratumoral chemotherapy (using, e.g., a 1.6 mm gauge needle), and RFA (using, e.g., a 2 mm gauge needle). All these procedures are performed percutaneously by inserting a specific needle through the skin, on a linear trajectory, inside the tumor. In order to enable safe and accurate needle positioning, the authors propose the use of an intraoperative ultrasound probe that can monitor in real time the needle placement. Thus, the procedure has three main stages:

1.   Preoperative: the patient is investigated using non-invasive molecular imaging techniques that determine the tumor location and characteristics (size, density, proximity to blood vessels), the most efficient treatment option, and possible safe needle trajectories;
2.   Intraoperative: the therapy is performed in the operating room, by surgeons, using a total of three medical instruments [13]: (1) an endoscopic camera, guided manually by a surgeon, that enables the fast transfer of the intraoperative ultrasound (I-US) probe in the targeted area of the liver and the continuous evaluation of the surgical field; (2) the I-US probe, guided by a robotic arm, used to locate the tumor and monitor in real time the needle insertion in the liver parenchyma; (3) the therapeutic needle(s), guided by a second robotic arm used to insert the needle on a linear safe trajectory from outside the body into the live tumor;
3.   Postoperative: the patient evolution is monitored by assessing the procedure results and the patient evolution.

The procedure takes place in an operating room (Figure 2) where the necessary equipment is provided: an operating table adjustable on three axes, a vital signals monitoring system (respiration, cardiac rhythm), and anesthesiology equipment. On the lateral side the laparoscopic tower is positioned, and in its proximity, the ultrasound tower.

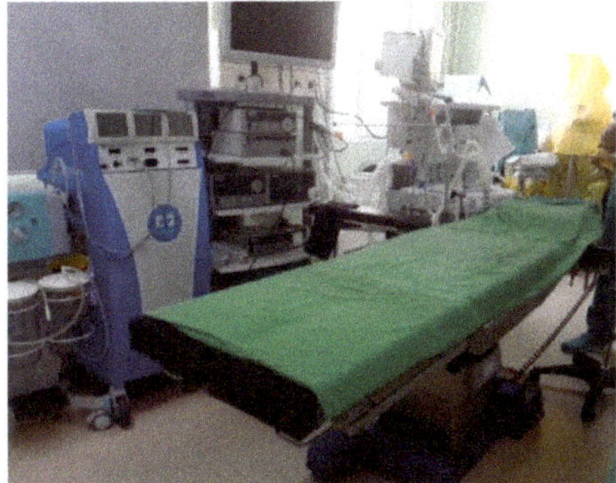

Figure 2. Typical operating room with the necessary medical equipment for the procedure.

2.1.1. Remote Center of Motion Concept for the Guiding of the I-US Probe

In 1995, Russell Taylor [14] introduced the concept of the remote center of motion (RCM), defining it as the point of entrance in the abdominal cavity, a fixed point which should not be displaced during the medical task. In minimally invasive surgery (MIS), it is used for instrument insertion into the body, and for our procedure this concept is used for the manipulation of the I-US probe. Based on Figure 3, it can be stated that with respect to point B (RCM), the instrument can achieve four independent motions:

- In spherical coordinates: two rotations that would position the point E on a surface of a "sphere" with radius BE, one translation along the A-B-E segment or the longitudinal axis of the instrument, and one rotation around the same axis;
- In Cartesian coordinates: three translations which enable the positioning of the point E in space with respect to the point B and one rotation around the longitudinal axis of the instrument.

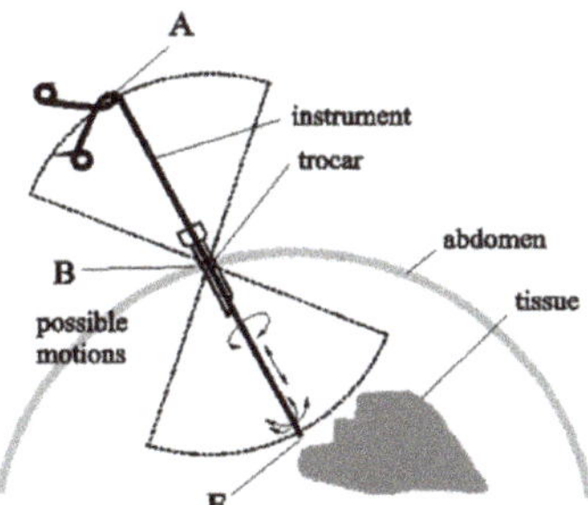

Figure 3. The remote center of motion (RCM) concept [14].

When it comes to technical solutions based on robotic systems, the RCM can be achieved in two ways: by using the tissue around the insertion point as guidance or by mechanically constraining that point in space. Additionally, a third category of RCM can be added that combines the simplicity of the first with the capabilities of the second, namely, architecturally constrained. The first approach imposes a simple mechanical construction at the anchor point of the instrument to the robotic guiding device (in the form of a 2-DOF, degrees of freedom, passive Cardan joint), but its usability is limited to the manipulation of instruments that do not come in intimate contact with the internal tissues (e.g., a laparoscopic camera). The second approach, where the RCM is mechanically constrained, imposes the use of an active 2-DOF joint at the anchor point of the instrument. This, in turn, determines a much more complicated construction of the robotic device but it enables the manipulation of instruments that come into direct contact with the internal tissues (e.g., the surgical instruments). The third approach imposes the development of the robotic device in such a way that without the addition of any supplementary joint it would keep the location of the RCM fixed. Even though such an approach will impose the use of a positioning mechanism that would adjust the position of the RCM in space, an architectural constraint mechanism can be limited to only 2 DOF. This approach also has an increased safety aspect because after the insertion of the instrument inside the patient, the positioning mechanism will remain fixed, reducing to a minimum the pressure exerted on the tissue walls and eliminating the risk of unwanted motions that could harm the patient.

2.2. Definition of Possible Hazards and Associated Risks

Steps 2 and 3 in the risk management process (see Figure 1) are intended to evaluate all the possible hazards and estimate their occurrence risk. The possible hazards were determined (and detailed in Table 1) with the help of nine clinicians from the "Iuliu Hatieganu" University of Medicine and Pharmacy, Cluj-Napoca. Furthermore, each clinician filled in a questionnaire (see Appendix A) regarding the severity and occurrence probability of each hazard, and the mean values are also presented in Table 1 (in Section 3, the risks associated with the robotic assisted procedure are reevaluated based on the proposed technical solution). The severity and probability scales were derived from a typical risk assessment matrix and the overall score definition was chosen to allow a simple yet comprehensive evaluation of the risks (which was also used in [15]).

Table 1. Identified hazards and associated risks for the targeted therapy of hepatocellular carcinoma (HCC).

No.	Hazard	Associated Risk	Severity *	Probability **	Overall Score ***
H1	Needle positioning error beyond acceptable value	Linear positioning error (the error remains constant at any depth)	85	60	150
		Orientation positioning error (the error multiplies with the depth)	90	95	195
		Insertion depth error	80	50	130
H2	I-US (intraoperatory ultrasound) probe positioning error	I-US probe positioning does not respect the predefined trajectory	50	50	100
		I-US probe is pressing too hard on the liver	90	70	160
		I-US cannot reach the targeted area	40	40	80
H3	I-US entry port (RCM) is not preserved	The I-US probe applies pressure on the abdominal wall	100	80	180
		The I-US probe movements are different than intended	90	80	170
H4	The targeted area cannot be reached	Incapability of performing the procedure (the instruments cannot reach the specific liver area)	100	50	150
		Incapability of achieving the necessary orientation for the I-US probe	60	45	105
		Incapability of achieving the necessary orientation for the therapeutic needle	95	45	140
H5	The operating table has to be repositioned to ensure better access to the targeted area of the liver	The relative patient–robot position is changed	100	60	160
H6	The patient state deteriorates during the procedure	The patient cannot be resuscitated in due time	100	40	140
H7	The needle cannot be viewed inside the liver parenchyma	The needle insertion cannot be monitored	90	90	180
H8	Improper sterilization	Infection risk	80	10	90

* Severity scale: (0–29) minor, (30–89) moderate, (90–99) serious, (100) catastrophic; ** Probability scale: (0–29) remote, (30–69) unlikely, (70–99) likely, (100) very Likely; *** Overall score: Severity + Probability.

2.3. Identifying Risk Control Measures for the Treatment of Non-resectable HCC Tumors

The necessity of a technical solution that enables targeted treatments such as HDR brachytherapy for HCC in a safe manner emerges from the relatively high probabilities of hazards (see Table 1). A parallel robotic system is proposed as the technical solution, and to proceed further into its design, an integrated medical protocol was developed with the aim to achieve the targeted treatment of non-resectable tumors performed in the operating room as a minimally invasive procedure. This also enables the definition of the technical characteristics of the robotic system and the specific motions for the guidance of instruments involved.

The robotic assisted *medical protocol* for the MIS procedure is as follows:

A. **Diagnosis and preplanning** stage:

1. **Diagnostic:** This step refers to the patient's preliminary analysis where it is determined whether or not a hepatic tumor is non-resectable (the inclusion criterion). This step may require various diagnosis methods such as medical imaging, biopsies, etc.;

2. **Establishing the optimal treatment:** The optimal treatment approach is established, which may be targeted brachytherapy, targeted chemotherapeutic agent delivery, or RFA;

3. **Trajectory definition:** The coordinate points of interest (e.g., insertion and target points) are established relative to an external system of markers (fixed on bone mark). The relative positioning of the liver and the markers will not change once the carbon dioxide is insufflated. Based on previously obtained imaging data, the I-US probe insertion coordinates are determined together with the insertion coordinates of the needles, the needle trajectories, and the target points. *Note:* Due to possible displacement of the liver, most likely from the pneumoperitoneum described in point B.2. below (the difference in position from the volumetric information and the position in the real-time procedure), in most of the cases the predefined trajectories are only an idealization (some corrections are required during the actual medical procedure).

B. **Procedure Preparation** stage:

1. **Patient and robotic system registration:** The patient is positioned on the operating table, the markers are identified, and the robotic system is fixed such that the trajectories of interest are in the central zone of its workspace. The mathematical correlation between the robot coordinate system and the patient coordinate system is determined to achieve patient–robot registration. *Note:* The robot–patient registration serves only as an initial guideline; during the procedure, the clinician will actively search the tumor using the I-US probe.

2. **Pneumoperitoneum creation:** The patient is prepared for the intervention by insufflating CO_2 up to a certain pressure that is maintained constant through the whole procedure. The pressure will create an empty volume inside the abdominal cavity, enabling instrument manipulation inside the patient body. The first incision is created to allow the insertion of a 10 mm trocar for the insertion of the laparoscopic camera (manually guided) which will monitor the whole procedure.

C. **The robotic assisted procedure:**

1. **Ultrasound probe insertion and tumor location:** After the patient is prepared and the laparoscopic camera is introduced, the robotic system guides the tip of the I-US probe such that it touches the patient's skin. At this point, the surgeons fix a 10 mm trocar, this being registered as the RCM point for the I-US probe guiding module, and the US probe is inserted until it makes contact with the hepatic parenchyma. After contact, the clinician actively searches and locates the tumor, and by knowing the approach plane of the needle

trajectory and the tumor depth, the US probe is fixed in the same plane to enable needle monitoring during its insertion into the hepatic tissue.

2. **Therapeutic needle insertion:** The needle guiding module positions the needle on the defined trajectory (confirmed with possible minor corrections after tumor location with the I-US probe). The needle is then positioned on the desired trajectory above the patient's skin, but as close as possible to it. After the trajectory validation, the robot will remain in the current position and the needle will be inserted using the insertion module, monitoring in real time the resistive force to avoid needle deflection;

3. **Therapeutic needle insertion into the hepatic parenchyma:** The needle is inserted continuously up to the proximity of the hepatic parenchyma. When the needle touches it, the trajectory is once more validated and the needle insertion into the hepatic tissue is initiated. From this point on, the needle can be seen by means of the US probe that confirms reaching of the target point.

   - **Multiple needle insertion:** When the treatment requires the use of multiple needles, those are inserted by following Steps C.2 and C.3 until all the required therapeutic needles are inserted. The insertion order is established such that the I-US probe may be repositioned without affecting the already inserted needles. *Note:* Since multiple needle insertions increase the risk of hepatorrhagia, further research (in vivo) is intended to determine the maximum number of needles allowed.

4. **Delivery of the treatment:** After all needles are inserted, their position is confirmed by the medical personnel and the treatment is delivered. For brachytherapy treatment, the patient should be transferred to a controlled room specialized for brachytherapy treatment;

5. **Needle extraction and operating field check:** After completing the procedure, the needles are extracted either by the robot or manually (depending on the procedure). After extraction, the operating field is checked for eventual hemorrhages;

6. **Ultrasound probe extraction:** After validating the position of all needles, the US probe is retracted, followed by extracting the laparoscopic camera and suturing the incision points of the two trocars.

The analysis of the possible hazards and their estimated occurrence, together with the procedure protocol, led to the definition of the main design constraints for the new robotic system which should be accounted for and implemented as risk control measures:

i. The needle guiding module must operate with **high precision**, due to the fact that the robot must insert therapeutic needles within the tumors (a maximum error of 2 mm is accepted [3,4]);

ii. The robot workspace should have no singularities (robot configurations where the mechanism is not well behaved, losing or gaining degrees of freedom). A singularity-free workspace may be achieved through mechanism design;

iii. Since the medical procedure requires the manipulation of two medical tools (therapeutic needles and I-US) in a distinct manner, the robotic system must have two independent modules. The first module is designed to insert and manipulate the **US probe**, whereas the **second module** is designed to **insert the therapeutic needles**. The relative positioning of the two guiding modules is always known; therefore, while the clinician actively locates the tumor, the needle will also adjust its position such that the needle targets the tumor;

iv. Because the procedure is to be minimally invasive, the **US probe guiding module** needs to work based on the **RCM concept;**

v. Based on the liver size and the multiple trajectories that need to be used to target tumors located in all areas of the liver, the robot should have workspace of $500(X) \times 300(Y) \times 500(Z)$ and orientation capability of $90°$ around Y and $150°$ around Z;

vi.　A low weight is desired since the robot needs to be mounted on the lateral sides of the operating table;

vii.　The robotic system mounting needs to be adapted to the surgical environment as the procedure takes place in an operating room;

viii.　Due to the intimate interaction between the robot and the patient, **safety** is a critical aspect, including also the required ability to quickly remove the instruments from the patient's body;

ix.　The robotic system should occupy a volume as small as possible in the operating room.

2.4. Implementing the Risk Control Measures for the New Technical Solution

In order to achieve the design of a safe robotic system for the treatment of non-resectable HCC, a roadmap was developed and is presented in Figure 4. Starting with the *design constraints*, an AHP (analytic hierarchy process) was performed in order to prioritize the importance of the technical characteristics of the robot (with respect to enhancing safety and decreasing the associated risks).

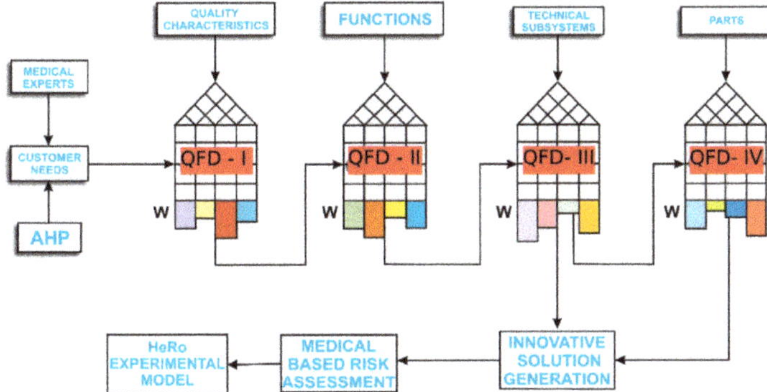

Figure 4. Roadmap for the design of a new robotic system of the treatment of HCC.

In a four-phase QFD, an importance analysis of the measurable quality characteristics, functions, technical subsystems, and individual robot components is performed along with the identification of the unique selling points of the proposed solution. The design methods were implemented using Qualica software [16]. Usually there are three scales for the numerical correlations among the analyzed parameters in the QFD (1, 3, 9). However, the authors chose an extended scale (1, 3, 9, 27, 81) to allow a better "resolution" of the analysis. Furthermore, the correlation number was chosen by clinicians and engineers based on their professional experience: for QFD-I, the clinicians attributed the correlation values based on the previously presented risk assessment (Table 1); for QFD-II, -III, and -IV, the engineers attributed correlation values by closely considering fundamental theoretical aspects (from mechanism science, such as kinematics and singularity analysis) and the predefined robotic assisted medical protocol (see Section 2.4).

3. Results

The medical protocol and the design constraints led to the definition of 10 critical technical characteristics of the robotic system which, through adequate implementation, should lead to a reliable technical solution minimizing the risks associated with the medical procedure. An analytical hierarchy process (AHP) analysis was performed to determine the relative importance and criticality of these characteristics with respect to the medical task and is presented in Figure 5.

AHP imposes the completion of a comparison matrix where each of the technical characteristics is compared in terms of importance with the others based on the specific requirements of the procedure. A five-level comparison scale was used, while for the final sorted results, the most important item was made to be 3 (three) times more important than the least important one. The analysis revealed that the procedure accuracy is the most critical characteristic of the robotic system, with weight 15.9% in the final importance (see Figure 6). This is due to the fact that accuracy is actually the main technical challenge of the medical procedure (i.e., the accuracy was the reason for appealing to robotic systems for the medical procedure in the first place). The patient safety is the second most important characteristic, with a weight of 15.1%. Safety is strongly correlated with the accuracy of the robotic system (since high accuracy reduces the risk associated with the needle penetrating unwanted tissue), the mechanical design of the robotic system (e.g., if the mechanism has no singularities, the safety in operation of the robotic system is in turn increased), and the control reliability (e.g., fine-tuned intelligent control may have faster reaction times than a human when anomalies are encountered during the medical procedure). The stiffness of the structure had weight 14.4%, and it is again correlated with the accuracy and safety (a perfectly stiff robotic device will have no parasite motion in its mechanical joints). The motion repeatability of the robotic system had weight 13%, and it is correlated with the robotic system accuracy and stiffness and, in turn, influences the safety in operation. The workspace of the robotic system has a total weight of 10%, and it determines the spectrum of insertion trajectories and RCM manipulation of the medical tools. All other factors (dimensions, number of components, etc.) have less than 10% weight (individually); therefore, in simple terms, the robot design should focus more on attaining the imposed accuracy level and safety rather than reducing its weight to a minimum or having a high degree of universality.

How Important is the Left Item (Row) as Compared to the Top Item (Column) ?

Group:	Top Level ITEMS	1 Minimum number of components	2 Patient safety	3 Workspace	4 Accuracy	5 Repeatability	6 Stiffness	7 Weight	8 Dimensions	9 Modularization	10 Degree of universality	Importance in group
	1 Minimum number of components		2	½	½	½	½	○	-	½	½	7.5%
	2 Patient safety			2	○	○	+	2	2	2	2	13.5%
	3 Workspace				-	-	-	+	+	+	2	10.0%
	4 Accuracy					+	○	2	2	2	2	14.1%
Input	5 Repeatability						○	2	2	2	2	13.0%
	6 Stiffness							2	2	2	2	13.1%
	7 Weight								-	+	○	6.8%
	8 Dimensions									+	+	8.0%
	9 Modularization										+	7.2%
	10 Degree of universality											6.7%

AHP Toplevel Matrix legend: 2 = 2,00 twice as important; + = 1,50 somewhat more important; ○ = 1,00 equally important; - = 0,67 somewhat less important; ½ = 0,50 half as important.

Figure 5. AHP comparison matrix between the technical characteristics.

The technical characteristics defined for the robotic system were used as input data in the first QFD matrix (Figure 7) and compared with the quantifiable quality characteristics (CTQs) which are imposed by the design constraints. For a reliable solution, the patient safety, robotic system accuracy, and stiffness have a combined weight of above 50% (in the Phase 1 QFD analysis of relative importance). Following the critical correlations between these three characteristics and the design constraints shows that the development of the robotic system should mostly focus on a kinematic design that ensures the RCM architecturally, a kinematic design that has no singularities in the workspace, a technical solution that ensures stiffness such that the targeting needle positioning error is less than 2 mm, and a fail-safe control to facilitate the robotic system operation.

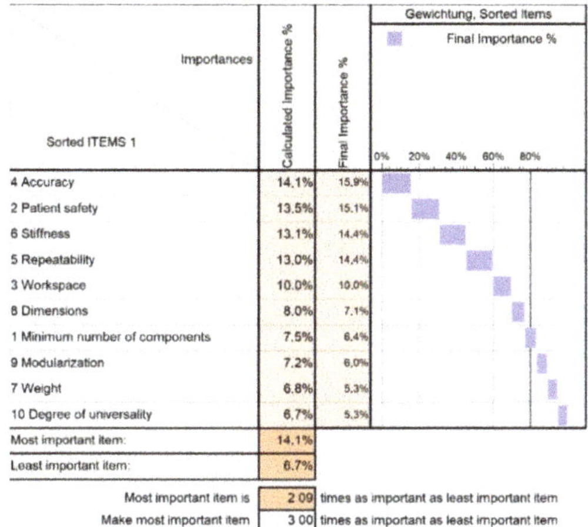

Figure 6. AHP sorted results.

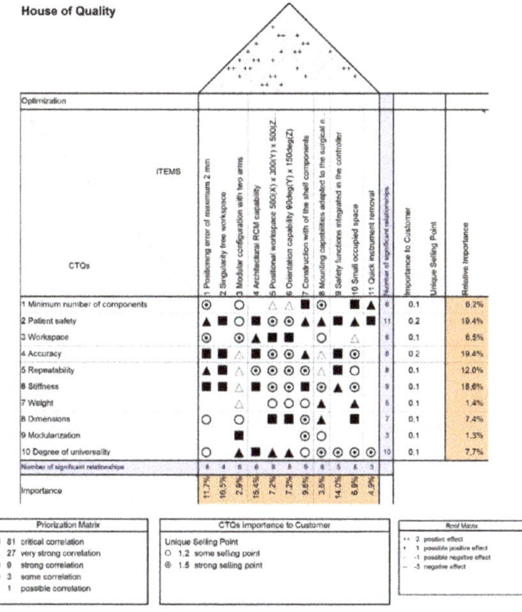

Figure 7. QFD Phase 1—Analysis of technical characteristics with quantifiable quality ones.

The next step in the analysis refers to the evaluation of the functions that the robotic system has to achieve with respect to the quantifiable characteristics defined in the first step. The functions are elaborated in Table 2 to underline their relevance in the robotic system design. The second QFD phase is illustrated in Figure 8, and among the critical functions identified are those associated with the I-US probe motion inside the body (RCM principle) and the needle insertion on linear trajectories (pair of points), followed by decoupled positioning and orientation motions. For the medical procedure, decoupled motions have the advantage that they increase the precision (e.g., while the medical tool is orientated the position of the RCM is fixed, which, in turn, may reduce positioning errors).

Table 2. Defined functions for the robotic system.

Function	Observations
Mounting on the side of the table	Table mounting is preferred (instead of ground mounting) to maintain a constant position (robot/patient) when the table is moved.
I-US probe attachment to the robot Needle module attachment to the robot	The commercially available medical tools (I-US probe and therapeutic needles) must be mounted/demounted to/from the robot.
I-US probe positioning and orientation outside the body Needle positioning and orientation outside the body	Before manipulating the medical tools inside the body during the medical procedure, they must be positioned correctly outside the body for the medical tool insertion.
I-US probe motion inside the body Needle insertion on linear trajectory	The I-US must be manipulated using the RCM concept, whereas the needles are always inserted on linear trajectories.
Real-time tissue resistance force monitoring	Monitoring the tissue resistance reduces the risk of needle bending during the needle insertion.
Automated safety function	Using an intelligent algorithm which takes multiple real-time parameters as inputs increases the safety of the procedure and allows fast reactions for situations where trauma risk increases.
Decoupled positioning and orientation motions	Decoupling the motions may increase the precision since while the orientation is changed the Cartesian robot is stiff (there are fewer overall errors during the medical instrument manipulation).
Emergency instrument detachment	In case of hemorrhage or other complications, the surgeons must be able to intervene promptly. Easy-to-use elements that allow fast instrument demounting are required.
Sterilization	The components that are in close proximity to the patient must be detachable for sterilization purposes.
Preprocedural positional optimization	The robotic system must allow positional corrections since the target points may change due to errors and tissue elasticity.

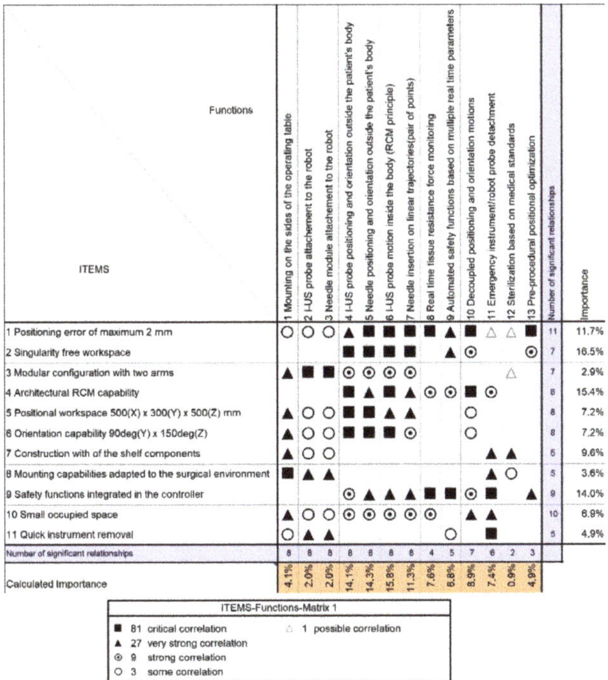

Figure 8. QFD Phase 2—The evaluation of the robotic system functions.

The next step in the analysis refers to the evaluation of the functions that the robotic system has to achieve and the technical subsystems of the robotic system. Table 3 describes the subsystems which compose the robotic system, whereas Figure 9 illustrates the Phase 3 QFD. The technical subsystems which have the most influence in achieving the predefined accuracy and safety aspects are the instrument mounting subsystems and the actuation subsystem, followed closely by the positioning subsystems (for the XYZ positioning and YZ orientation).

Table 3. Defined subsystems for the robotic system.

Subsystems	Observations
Operating table mounting solution	The subsystem fixes the robot base to the lateral side of the operation table such that the two modules operate in symmetry to each other.
Modular XYZ positioning subsystem	Due to the decoupled motion, the subsystem positions the insertion point or the RCM point in Cartesian space.
Modular YZ orientation subsystem	Due to the decoupled motion, the subsystem provides the orientation of the medical tool (I-US probe or therapeutic needle).
Instrument mounting subsystem	The subsystem provides the ability to mount the needle or the I-US probe instrument on the robotic system, increasing modularity.
Actuation subsystem	The subsystem provides the motions (the DOFs) of the robotic system.
Sensorial subsystem	The system that allows fine control of the robotic system.

The final step in the analysis refers to the evaluation of the technical subsystems of the robotic system and its component parts. Among the most important parts of the robotic structure (see Figure 10) resulting from the analysis are the vertical and horizontal pulley boxes, the linear modules, and the circular rails. All these components are closely related with the robot architecture and they affect the accuracy of the robotic system as a whole.

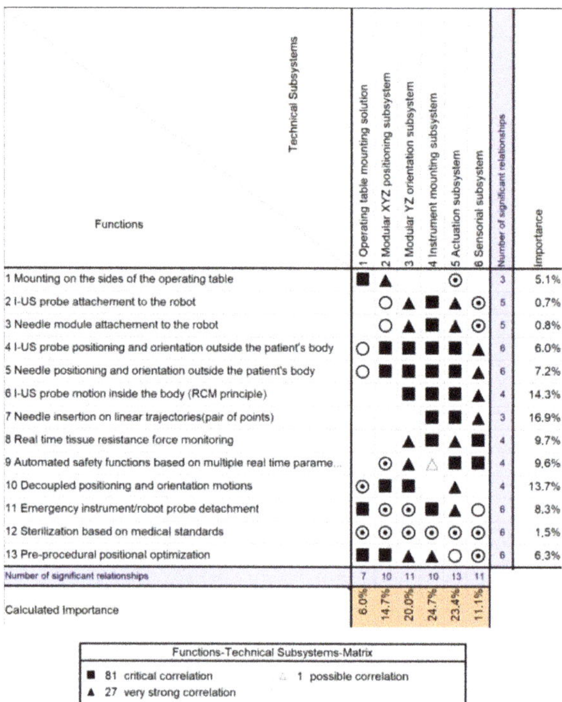

Figure 9. QFD Phase 3—Robot subsystems analysis.

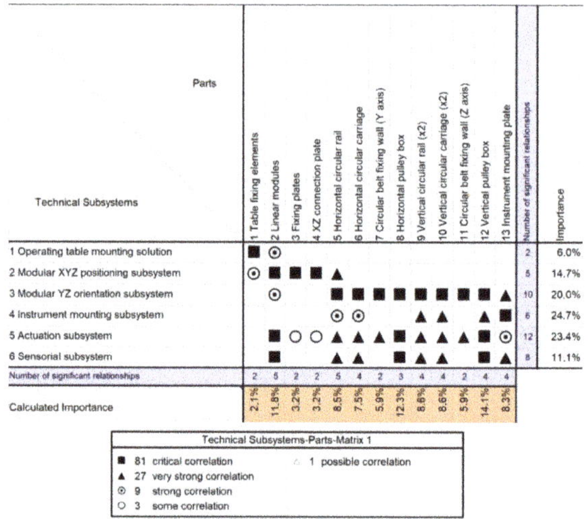

Figure 10. QFD Phase 4—Component analysis.

3.1. HeRo Design Concept

Based on the medical protocol and the design constraints (the main purpose of which was to minimize the risk while increasing the therapeutic efficiency and reliability), a set of QFD analyses was conducted and the design of the HeRo parallel robotic system [17] emerged. Figure 11a illustrates the HeRo concept augmented into the relevant medical environment (in the operating room), whereas Figure 11b illustrates the CAD (computer aided design) of one guiding module. The following components with their technical characteristics are highlighted:

- The parallel robotic system is composed of two modules, Module 1 and Module 2, which are attachable by the operating bed (via modular attaching mechanisms) with each module being capable of guiding the needle instrument or the I-US instrument;
- The main components of each module have a total of 5 DOFs and are composed of two main mechanisms: one **gantry mechanism** with 3 Cartesian DOFs for the insertion point/RCM positioning (having three linear drives) and one **spherical mechanism** with 2 orientation DOFs for the needle trajectory or I-US probe manipulation (having two circular drives);
- The characteristics of the main components are defined by reliable motion with low friction due to the **3 linear guides** and **2 circular guides** (with rails based on bearing balls and carriages) and **5 servo motors** with high rotation motion resolution. Each module allows a quick plug/unplug of the medical tools (needles, I-US probe) to facilitate fast intervention times, for the clinicians, in the case of unexpected patient trauma.

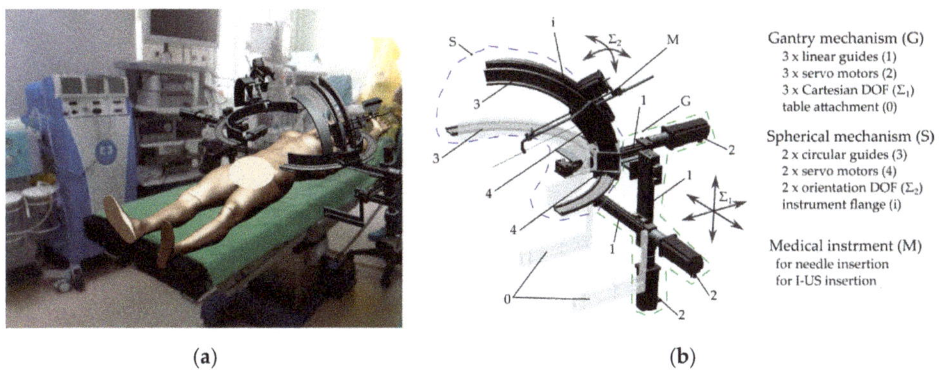

(a)   (b)

Figure 11. The concept of the HeRo parallel robotic system: (**a**) augmented into the medical environment; (**b**) CAD (computer aided design).

3.2. HeRo Concept Residual Risk Estimation

After the HeRo parallel robotic system design, the residual risks were evaluated together with other forms of risk which may be introduced by the robotic system. Table 4 defines the risks as well as their severity and probability.

Table 4. Residual risk estimation.

No.	Associated Risk	Comments	Severity	Probability	Overall Score
RR1	Linear positioning error	The risk is reduced as the position is achieved using actuators which have known positions, and furthermore, the I-US probe monitors in real time the needle insertion	50	50	100
RR2	Orientation positioning error	As the orientation error increases with depth, the severity is higher	70	60	130
RR3	Insertion depth error	The insertion depth is achieved by a dedicated actuator and monitored with the I-US probe	60	10	70
RR4	I-US probe positioning does not respect the predefined trajectory	The I-US probe positioning is controlled by actuators and can be corrected if needed	30	20	50
RR5	I-US probe is pressing too hard on the liver	The I-US module must use a force sensor to ensure the necessary safety	50	10	60
RR6	I-US cannot reach the targeted area	The robot is attached at the beginning of the procedure in a convenient way with respect to the patient	40	5	45
RR7	The I-US probe applies pressure on the abdominal wall	An efficient solution to ensure the RCM has to be used	100	40	140
RR8	The I-US probe movements are different than intended	This can occur when the RCM is improperly defined	90	30	120
RR9	Incapability of performing the procedure (the instruments cannot reach the specific liver area)	The robot is attached at the beginning of the procedure in a convenient way with respect to the patient	50	20	70
RR10	Incapability of achieving the necessary orientation for the I-US probe	Based on the defined workspace, this risk is eliminated	-	-	-
RR11	Incapability of achieving the necessary orientation for the therapeutic needle	Same as above	-	-	-
RR12	The relative patient–robot position is changed	As the robot is attached to the operating table, this risk is eliminated	-	-	-
RR13	The patient cannot be resuscitated in due time	The robot must have a quick removal protocol	100	10	110
RR14	The needle insertion cannot be monitored	The two modules should use the same controller which ensures that the pose of the two instruments is known	90	10	180
RR15	Infection risk	A robot is generally covered with a sterile plastic sheet, so this risk is eliminated	-	-	-

Appl. Sci. **2020**, *10*, 52

Table 4. *Cont.*

No.	Associated Risk	Comments	Severity	Probability	Overall Score
		New identified risks			
NR1	The robot has an electrical malfunction	Fail-safe state-of-the-art automation systems are highly reliable and diminish this risk	100	30	130
NR2	The robot has a mechanical malfunction	Periodical optimal maintenance is facilitated by the robotic system's simple design and implementation	100	50	150
NR3	The robotic system guiding modules may collide	Avoided with an advanced control which is used to correlate the motion between the two modules	80	50	130

4. Discussion

Since targeted HDR brachytherapy and targeted chemotherapeutic agent delivery do not offer a therapeutic solution by today's standards, the implementation of a technical solution (which minimizes the drawbacks of the medical procedure) may be valuable for the medical community. Based on the estimated hazards of the medical task and their occurrence risks (see Table 1), the authors proposed the HeRo concept (which resulted from an AHP and a multiphase QFD) and estimated the residual risks thereafter (see Table 4). One question still stands, which is "do the benefits of the medical procedure outweigh the residual risks?" The authors attempt to answer this from a technical point of view (since there are no relevant medical data about the therapeutic index of HDR brachytherapy and targeted chemotherapy for non-resectable HCC, one can only assume that these therapies would do more good than harm due to their positive results in other areas of the body).

The safety aspect regarding HeRo robotic system exploitation refers to the mechanical aspects of the robot and the control of the robot. Since the HeRo robotic system is composed of one gantry mechanism and one spherical mechanism, it follows that the robotic system has no singularities in the workspace. Moreover, throughout the QFD analysis, the correlation between various characteristics (e.g., accuracy, stiffness, etc.) was emphasized. Consequently, choosing technical solutions that increase accuracy (such as linear and circular guides and high-quality actuation solutions) has a positive impact on the safety as well. The mechanical solutions together with state-of-the-art automation solutions and sensors should lead to the development of a reliable experimental model for the HeRo concept.

The two modules of the HeRo parallel robotic system operate "mirrored" relative to each other, a fact that provides multiple advantages. A variety of insertion points and trajectories may be achieved since the insertion instruments (for the needle and I-US probe) may be mounted on either Module 1 or Module 2 of the robotic system. Furthermore, the simple design with decoupled motions has the advantage that it allows optimal technical maintenance of the robotic system since the robot becomes easy to assemble or disassemble.

From the three ways in which RCM manipulation may be achieved (see Section 2.1.1), following the systematic development of HeRo, the authors chose a mechanical constraint using a spherical mechanism which is located outside the patient body. Consequently, the RCM will be fully constrained, and using this approach should imply lower overall errors (with respect to other technical approaches). The errors for this case are correlated with the radius of the sphere. The advantage of a larger radius is that it increases the orientation precision. The disadvantage is that the mechanism may suffer from component elasticity, vibrations, and mechanical stress. However, the circular guides used for the HeRo design are commercially available (well calibrated and with well-known mechanical characteristics), and the elasticity of the material is insignificant.

As previously stated, assuming a $1°$ angular error and 1 mm linear error results in (combined) errors of 2.5 mm and 5 mm at 50 mm depth and 200 mm depth, respectively (not taking into account any needle deflection) [5], and according to [4], these values are not acceptable for the targeted treatments. Figure 12 illustrates point clouds to show the error distribution for the mentioned values in Cartesian coordinates (where the points within the point clouds are not due to a random distribution but rather computed with incrementally numerical data for the input). To evaluate the accuracy of the gantry mechanism is straightforward since there is a one-to-one dependency between the Cartesian position of the RCM point and the values of the gantry mechanism actuators. Each actuator changes only one coordinate in the Cartesian space; therefore, the error in this case will be mainly due to the quality of the mechanical design of the gantry subsystem. An example of this error propagation is illustrated in Figure 13a, where a 0.7 mm error is assumed (due to high-resolution motors and $2°$ maximum backlash due to the gearheads) for the actuators of the gantry mechanism. These errors are, however, minimized by using the linear guides in the robotic system design (which are well calibrated). The results are different when the errors of the spherical mechanism are considered. Assuming a maximum $\pm 2°$ (angular degrees) error as the actuator backlash (which is appropriated for gearheads nowadays—see, for example, gearheads from Maxon Motors [18]) the associated error (using belts and pulleys for the motion of the carriage on the circular guides, which inherently have no backlash) is approximately 0.34 mm (due to a radius of the rails of 200 mm). This, in turn, will translate to an angular error at the center of the sphere of about 0.076° (which, by using the similar triangle rule, may double at a 200 mm insertion depth). Figure 13 shows this error distribution at two insertion depths (100 mm and 200 mm), showing the accuracy of the HeRo parallel robotic system (e.g., at 200 mm depth, the error is less than 1 mm).

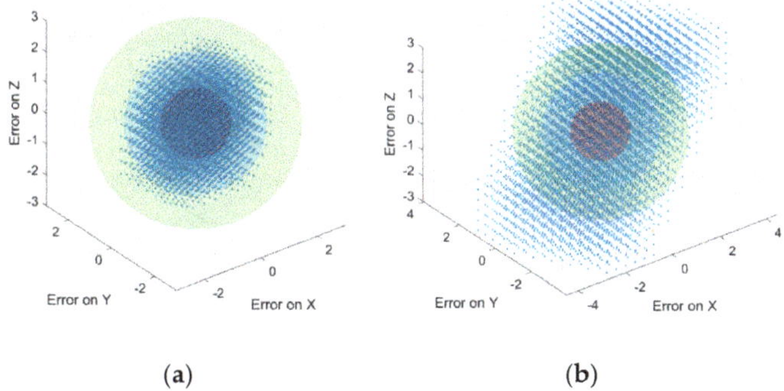

(**a**)   (**b**)

Figure 12. Error dispersion depending on the targeted depth for the needle target point for different insertion depths: (**a**) 50 mm depth; (**b**) 200 mm depth.

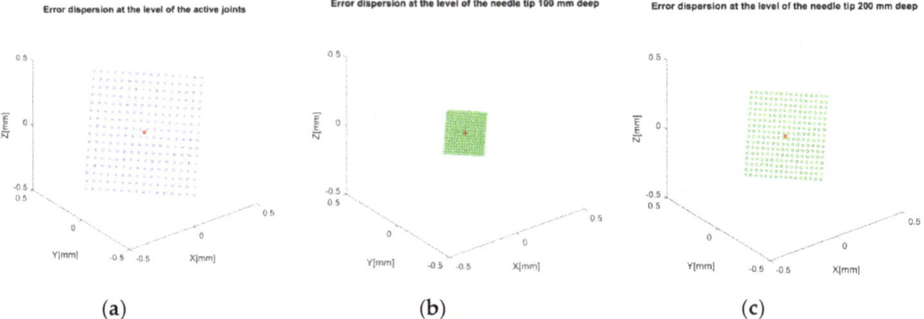

(**a**)   (**b**)   (**c**)

Figure 13. Estimated error distributions for the needle tip using the HeRo robotic system: (**a**) linear error; (**b**) angular error at 50 mm insertion depth; (**c**) angular error at 200 mm depth.

An interesting solution for needle insertion was presented in [19], where the authors used US images to determine the target volume for the needle tip directly from the graphical user interface which, in turn, served as input for the robotic system actuators. This strategy is also being considered for the further development of the HeRo robotic system, specifically for the needle insertion, since it may be valuable for the accuracy of the procedure. Consequently, the needle insertion should be automated, as opposed to other medical robotic systems where the insertion of the needle is manual (see, for example, the 7-DOF robotic system found in [20]). Moreover, the authors intend to use robust control solutions (e.g., B&R automation [21]) to ensure high reliability, even though it has been proven in the past that cheaper controlling solutions may also be used (see, for example, the medical robot proposed in [22] which is controlled by a PI controller). It is also important to note that the HeRo is designed to use commercially available medical tools (such as needles and the I-US probe), which were also common in [23]. This approach (in contrast to implementing medical tools directly in the robotic system) has the advantage of reduced development cost and increased modularity (since the robotic system may be designed to work with multiple variants of the medical tools).

5. Conclusions

Due to today's medical standards, targeted therapies such as HDR brachytherapy and chemotherapeutic agent delivery are not viable for the treatment of non-resectable HCC since the procedure involves large amounts of risks. Through a process of risk management in accordance with the ISO 14,971 standard for healthcare devices, the authors identified the risk associated with the medical procedure and proposed a new technical solution in the form of a parallel robotic system (the HeRo concept) which has the potential to facilitate patient safety during the targeted therapy of HCC. For the design of the HeRo parallel robotic system the medical protocol was defined, which, in turn, helped to establish the design constraints for the new robotic system. By considering the design constraints (with respect to the medical requirements), an AHP analysis was conducted which led to a four-step QFD. The resulting robotic system is composed of two identical independent modules (for needle and I-US probe guidance) both having RCM manipulation imbedded due to the spherical mechanism of the robotic system architecture, which has better precision of insertion (with respect to other modalities of obtaining the RCM which are not fully constrained). Moreover, the robotic system has no singularities and has decoupled motions between the positioning of the RCM and/or insertion point and the orientation of the medical tool, and these two aspects should also contribute to the overall risk reduction. The residual risks introduced by the robotic system were also evaluated, showing a reduction in hazard occurrence and severity which validates the HeRo concept as a possible technical solution for the treatment of un-resectable HCC. Future work aims to develop the first experimental model of HeRo and continue the risk management through risk evaluation (e.g., defining foreseeable sequences of events that lead to hazards) and optimization of the robotic system until it reaches the maturity level of TRL5 (technology readiness level where the prototype is evaluated in relevant environments), which will prepare the technical solution for technological transfer.

Author Contributions: Conceptualization, D.P., V.C., B.G. and N.P.; Data curation, V.C., I.B. and B.G.; Formal analysis, V.C. and N.P.; Funding acquisition, D.P.; Investigation, I.B., B.G. and N.P.; Methodology, D.P., V.C., N.A.H., B.G. and C.R.; Project administration, D.P.; Supervision, D.P.; Validation, V.C., N.A.H., N.P. and C.R.; Writing—original draft, V.C. and I.B.; Writing—review and editing, D.P., V.C. and I.B. All authors have read and agreed to the published version of the manuscript.

Funding: This work was supported by a grant from the Romanian Ministry of Research and Innovation, PCCDI—UEFISCDI, project number PN-III-P1-1.2-PCCDI-2017-0221/59PCCDI/2018 (IMPROVE), within PNCDI III.

Conflicts of Interest: The authors declare no conflict of interest.

Appendix A

The following questionnaire aims to assess the risks associated with the therapy of non-resectable HCC by means of HDR brachytherapy or targeted chemotherapeutic agent delivery. The subject that fills this questionnaire will not be required to provide any sensitive personal data; thus, confidentiality is a priority.

1) Please state your medical expertise (surgeon, oncologist, etc.):
2) Please state your experience in the medical field (months/years):

Please fill the following table (according to your opinion strictly based on your experience in the medical field) which describes possible hazards, their associated risks, and the severity of each hazard, and please estimate their probability of occurrence. The following scales should be used as guidelines: **Severity scale:** (0–29) minor, (30–89) moderate, (90–99) serious, (100) catastrophic. **Probability scale:** (0–29) remote, (30–69) unlikely, (70–99) likely, (100) very likely.

Hazard	Associated Risk	Severity *	Probability **
Needle positioning error beyond acceptable value	Linear positioning error (the error remains constant at any depth) Orientation positioning error (the error multiplies with the depth) Insertion depth error		
I-US probe positioning error	I-US probe positioning does not respect the predefined trajectory I-US probe is pressing too hard on the liver I-US cannot reach the targeted area		
I-US entry port (RCM) is not preserved	The I-US probe applies pressure on the abdominal wall The I-US probe movements are different than intended		
The targeted area cannot be reached	Incapability of performing the procedure (the instruments cannot reach the specific liver area) Incapability of achieving the necessary orientation for the I-US probe Incapability of achieving the necessary orientation for the therapeutic needle		
The operating table has to be repositioned to ensure better access to the targeted area of the liver	The relative patient–robot position is changed		
The patient state deteriorates during the procedure	The patient cannot be resuscitated in due time		
The needle cannot be viewed inside the liver parenchyma	The needle insertion cannot be monitored		
Improper sterilization	Infection risk		

Are there other hazards which are not mentioned in the above table? (If yes, please fill in the table below.)

Hazard	Associated Risk	Severity *	Probability **

References

1. Couri, T.; Pillai, A. Goals and targets for personalized therapy for HCC. *Hepatol. Int.* **2019**, *13*, 125–137. [CrossRef] [PubMed]
2. European Association for the Study of the Liver. EASL Clinical Practice Guidelines: Management of hepatocellular carcinoma. *J. Hepatol.* **2018**, *69*, 182–236. [CrossRef] [PubMed]
3. Inchingolo, R.; Posa, A.; Mariappan, M.; Spiliopoulos, S. Locoregional treatments for hepatocellular carcinoma: Current evidence and future directions. *World J. Gastroenterol.* **2019**, *25*, 4614–4628. [CrossRef] [PubMed]
4. Podder, T.K.; Beaulieu, L.; Caldwell, B.; Cormack, R.A.; Crass, J.B.; Dicker, A.P.; Fenster, A.; Fichtinger, G.; Meltsner, M.A.; Moerland, M.A.; et al. AAPM and GEC-ESTRO guidelines for image-guided robotic brachytherapy: Report of Task Group 192. *Med. Phys.* **2014**, *41*, 101501. [CrossRef] [PubMed]
5. Birlescu, I. Studies Regarding the Safe Operation of Innovative Medical Parallel Robots. Ph.D. Thesis, Technical University of Cluj-Napoca, Cluj-Napoca, Romania, 20 September 2019; pp. 31–32.

6. Pisla, D.; Galdau, B.; Covaciu, F.; Vaida, C.; Popescu, D.; Plitea, N. Safety Issues in the Development of the Experimental Model for an Innovative Medical Parallel Robot used in Brachytherapy. *Int. J. Prod. Res.* **2016**, *55*, 684–699. [CrossRef]

7. de Battisti, M.B.; de Senneville, B.D.; Hautvast, G.; Binnekamp, D.; Lagendijk, J.J.W.; Maenhout, M.; Moerland, M.A. A novel adaptive needle insertion sequencing for robotic, single needle MR-guided high-dose-rate prostate brachytherapy. *Phys. Med. Biol.* **2017**, *62*, 4031–4045. [CrossRef] [PubMed]

8. Dou, H.; Jiang, S.; Yang, Z.; Sun, L.; Ma, X.; Huo, B. Design and validation of a CT-guided robotic system for lung cancer brachytherapy. *Med. Phys.* **2017**, *44*, 4828–4837. [CrossRef] [PubMed]

9. Poulin, E.; Gardi, L.; Barker, K.; Montreuil, J.; Fenster, A.; Beaulieu, L. Validation of a novel robot-assisted 3DUS system for real-time planning and guidance of breast interstitial HDR brachytherapy. *Med. Phys.* **2015**, *42*, 6830–6839. [CrossRef] [PubMed]

10. Tractica. *Robots Market Forecasts*; Tractica: Boulder, CO, USA, 2018.

11. Multi Annual Roadmap, euRobotics AISBL. 2017. Available online: https://www.eu-robotics.net/cms/upload/topic_groups/H2020_Robotics_Multi-Annual_Roadmap_ICT-2017B.pdf (accessed on 26 November 2019).

12. ISO 14971—Medical Devices—Application of Risk Management to Medical Devices. Available online: https://www.iso.org/standard/72704.html (accessed on 26 November 2019).

13. Vaida, C.; Al Hajjar, N.; Lazar, V.; Graur, F.; Burz, A.; Elisei, R.; Mois, E.; Pisla, D. Robotics in Minimally Invasive Procedures: History, Current Trends and Future Challenges. In *6th International Conference on Advancements of Medicine and Health Care through Technology—IFMBE Proceedings*; Vlad, S., Roman, N., Eds.; Springer: Singapore, 2019; pp. 267–273.

14. Taylor, R.H.; Funda, J.; Eldridge, B.; Gomory, S.; Gruben, K.; LaRose, D.; Talamini, M.; Kavoussi, L.; Anderson, J. A Telerobotic Assistant for Laparoscopic Surgery. *IEEE Eng. Med. Biol.* **1995**, *14*, 279–287. [CrossRef]

15. Tucan, P.; Vaida, C.; Plitea, N.; Pisla, A.; Carbone, G.; Pisla, D. A Risk-Based Assessment Engineering of a Parallel Robot Used in Post-Stroke Upper Limb Rehabilitation. *Sustainability* **2019**, *11*, 2893. [CrossRef]

16. Qualica. Available online: https://www.qualica.net/software.html (accessed on 24 November 2019).

17. Vaida, C.; Pisla, D.; Plitea, N.; Gherman, B.; Tucan, P. Parallel Modular Robotic System for the Ultrasound Intraoperatory Probe Guidance and the Manipulation of Instruments for the Treatment of Hepatic Tumors. Patent Pending No. A01143/24.12.2018, 24 December 2018.

18. Maxon Motors. Available online: https://www.maxongroup.com/ (accessed on 24 November 2019).

19. Megali, G.; Tonet, O.; Stefanini, C.; Boccadoro, M.; Papaspyropoulos, V.; Angelini, L.; Dario, P. A Computer-Assisted Robotic Ultrasound-Guided Biopsy System for Video-Assisted Surgery. In *Medical Image Computing and Computer-Assisted Intervention—MICCAI, 2001*; Springer: Berlin/Heidelberg, Germany, 2001; Volume 2208, pp. 343–350.

20. Kettenbach, J.; Kronreif, G.; Figl, M.; Fürst, M.; Birkfellner, W.; Hanel, R.; Bergmann, H. Robot-assisted biopsy using ultrasound guidance: Initial results from in vitro tests. *Eur. Radiol.* **2005**, *15*, 765–771. [CrossRef] [PubMed]

21. B&R Automation. Available online: https://www.br-automation.com/ (accessed on 24 November 2019).

22. Mallapragada, V.G.; Sarkar, N.; Podder, T.K. Robot assisted real-time tumor manipulation for breast biopsy. *IEEE Trans. Robot.* **2009**, *25*, 316–324. [CrossRef]

23. Fichtinger, G.; Fiene, J.P.; Kennedy, C.W.; Kronreif, G.; Iordachita, I.; Song, D.Y.; Burdette, E.C.; Kazanzides, P. Robotic assistance for ultrasound-guided prostate brachytherapy. *Med Image Anal.* **2008**, *12*, 535–545. [CrossRef] [PubMed]

Article

Approach to Risk Performance Reasoning with Hidden Markov Model for Bauxite Shipping Process Safety by Handy Carriers

Jianjun Wu [1], Yongxing Jin [1,*], Shenping Hu [1,*], Jiangang Fei [2] and Yuanqiang Zhang [3]

[1] Merchant Marine College, Shanghai Maritime University, Shanghai 201306, China; jjwu@shmtu.edu.cn
[2] Australia Maritime College, University of Tasmania, Launceston 7248, Australia; jiangang.fei@utas.edu.au
[3] Faculty of Maritime and Transportation, Ningbo University, Zhejiang 315211, China; zhangyuanqiang@nbu.edu.cn
* Correspondence: yxjin@shmtu.edu.cn (Y.J.); sphu@shmtu.edu.cn (S.H.)

Received: 31 December 2019; Accepted: 11 February 2020; Published: 13 February 2020

Abstract: An approach based on the hidden Markov model (HMM) is proposed for risk performance reasoning (RPR) for the bauxite shipping process by Handy carriers. The unobservable (hidden) state process in the approach aims to model the underlying risk performance, while the observation process was formed from the time series of risk factors. Within the framework, the log-likelihood probability was used as the measure of similarity between historical and current data of risk reasoning factors. Based on scalar quantization regulation and risk performance quantization regulation, the RPR approach with different step sizes was conducted on the operational case, the performance of which was evaluated in terms of effectiveness and accuracy. The reasoning performance of the HMM was tested during the validation period using three simulated scenarios and one accident scenario. The results showed significant improvement in the reasoning capacity, and satisfactory performance for numerical risk reasoning and categorical performance reasoning. The proposed model is able to provide a reference for risk performance monitoring and threat pre-warning during the bauxite shipping process.

Keywords: risk performance reasoning; hidden Markov model; Handy bauxite carrier; process safety; performance evaluation

1. Introduction

Bauxite is abundant, totaling 30 billion tons globally in 2018, according to the data from the United States Geological Survey (USGS). The natural distribution of bauxite is extremely uneven, mainly concentrated in Africa, Oceania, South America, and Southeast Asia. China's demand for imported bauxite increased sharply from about 2.3 million tons in 2007 to 82.62 million tons in 2018 [1]. Panamax and Handy carriers transport 90% of the bauxite via shipping [2]. Here, Handy carrier is the collective term that refers to Handysize and Handymax bulk carriers. Handy carriers play an important role in bauxite shipping, accounting for 48% of the total industry. Meanwhile, 74% of deaths in the industry were linked to accidents involving Handy bulk carriers [3–6]. The liquefaction of bauxite during transportation is an important cause of ship accidents, directly responsible for more than 80 casualties of seafarers [7].

The basic reason for liquefaction is that the moisture content of bauxite exceeds the transportable moisture limit (TML). Influenced by the effect of the ship's stability and its cargo properties in a complex shipping process, bauxite with a high moisture content carried by Handy carriers tends to liquefy, which threatens the stability and safety of the ship. The bauxite performance, dynamic ship stability, and maritime environment have important effects on the risk level of the carrier during the transportation process. Risk identification and monitoring can improve the capacity for risk prevention on bauxite carriers. The study of risk reasoning of the transportation process based on cargo information and uncertain weather and sea conditions may allow more time for an emergency response, thus reducing the risk of loss or damage. A hidden Markov model (HMM)-based approach is introduced here to reveal the factor and time correlations of the observation index and hidden risk, thereby achieving risk reasoning for the bauxite transportation process.

The organization of this paper is as follows: recent studies related to bauxite liquefaction and its risk reasoning are reviewed in Section 2. The research theory and the model for reasoning are presented in Section 3. The model is applied to specific cases in Section 4, where the results are analyzed for effectiveness and accuracy. Section 5 presents the analysis and discussion of this study, encompassing scenario planning. Conclusions are drawn in Section 6.

2. Literature Review

2.1. System of Maritime Transportation

Maritime transportation is a complex process, which involves many factors such as human, ship, environment, management, and cargo. In order to carry out risk reasoning for the transportation process, it is necessary to determine any accident mechanisms related to cargo. Cargo has a complex correlation with the other subsystems in the operation safety of a ship. For example, Li [8] studied the safety evolution of seaborne dangerous chemicals under various uncertain conditions. The aforementioned research paved the way for a new mode of operation safety research for specific cargo ships. Unlike dangerous chemicals, cargoes that may liquefy are not inherently dangerous. Nevertheless, danger can occur when cargoes start moving on the carrier. Ma [9] studied the shipping risk of ore concentrate powder and revealed the accident mechanisms through risk identification of the system, with factors including human, ship, environment, management, and cargo. Bauxite is different from ore concentrate powder. The potential risk of bauxite liquefaction during shipping is more prominent. Seaborne bauxite presents potential danger when interacting with the carrier in a specific environment, and it is necessary to develop an approach to study the transportation safety of bauxite on the basis of a safety system engineering method.

2.2. Risk of Cargo Liquefaction

- Effect of moisture content

There are many influencing factors for transportation accidents involving cargoes that may liquefy. It is necessary to identify and monitor the accident factors according to the mechanism of cargo liquefaction, the ship's stability, and the marine environment. Shen [10] found that the actual moisture content of the cargo must not exceed the transportable moisture limit (TML) in order to prevent liquefaction. However, bauxite with an initial moisture content lower than the TML may still exceed the TML and liquefy due to changes in temperature and humidity during the transportation process. Wang [11] found that observable indexes such as saturation and compactness can be used as key indexes to measure the degree of liquefaction. A higher moisture content or saturation increases the risk of liquefaction. The initial saturated or unsaturated state of cargo is disturbed by internal and external factors; thus, the actual moisture content of cargo exhibits temporal fluctuation. It is critical to avoid an increase in moisture content during the process of cargo production, storage, loading, and navigation of the ship [2]. In order to strengthen the control of moisture content in the loading and post-loading stages, Popek [12,13] proposed that biodegradable thermoplastic polymer material be added to the concentrate to absorb moisture from granular pores, thereby preventing slippage and transfer of concentrate during storage and transportation. Altun [14] proposed that the application of suitable chemical filter aids in the filtration process of concentrate production could effectively reduce the water content of different mineral products to be 10%–15% lower than the TML. By reducing the initial moisture content of the cargo, the liquefaction resistance of the cargo during shipping can be improved.

- Effect of weather or sea

Heavy weather and adverse sea conditions are the main cause of many accidents involving Handy bauxite carriers. Once the impact of the external environment on the ship and cargo deteriorates into force majeure beyond the ship's disaster resistance [15], disaster eventually occurs. Therefore, special attention should also be paid to the complexity and variability of the environment of a sea route [16,17]. Air humidity increases the risk of liquefaction and movement of highly absorbent solid bulk cargo [18]. Furthermore, seawater often sweeps over the ship deck in heavy weather conditions, resulting in water penetrating cargo holds, which may increase the cargo moisture content and affect the safety of the ship's operation. Moreover, wind affects the speed and rolling angle of the ship [19]. In the case of random waves, especially on heavy seas, ships roll at large angles, which can easily lead to capsizing [20,21]. At the same time, ship rolling leads to cargo shifting. The initial shifting of cargo after liquefaction and the heeling moment of external wind and wave eventually lead to the ship capsizing [22]. Ship acceleration and kinematic waves affect cargo stability [23]. The hull vibration caused by rolling and machinery operation is not only harmful to the safety of the ship structure [24], but it can also change the characteristics of the cargo on board [25], even aggravating the liquefaction of cargo [26].

2.3. Risk Response for Shipping Process

At present, some achievements were obtained in the monitoring and reasoning of cargo liquefaction. Ju [27] quantitatively assessed the risk of liquefaction and its impact on ship stability by analyzing time-domain characteristics for different amplitudes and frequencies and initial saturations of cargo. Based on the effect of liquefaction on the ship's intact stability, Andrei [28] proposed a method to measure the heeling moment and the probability of cargo shifting caused by liquefaction. Munro [29] investigated the relationship between resistivity changes and pore pressure in an equivalent cargo hold model to monitor cargo liquefaction risks. Daoud [30] established a dynamic model through a static numerical simulation to monitor the ship movement posture and cargo state at all times, and studied the nickel liquefaction mechanism under swell using a nonlinear model [31]. Liu [32] developed a transport risk system framework for navigation safety in heavy weather in collaboration with the China Meteorological Administration for ships carrying cargoes that may liquefy in different seas. However, there are still gaps in risk monitoring and reasoning for bauxite carriers in current research and practice. For bauxite shipping, strengthening risk management in the whole process of shipping is critical.

2.4. Accident History of Bauxite Carriers

At present, bauxite is not formally listed in Group A (cargo that may liquefy) by the International Maritime Organization (IMO). The international maritime community still has doubts about its liquefaction characteristics. Depending on the particular circumstances of any given shipment, it would appear that bauxite may come with the risk of liquefaction and shift during shipping, which can cause a vessel to capsize at a moment's notice. Fortunately, up until 2 July 2013, none of the incidents resulted in losses to vessels or crew members, according to data from the North P&I Club. However, on 2 January 2015, M.V. Bulk Jupiter with 46,400 tons of bauxite capsized and sank in strong winds and swells off the coast of Vietnam, killing 18 crew members. The disaster of Bulk Jupiter, a Handy bauxite carrier, aroused a series of responses related to risks of the bauxite shipping process.

IMO requested that the global bauxite industry undertake research into the behavior and characteristics of bauxite cargoes during ocean transportation. From 14 to 18 September 2015, the second meeting of the Subcommittee on Cargo and Container Transport (CCC) of the Maritime Safety Committee approved CCC.1/Circ.2 to remind people of the potential risks of bauxite in maritime transport [33]. Considering that Handy carriers do not have a special structural design for cargoes that may liquefy, IMO recommends that the captain, on the basis of experience and relevant certifications, may refuse to carry the cargo if the carrying of such cargo may fail to ensure the absolute safety of the voyage; if the captain decides to carry it, necessary measures must be taken to ensure the safety of the vessel.

In recent years, the Global Bauxite Working Group (GBWG) designated by the IMO carried out a series of research studies on bauxite properties. From 11 to 15 September 2017, the fourth meeting of the CCC Subcommittee adopted CCC.1/Circ.2/Rev.1 in the draft amendment to classify certain bauxite as cargo that may liquefy, submitting it to the Maritime Safety Committee for consideration [34]. The 2019 amendments of the International Maritime Solid Bulk Cargoes (IMSBC) code was adopted by the 101st session of the Maritime Safety Committee. The new individual schedule for bauxite fines as a Group A cargo is expected to be implemented on 1 January 2021.

2.5. Risk Performance Reasoning for Bauxite Shipping Process

According to the research of the GBWG and the authors, an atypical motion of the ship (wobbling) may also be indicative of cargo instability. Extreme care and appropriate action must be taken, taking into account the provisions of relevant IMO instruments when handling and carrying bauxite in bulk. Bauxite may suffer instability due to its moisture content, resulting in dynamic separation and formation of liquid slurry (water and fine solids) above the solid material, resulting in a surface effect which may significantly affect the ship's stability. If left unchecked, this movement of cargo has the potential to further reduce the stability of the ship, and the risk of capsizing will significantly increase. Based on the knowledge of bauxite and its carrier, Wu [35] carried out a risk simulation on the first stage of the bauxite maritime transportation process using the Markov chain cloud model, and obtained spatial correlation between transportation risk and ship positions. This allowed risk reasoning of the transportation process of bauxite carriers to be achieved through combining with weather and sea forecast information.

The hidden Markov model has high applicability in reasoning. Chen [36] proposed a hidden Markov model (HMM) framework for modified analogue forecasting (MAF) of meteorological droughts to improve reasoning capacity and performance for a time series of the standardized precipitation index. Joshi [37] used the Baum–Welch algorithm to optimize the parameters of a hidden Markov model for temperature forecasts to reduce root-mean-square errors and improve reliability. Wu [15] introduced a hidden Markov model to analyze the causes of accidents involving ships carrying liquefiable cargoes and found that environmental deterioration was a direct cause and cargo liquefaction was a fundamental cause. This study laid the foundation for risk reasoning of the bauxite shipping process. Fabbri [38] carried out navigation risk assessment using meteorological and oceanographic (METOC) methods, which provided a useful reference for the risk reasoning of Handy bauxite carriers, embodying the performance of cargo.

Based on the initial state of bauxite, the carrier, and the meteorological dynamics of the routing, the risk reasoning of bauxite carriers can be realized using real-time maneuvering data as input. This paper attempts to establish an HMM-based approach for risk performance reasoning, which aims to determine cargo performance and ship posture.

3. Methods

3.1. Theroy of Hidden Markov Model

3.1.1. Hidden Markov Model

An HMM is a probabilistic model describing double stochastic processes [39] with parameters, which include Markov processes of hidden states and observation processes associated with hidden states. The process of hidden state transition is not observed directly. The change in hidden state can be inferred by observing the sequence of indexes. Objective indexes can be divided into observed variables which are convenient for direct measurement and hidden variables which cannot be directly observed. That is to say, the risk state of the hidden variable needs to be judged using observed indexes. The application of HMM can reduce the dependence on experts' subjective experience. There is a hierarchical independent mapping relationship between these observed indexes and hidden variables. The correlation process between the hidden risk state and the observed state in the model is shown in Figure 1.

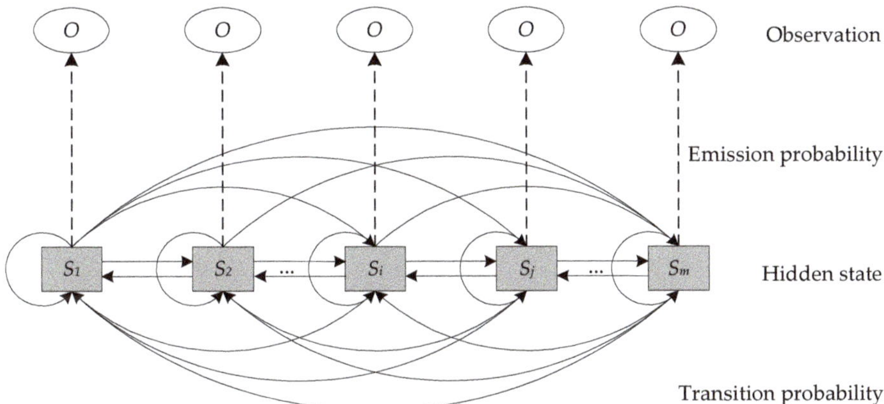

Figure 1. Relationship between hidden states and observations in the hidden Markov model (HMM).

3.1.2. HMM Parameter Learning

The parameter learning algorithm is called the Baum–Welch algorithm [40], which iteratively optimizes the parameters of the HMM. Let (Ω, F, P) be the probability space and $\{Y_t\}_{1 \le t \le T}$ and $\{X_t\}_{1 \le t \le T}$ be sequences of random variables of observable and hidden states, where $X_t : \Omega \to S_1, S_2, \ldots, S_m$ and $Y_t : F \to N$ or any set of possible states.

The specific implementation steps are as follows:

Step 1: In the given sample training space, the first observation sequence $y = (y_1, y_2, \ldots, y_T)$ is trained and the initial model parameters are re-estimated to obtain the model parameters $\lambda_1 = (\pi_1, A_1, B_1)$, where π_1 stands for the distribution of hidden states, A_1 is the transition probability matrix, and B_1 represents the distribution of observable states.

Step 2: The observation sequence $y = (y_1, y_2, \ldots, y_T)$ is trained using the new re-estimated parameters $\lambda_1 = (\pi_1, A_1, B_1)$ to obtain the next new model parameters $\lambda_2 = (\pi_2, A_2, B_2)$.

Step 3: Step 2 is repeated until the model converges. Using the three re-estimation equations mentioned below, the initial parameters of the HMM model are updated from $\lambda_0 = (\pi_0, A_0, B_0)$ to $\lambda = (\pi, A, B)$, satisfying $P(y|\lambda) \ge P(y|\lambda_0)$.

The parameter re-estimation equations are as follows.

$$\pi' = \gamma_1(i) = \frac{\alpha_1(i) \cdot \beta_1(i)}{\sum_{i=1}^{m} \alpha_1(i) \cdot \beta_1(i)}. \tag{1}$$

$$a_t(i,j)' = \frac{\xi_t(i,j)}{\gamma_t(i)}. \tag{2}$$

$$b_i(y_t)' = \frac{\alpha_t(i) \cdot \beta_t(i)}{\sum_{i=1}^{m} \alpha_t(i) \cdot \beta_t(i)}. \tag{3}$$

In Equation (1), π' is the estimation of initial probability π, and it stands for the probability of the hidden risk state S_i at time $t = 1$; $\alpha_t(i)$ represents the forward probability function of the observation $O = (y_1, y_2, \ldots y_t)$ at time t in state i, and $\beta_t(i)$ stands for the backward probability of the partial observation sequence from time step $t + 1$ to the end. In Equation (2), $a_t(i, j)\prime$ is the estimation of state transition probability $a_t(i, j)$, and it stands for the quotient of frequency of the risk state transition from state S_i to S_j divided by the frequency of the hidden state transition from state s_i to others; $\xi_t(i, j) = P\big(X_t = S_i \big| X_{t+1} = S_j\big)$ is the probability of being in state S_i at time t and state S_j at time $t + 1$, while $\gamma_t(i) = P(X_t = S_i)$ is the probability of being in state S_i at time t. In Equation (3), $b_i(y_t)\prime$ is the estimation of observation probability $b_i(y_t)$, and it stands for the quotient of frequency of the observed state O_i from hidden state S_i divided by the frequency of observation from the hidden state S_i.

3.2. The Application of Hidden Markov Model

3.2.1. Description of Bauxite Shipping

- Process Risk

The process risk is a dynamic characterization of the risk state at any time during the system's operation. It is the output of the coupling effect of uncertain (random) events under the influence of risk factors [41]. The process risk of bauxite shipping describes the development and evolution of the bauxite shipping system between the safety and accident subsystems of Handy bauxite carriers, where bauxite continuously interacts with the traffic environment over time.

- Risk performance

The risk performance indicates the general status of the risk at a particular time, as well as the properties and characterization of the mechanism of the risk variation. The performance introduced in risk research highlights the temporal processes and spatial spread. Based on the severity of consequence and response, the risk performance of a process is quantified and classified as normal, medium, high, or uncontrolled. The responses of the four-state sequence are undesired intervention, partial intervention, consistent intervention, and invalid intervention, respectively.

3.2.2. Risk Performance Transition of Bauxite Shipping

The factors of risk reasoning include static variables, dynamic variables, and voyage variables. Once a bauxite carrier is identified as a convenient bulk carrier, its ship parameters mostly represent static variables, such as ship age and ship technical status, while dynamic variables are environmental, such as relative length of ship (Length overall (LOA)/wavelength), wave, current, wind, and temperature. Voyage variables refer to bauxite attributes and ship maneuvering. These variables are the components of the risk evaluation system, and a change in their index values is related to the degree of system risk. However, this degree of risk cannot represent the degree of risk of the whole system. Through the risk evaluation index (observation variable), the expression of the risk state is established to indirectly show the level of total risk.

The elements in the structural model of risk reasoning for the bauxite shipping process involve objective indexes of the cargo, ship, and environment. The transition process between the risk state of the Handy bauxite carrier and the reasoning indexes constitutes an HMM. The requirements of parameter input and output in the hidden Markov model and its operation mechanism are shown in Figure 2.

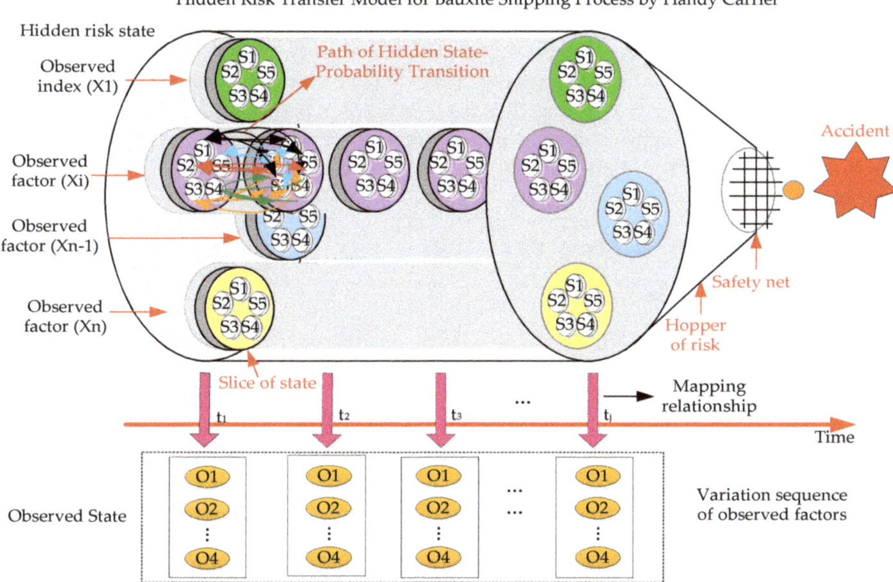

Figure 2. HMM-based mapping of risk transfer for bauxite shipping process.

3.2.3. HMM-Based Approach to Risk Performance Reasoning

The risk reasoning approach in an HMM framework can be achieved in several steps: data learning, modeling of HMM for risk reasoning, reasoning of risk performance, and performance evaluation of reasoning. The flowchart of the proposed model is shown in Figure 3.

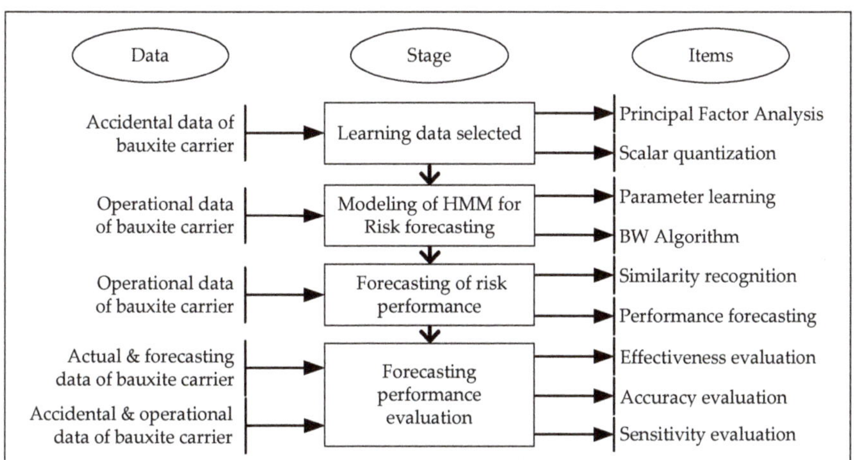

Figure 3. Flowchart of the risk performance reasoning for bauxite shipping process.

Firstly, the risk factors of the bauxite shipping process were identified based on the accident data of "M.V Bulk Jupiter", i.e., the Report of the Marine Safety Investigation into the Loss of a Bulk Carrier. The principal factors were obtained using principal factor analysis (PFA) from a dimensionality reduction of the risk factor set. The risk classification criteria were established after scalar quantification of the principal factors. Based on the first 50 groups of operational data of a Handy bauxite carrier, the HMM was trained using the Baum–Welch algorithm. After the steps of feature extraction, dimensionality reduction using PFA, and scalar quantization, the last nine groups of operational data were input as test data into the HMM to obtain the log-likelihood probability, which was then used for similarity recognition and risk performance reasoning. Then, the reasoned risk performance of each factor was compared with the test data from the last nine groups to evaluate the performance of risk reasoning in terms of classification effectiveness, measurement accuracy, and reasoning sensitivity.

3.3. Modeling of HMM for Risk Performance Reasoning

3.3.1. Principal Factor Analysis

Based on the risk identification of the risk of bauxite shipping process, a total of 15 risk factors (RFs) were obtained for the three types of factors: bauxite, Handy carrier, and environment. After experimental analysis of the coupling effect of the bauxite and Handy carrier under complicated marine conditions, seven principal factors (PF) were selected from the RFs, including static factors, voyage factors, and dynamic factors, for risk reasoning of the bauxite shipping process. System of principal factors is present in Figure 4.

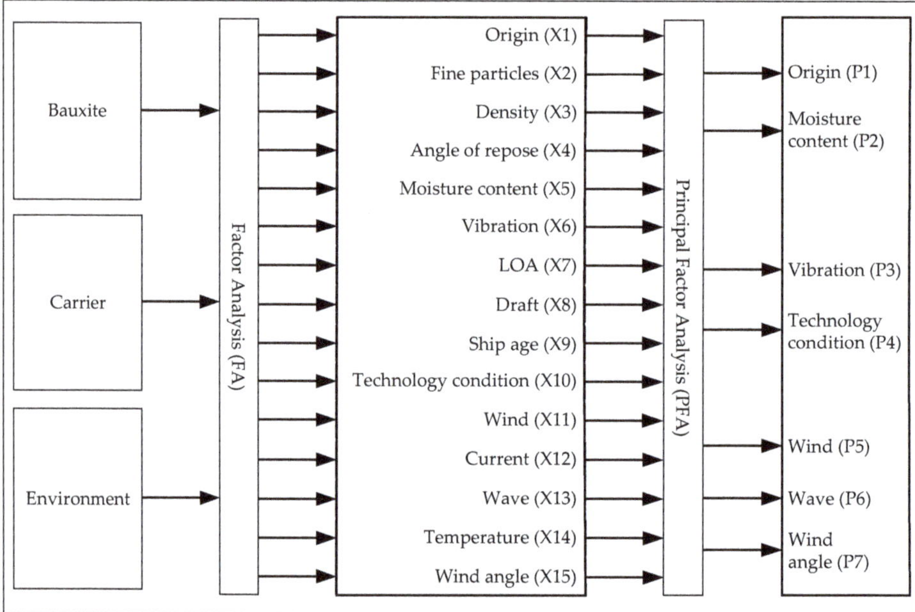

Figure 4. System of principal factors for risk performance reasoning.

3.3.2. Scalar Quantization

- Factors and classification criteria

With the rapid development of big data, much more information in maritime safety needs to be processed quickly. Studies on the quantitative risk analysis of ships are becoming more important [42]. Quantifying the factor values that can characterize the risk performance is helpful to explain their impact on the bauxite shipping process. In order to facilitate the application of the original data in the hidden Markov reasoning model, continuous values are discretized [43] and the risk grade is divided by the interval value for the quantitative factors. The other qualitative factors can be transferred into grades, as shown in Table 1.

Table 1. Factors and criteria of risk reasoning for Handy bauxite carrier. TML—transportable moisture limit, FMP—Flow moisture point.

Observed Factor	Factor Value	Normal Risk	Low Risk	High Risk	Uncontrolled
Origin (P1)	Risk property	Better	Good	Bad	Worse
Moisture (P2)	Absolute value	<0.8 × TML	0.8 × TML–TML	TML–FMP	>FMP
Vibration (P3)	Intensity scales	Weak	General	Strong	Violent
Technology (P4)	Reliability	Better	Good	Bad	Worse
Wind (P5)	Beaufort Scale	1–3	3–6	7–9	>9
Wave (P6)	Wave scale	1–2	3–4	5–6	>7
Wind angle (P7)	Intersection angle	0–30 or 150–180	30–60 or 120–150	60–80 or 100–120	80–100

Here, the range of risk grade values is based on the consideration of facilitating seafarers to classify and control the aforementioned risk indexes. Considering the complexity of ship maneuvering, it is difficult to accurately determine the risk level in the risk range of continuity for crews. For example, clear risk guidance is necessary for emergencies in Beaufort 8 wind. Although the risk level of the index changes when the actual index value is at the critical value of the adjacent risk level, the variables are still insufficient to achieve the degree of mutation in the total risk reasoning. When the risk level is very high, a sudden change in some key indexes may lead to a significant increase in the total risk of the ship, which highlights the need to control disaster-causing factors in bauxite shipping.

- Quantization regulation of factors of risk performance

Based on the risk scale criteria of principal factors, each grade risk is defined as a standard value of 1–4, representing normal, low risk, high risk, and uncontrolled risk, respectively. After the total risk value is obtained by quantization and combination of the risk performance of principal factors, the total risk scale can be obtained by scalar quantization. The risk value belonging to any risk scale interval can be defined as the standard value of risk. When the total risk value $R_t \epsilon (i, \ i+1]$, then we accept its risk scale $S_t = i + 1$ at moment t, where $0 < i < N, 1 \leq t \leq T$ (see Figure 5).

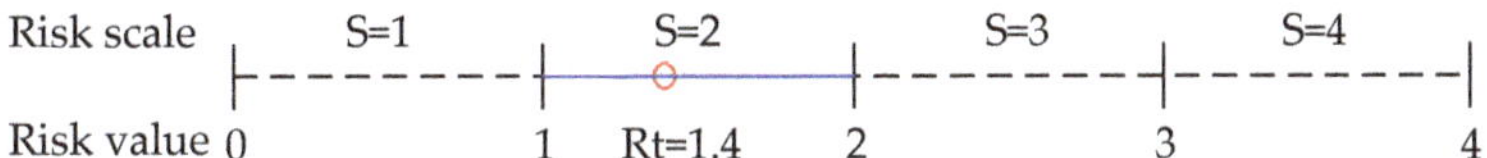

Figure 5. Quantization regulation of factors of risk performance.

3.4. Risk Performance Reasoning

3.4.1. Similarity Recognition

HMM parameters were trained using actual data of Handy bauxite carriers under normal conditions, representing a better cargo state, satisfactory ship conditions, and a good environment. Similarity recognition was performed using the Viterbi algorithm to obtain the maximum of log-likelihood probability, which is expressed as follows:

$$LLP_t = \log P(Y_t = O|\lambda), \tag{4}$$

where LLP_t is the log-likelihood probability of the observations $\{Y_t\}$ at the current time t under the HMM of λ.

The next step is to detect the closest LLP_{t0} from the LLP_{t-max} of the historical data and obtain the risk performance at time $t0$. This function is expressed as follows:

$$Diff(LLP_t) = \left| \log P(Y_t = O_{Ft}|\lambda) - \log P(Y_t = O_H|\lambda) \right|, \tag{5}$$

where $Diff(LLP_t)$ is the difference between $\log P(Y_t = O_{Ft}|\lambda)$ and $\log P(Y_t = O_H|\lambda)$, $\log P(Y_t = O_{Ft}|\lambda)$ indicates the max log-likelihood probability of the forecasted O at the current time t under the HMM of λ, and $\log P(Y_t = O_H|\lambda)$ expresses the log-likelihood probability of the historical O at the under the HMM of λ.

The series of $Diff(LLP_t)$ was sorted in ascending order by the MATLAB function "[LLsort, LLpos] = sort []", from which the original element position was returned. Given n as the number of steps, the n closest LLP_{t0} was obtained based on the n smallest $Diff(LLP_t)$.

The above approach is called similarity recognition.

3.4.2. Risk Performance Reasoning of Factors

For each principal factor, the difference between the current risk value and the next reasoned value is the same as the difference between the two adjacent risk values discovered using similarity recognition. The reasoned value at time $t + 1$ can be obtained from the former risk value. The approach can be expressed as follows:

$$
\begin{cases}
F_1 - H_K = \frac{1}{N} \sum_{n=1}^{N} \left(H_{k+1}^{(n)} - H_k^{(n)} \right), & k \in [1, K-1] \\
F_t - F_{t-1} = \frac{1}{N} \sum_{n=1}^{N} \left(H_{k+1}^{(n)} - H_k^{(n)} \right), & t \in [2, T]
\end{cases}
\tag{6}
$$

where T is the length of series reasoned, K is the length of historical series, N is the number of the closest LLP of HMM for the carrier at the current time, F_1 is the first value, H_K is the last historical risk value which is also the benchmark of risk reasoning, F_t is the risk value reasoned at time t, F_{t-1} is the risk value reasoned at time $t - 1$, $H_k^{(n)}$ is the n closest historical risk value compared with the LLP_t of the benchmark and F_{t-1}, $H_{k+1}^{(n)}$ is the next historical risk value of $H_k^{(n)}$.

Nevertheless, it must be noted that, if $F_1 - H_K > 0$, then the reasoned (forecasted) risk value F_t grows linearly at every moment based on Equation (6); therefore, the output of every step is normalized using the quantization regulation.

3.4.3. Risk Performance Reasoning of Ship

The reasoned risk value of principal factors and the total risk value were set as intervals of the risk state value ranging from 0–4. The risk performance of principal factors was reasoned through a quantization of the risk grade using scalar quantization regulation. The quantization regulation of the risk performance can be used to gain the total risk performance reasoning at any future time.

Reg. 1: when the risk grade of the moisture content (K2) is 1, that is, $S_{K2} = 1$, the total risk value of the bauxite carrier is $R_t = Average(S_{K1} : S_{K7})$, and then the total risk grade S_t can be obtained using quantization regulation.

Reg. 2: when the risk grade of K2 is 2, that is, $S_{K2} = 2$, and the risk grade of vibration (K3) or wind (K5) is between 2 and 3, that is, $2 \leq (S_{K3} \, or \, S_{K5}) \leq 3$, the total risk value of the bauxite carrier is $S_t = max(S_3, Average(S_{K1} : S_{K7}))$.

Reg. 3: when the risk grade of K2 is 2, that is, $S_{K2} = 2$, and the risk grade of K3 or K5 is between 2 and 3, that is, $S_{K3} \, or \, S_{K5} = 4$, the total risk value of the bauxite carrier is $S_t = max(S_4, Average(S_{K1} : S_{K7}))$.

Reg. 4: when the risk grade of K2 is 3, that is, $S_{K2} = 3$, the total risk value of the bauxite carrier is $S_t = max(S_3, Average(S_{K1} : S_{K7}))$.

Reg. 5: when the risk grade of K2 is 4, that is, $S_{K2} = 4$, the total risk value of the bauxite carrier is $S_t = max(S_4, Average(S_{K1} : S_{K7}))$.

3.5. Performance Evaluation of Reasoning

3.5.1. Effectiveness Evaluation

The effectiveness evaluation of the risk performance reasoning can be expressed using the degree of bias (DOB) and degree of detection (DOD), defined as follows:

$$\text{DOB} = F_{ij} - F_{ji} = \frac{\sum_{i=1}^{4}\left(f_{ij} - f_{ji}\right)}{T}, \tag{7}$$

$$\text{DOD} = F_{ii} - F_{ji} = \frac{\sum_{i=1}^{4}\left(f_{ij} - f_{ji}\right)}{T}, \tag{8}$$

where i and j denote the ordinal values of the risk category, satisfying i *and* $j \in (1,2,3,4)$; T is the length of series to be reasoned, and F_{ij} is the relative frequency of the forecast risk category i while the observed risk category is j (a non-i interger value), which can be calculated as the count of f_{ij} occasions in all four risk categories divided by the length of the reasoned series. F_{ji} can be similarly obtained. F_{ii} is the relative frequency of the forecast risk category equal to the observed risk category i which can be calculated as the count of f_{ii} occasions in all four risk categories divided by the length of the reasoned series.

The degree of bias compares the number of times a risk category was forecast to the number of times the risk category was observed. It indicates that the forecast categories were over-forecast or under-forecast with a value greater or less than 0, respectively, while a value of 0 describes unbiased forecasts. The degree of detection is the fraction of occasions when the risk forecast category occurred for occasions when it was also forecast. This value represents the success rate for detecting different risk categories and ranges from −1 as completely wrong to 1 as completely accurate.

3.5.2. Accuracy Evaluation

The accuracy evaluation of the risk performance reasoning is expressed using the root-mean-square error (RMSE) and modified Nash–Sutcliffe model efficiency coefficient (MNSE) [36], defined as follows:

$$\text{RMSE} = \sqrt{\frac{1}{T}\sum_{t=1}^{T}(O(t) - F(t))^2}, \tag{9}$$

$$\text{MNSE} = 1 - \frac{\sum_{t=1}^{T}(O(t) - F(t))^2}{\sum_{t=1}^{T}\left(O(t) - \overline{O}\right)^2}, \tag{10}$$

where $O(t)$ and $\overline{O}$ are the observations and the mean values of the observations, respectively; $F(t)$ stands for the reasoned values, while T is the length of series to be reasoned.

RMSE is mainly used to represent the standard deviation of the differences between observations and forecasts. The range of the RMSE lies between 0 and infinity. A value of 0 for the RMSE indicates that the forecasts are as accurate as the mean of the observations, while bigger values show that the reasoning model is worse than the observed expectation.

MNSE is used to measure the evaluation accuracy and is defined as one minus the sum of the absolute squared differences between the observed and reasoned values divided by the sum of the absolute squared differences between the observations and observed expectation. The range of the MNSE lies between negative infinity and 1.0 (perfect match). A value of 0 for MNSE denotes that the value reasoned is the same as the observed expectation, while negative values for MNSE show that the reasoning model is worse fitted to the observation or explains relationships poorly compared to the model defined using the mean value of observations.

3.5.3. Sensitivity Evaluation

The data from the simulation scenarios and the accident scenario were used for a sensitivity evaluation to judge the deviation of those scenarios from the normal state of bauxite shipping. The deviation of the HMM-based risk reasoning approach is defined as the degree of sensitivity (DOS), which is modified from the concept of the discrete degree of classification [44], in order to avoid invalid measurements due to some LLPs being minus infinity. DOS is expressed as follows:

$$\text{DOS}_i = \frac{1}{2} \times \left(\frac{\log P_{max}(O_i|\lambda) - \log P_{max}(O_0|\lambda)}{\log P_{max}(O_i|\lambda)} + \frac{\log P_{2nd\ max}(O_i|\lambda) - \log P_{2nd\ max}(O_0|\lambda)}{\log P_{2nd\ max}(O_i|\lambda)} \right) \times 100 \quad (11)$$

where $\log P_{max}(O_0|\lambda)$ and $\log P_{2nd\ max}(O_0|\lambda)$ are the maximum and the secondary maximum of *LLP* under the HMM of the normal scenario, respectively, while $\log P_{max}(O_i|\lambda)$ and $\log P_{2nd\ max}(O_i|\lambda)$ are the maximum and the secondary maximum of *LLP* under the HMM for the *i* scenario, respectively. Furthermore, the maximum and the secondary maximum of *LLP* are the first and second values of the *LLP* series listed in descending order.

DOS ranges from 0 to 100. As DOS approaches zero, the total performance becomes more normal. A greater DOS value denotes greater deviation of the risk performance from the normal.

4. Results

4.1. Data of Handy Bauxite Carrier

4.1.1. Ship Parameters

In order to verify the approach to risk performance reasoning, two Handy bauxite carriers were selected: one still in service, and another which previously sank. Their general parameters are listed in Table 2.

Table 2. Parameter of Handy bauxite carrier.

Ship Name	LOA	Breadth	Depth	Service Speed	Total Cargo Weight
M.V. Yuming	189.9 m	32.36 m	15.7 m	14.2 kn	42,700 t
M.V. Bulk Jupiter	189.99 m	32.26 m	17.9 m	14.5 kn	46,400 t

4.1.2. Operational Case

The No. 1509 voyage of M.V. Yuming carried bauxite from Guandan Port, Malaysia, on 28 July 2015. The weather at the loading port was fine a few days before arrival and during the loading period. It took 10 days to reach the destination port of Laizhou, China. There was no water leakage in the sewage well during the voyage of the ship, and the actual performance of the cargo was stable. The annual mean wave height is 1.62 m and the average wave period is 6.62 s in the waters of the Taiwan Strait [45]. Here, 94% of the wave lengths are less than 100 m, and 15% of them are between 70 and 100 m. During the voyage, the weather was good, the visibility was medium, the pressure was stable, the temperature was 20–32 °C, and the meteorological wind and wave levels were 3°–4°. Wind direction was mostly in the bow and stern direction. Except for the first day when the wind pressure difference was as high as 7°, the voyage flow pressure difference did not exceed 3°.

4.1.3. Accident Case

M.V. Bulk Jupiter sailed from Guantan Port, Malaysia, to Qingdao, China. During the loading period, the eastern coast of Malaysia suffered record-breaking rainy weather. The loading operation was delayed repeatedly due to heavy rain. Rainfall on 21 and 23 December during the subsequent loading periods was as high as 240 mm and 258 mm, respectively. After consulting the accident investigation report [46], it was found that the total weight of bauxite in Bulk Jupiter's voyage was

46,400 tons, with an average water content of 21.3%. After sailing on 30 December, the sea weather deteriorated gradually. On 31 December, the northeasterly wind was 6–7, the sea condition was 4–6, and the average wave height was 2.2 m. On 1 January 2015, the northeasterly wind was 8, while the wave height was 2.5–4 m. The swell direction was from the northeast (NE), the vessel's route placed the sea on the port bow as the vessel sailed east-northeast on a heading of 060°. In the final moments prior to sinking, the vessel's speed was 4.3 knots.

4.1.4. Data Collection

The factors value of time series with length of 59 were from the 4-hour interval records of No. 1509 voyage in Logbook onboard M.V. Yuming. According to the criteria of risk reasoning for Handy bauxite carriers, the risk performance rating of factors are obtained in style of PFA and FA, which is shown in Figures 6 and 7. The first 15 groups of data of principle factor analysis of M.V. Yuming is set as Scenario No.1. Scenarios No.2 to No.4 are simulated conditions based on the worse cargo condition, unsatisfied ship condition and terrible environment respectively, and Scenario No.5 is an accident condition based on the last voyage of M.V. Bulk Jupiter.

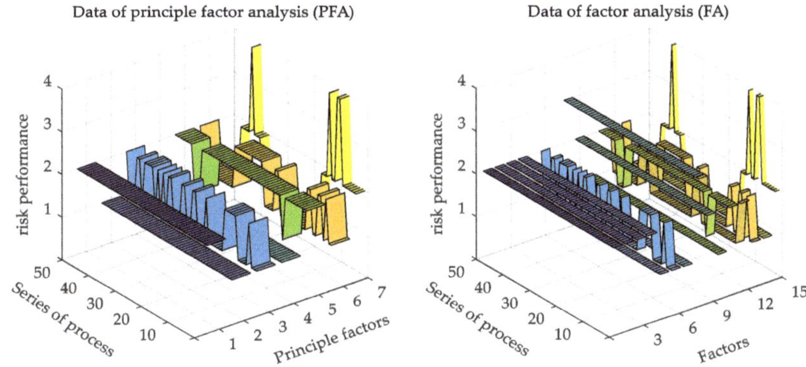

Figure 6. Data of normal scenario No. 1 (M.V. Yuming).

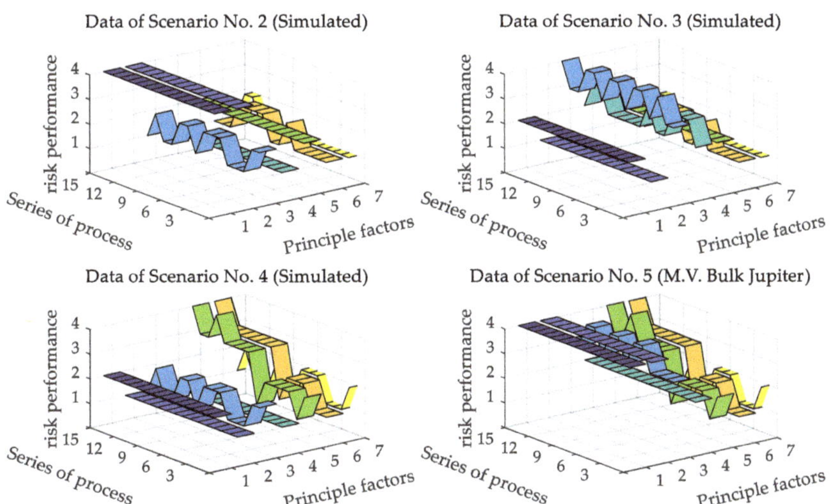

Figure 7. Data of scenario simulated and accident case (principle factor analysis (PFA)).

4.2. Parameter Training

The approach to selecting 15 factors for risk performance was factor analysis (FA). The approach to selecting seven principal factors for risk performance reasoning based on the bauxite model test and FA with 15 factors was principal factor analysis (PFA). Both FA and PFA were used to establish the HMM in order to reason the risk performance of bauxite shipping. In the process of parameter learning, logarithmic likelihood values were used to represent the matched degree between parameters and models. As the number of iterations increased, the matched degree tended to converge. The model parameters obtained from the training were valid because they satisfied the local optimal characteristics of the parameters.

Figure 8 indicates that the HMM tended to converge until the 37th and 34th iterations for PFA and FA, respectively. Nine groups of data were used to test the HMM and get a converged likelihood. Taking the PFA as an example, the trained and optimized HMM parameter was $\pi = [1.0\,0\,0\,0]$.

$$A = \begin{bmatrix} 0.0000 & 0.0000 & 0.0000 & 1.0000 \\ 0.2936 & 0.2711 & 0.1408 & 0.2945 \\ 0.0000 & 0.0019 & 0.0000 & 0.9981 \\ 0.1460 & 0.2830 & 0.5710 & 0.0000 \end{bmatrix}, \text{and } B = \begin{bmatrix} 0.0000 & 1.0000 & 0 & 0.0000 \\ 0.0006 & 0.9994 & 0 & 0.0000 \\ 0.6020 & 0.3474 & 0 & 0.0505 \\ 1.0000 & 0.0000 & 0 & 0.0000 \end{bmatrix}.$$

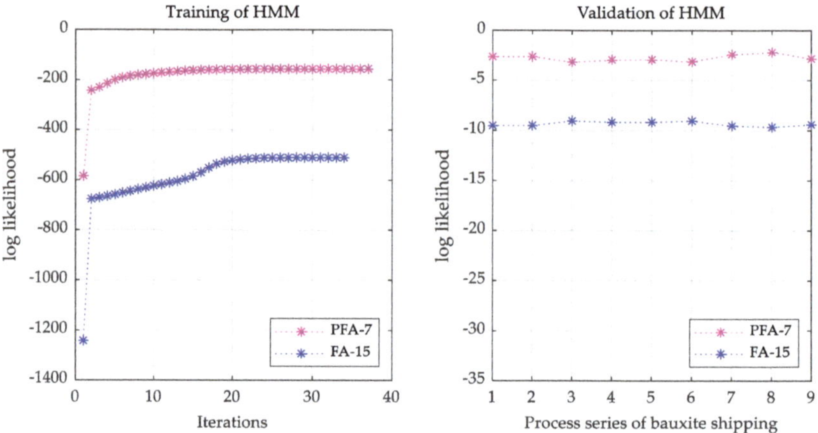

Figure 8. Parameter training and validation of HMM for risk reasoning.

4.3. Selection of Approach to Reasoning

Taking the risk reasoning of M.V. Yuming as an example, the best approach to reasoning was selected as shown in Figure 9.

The RMSE values indicate that the best forecast was performed with 3K (three steps and seven principal factors). The MNSE values indicate that the best forecast was performed with 2F (two steps and 15 factors).

4.4. Result of Reasoning

By using the reasoning approach for the risk performance of the factors, every factor was reasoned within a risk grade boundary of 1–4. Risk grades were obtained for 15 factors (FA approach) and seven principal factors (PFA approach). According to the abovementioned quantization regulation of risk performance, the time series of the total risk performance of the bauxite shipping process was obtained. Likewise, the same approach could be used to obtain the observed sequence of the total risk performance. Taking M.V. Bulk Jupiter as an example, the risk performance reasoning model based on a three-step PFA approach was adopted. The first 12 sets of data were used to train the model parameters, and the last three sets of data were used for reasoning and testing; the results are shown in Figure 10.

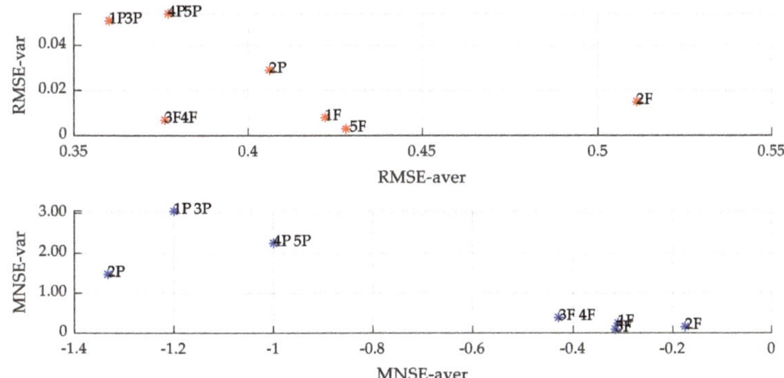

Figure 9. Root-mean-square error (RMSE) and modified Nash–Sutcliffe model efficiency coefficient (MNSE) with various combinations of numbers and step sizes.

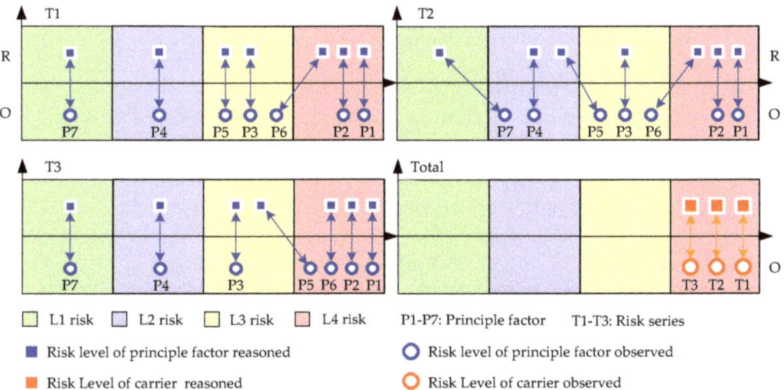

Figure 10. Comparison between calculated and observed risk performance of principal factors and the total process of M.V. Bulk Jupiter.

The comparison with the observed risk performance shows that the reasoned risk performance with seven principal factors was mostly accurate. The risk performance of the factors was transformed into an overall risk performance using the quantification regulation. The last three periods of the whole shipping process had the highest risk rating. In fact, the accident report showed that the actual moisture content of bauxite severely exceeded the transportable moisture limit (TML), and the carrier encountered strong winds and moderate swells along the coast of Vietnam. Consequently, capsizing occurred. Therefore, the calculated risk was consistent with the actual situation. Therefore, the model constructed is effective in forecasting risk performance for the bauxite shipping process.

4.5. Effectiveness and Accuracy

Figure 11 shows the comparison of the effectiveness evaluation of reasoning using the FA and PFA approaches, where the evaluation was conducted on M.V. Yuming for the risk performance of a bauxite shipping process. It can be seen that the calculated risk rating was completely consistent with the observed risk rating, and the effectiveness of the performance reasoning was ideal. The effectiveness of the risk reasoning based on FA or PFA was also quantitatively evaluated using DOB and DOD. The degree of deviation for the risk reasoning model was DOB $= 0$, indicating no deviation in the reasoning for the four risk ratings. The degree of detection was DOD $= 1$, showing that the reasoned risk rating was completely accurate.

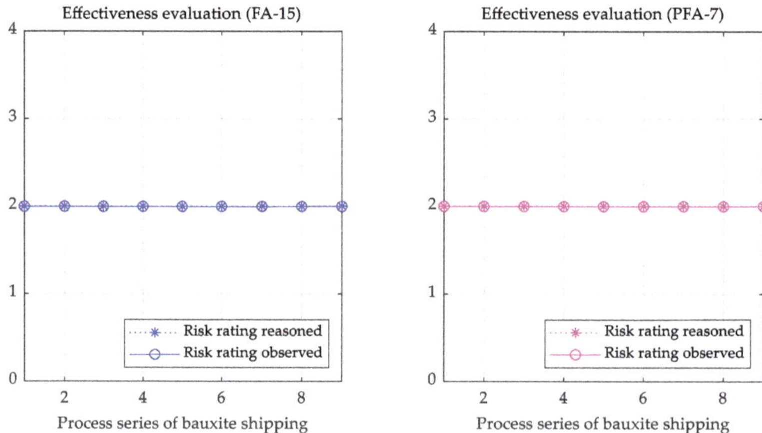

Figure 11. Effectiveness evaluation of reasoning for risk performance.

Accuracy evaluation was conducted on the calculated nine-step risk performance matrix and the actual observed risk performance matrix, as shown in Figure 12. The RMSE of reasoning based on the FA approach was 0.336 on average with a maximum of 0.516 and a minimum of 0.0, while the RMSE of the PFA approach was 0.36 on average with a maximum of 0.655 and a minimum of 0.0. The error of reasoning was small; thus, the accuracy was better evaluated. The average of the MNSE calculated using the FA approach was −0.459, ranging from 1.0 to −2.0. The average of the MNSE calculated using the PFA approach was −1.20 with a maximum of 1.0 and a minimum of −4.40. The MNSE values based on the two approaches were slightly smaller than zero; therefore, the reasoned performance was slightly conservative due to the values being lower than the actual risk performance.

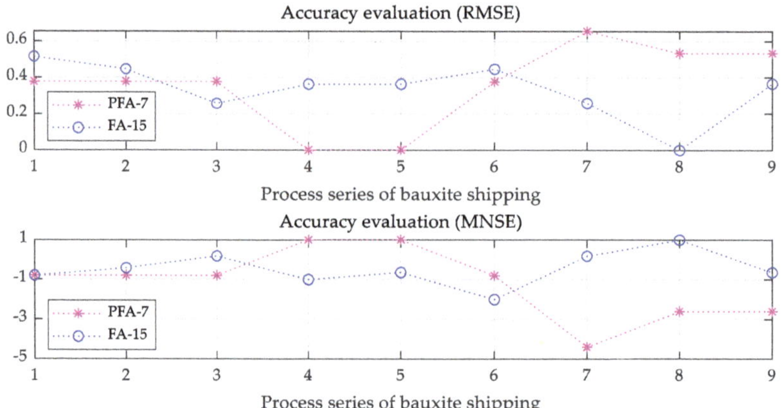

Figure 12. Accuracy evaluation of reasoning for risk performance.

5. Analysis and Discussion

5.1. Sensitivity Analysis

Five groups of test data of the bauxite shipping process were selected for sensitivity evaluation. Scenario No. 1 featured normal conditions based on the No. 1509 voyage of M.V. Yuming. Quantitative analysis was carried out on the conformity of risk performance under each scenario. The sensitivity function of the risk reasoning model established previously was used to obtain the degree of sensitivity (DOS) for each scenario relative to normal scenario No. 1.

Table 3 indicates that the effects of scenarios No. 2 and No. 5 were relatively close, representing the worst risk performance deviated from the normal scenario for bauxite shipping. Compared with the FA-based DOS, the PFA-based DOS was more sensitive. The risk assessment and classification of scenarios No. 2 and No. 5 were the best, with sensitivity values above 95, meaning that any abnormal risk performance of bauxite shipping can be detected more accurately. The sensitivity of scenario No. 4 was very small, showing that the model could not effectively distinguish the risk performance of bauxite shipping using the benchmark model. When the environment was poor, the risk performance of scenario No. 4 was consistent with that of normal scenario No. 1, and the risk was very low.

Table 3. Degree of sensitivity (DOS) of HMM for risk performance reasoning.

Scenario	Scenario No. 2	Scenario No. 3	Scenario No. 4	Scenario No. 5
FA-based DOS	88.962	35.789	3.118	89.567
PFA-based DOS	96.234	80.398	0.001	96.854

5.2. Pre-Warning of Threat

The risk performance reasoning model constructed in this paper is based on the normal conditions of cargo, ship, and environment for Handy bauxite carriers. The data of factors for scenarios No. 2 to No. 5 were input into the HMM for risk performance reasoning based on Scenario No. 1, where the log-likelihood probability of the output could provide a guide to abnormal risk monitor. A larger log-likelihood would denote a greater probability of low risk performance. Figure 13 demonstrates the log-likelihood value of 15 time series for scenarios No. 1 to No. 5.

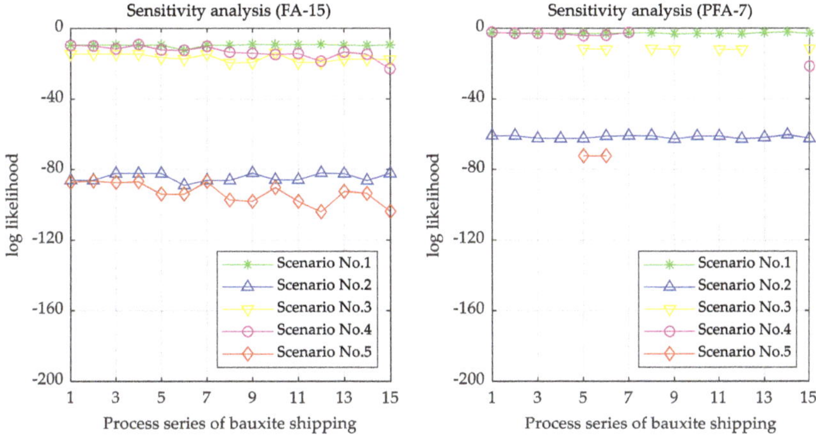

Figure 13. Sensitivity evaluation of reasoning for risk performance.

The logarithm likelihood values of scenario No. 3 with unsatisfactory ship conditions and of scenario No. 4 with a terrible environment were close to those of the normal scenario No. 1.

The log-likelihood value of scenario No. 2 with worse cargo conditions more obviously deviated from that of scenario No. 1 than that of Scenario No. 3 and Scenario No. 4, indicating that the risk performance of the Handy bauxite carrier in scenario No. 2 was worse, and that the deterioration of cargo with respect to ship conditions and the environment had a greater impact on the total safety of the bauxite shipping process.

Scenario No. 5 had a more serious deviation from the normal condition. This deviation was more significant than the worst scenario of single-category factors such as the cargo, ship, or environment, which all contribute to the total risk of bauxite shipping process. It was found that the coupling effect of risk factors produced a coupling risk, which aggravated the total risk performance during the shipping process.

When using the PFA-based risk reasoning model to detect the performance, the log-likelihood value of individual time points was negative infinite, indicating that the risk behavior at this time was seriously inconsistent with the normal scenario. The risk performance at this time point can be determined as the highest rating, which is unacceptable.

In scenarios No. 3, 4, and 5, the log-likelihood probability decreased gradually, showing that the total performance deviated from the normal condition and deteriorated gradually, thereby achieving the state monitoring and issuing a pre-warning threat during the shipping process.

5.3. Risk Performance Reasoning with Hidden Markov Model

- The influencing factors of the reasoning performance were obtained through the demonstration, thereby providing a reference for optimizing the model parameters and the reasoning approach. According to the results and analysis of the risk performance reasoning, increasing the training data and identifying principal factors can improve the reasoning performance.

- Due to the great influence of cargo performance on the total risk performance of Handy bauxite carriers, high-risk cargo factors should be avoided. In particular, when the size of a bauxite carrier and the shipping environment or ship routing cannot be changed, the cargo quality related to shipping safety must be closely monitored, and a moisture content audit must be performed before loading, while rainproof measures should be ensured during loading and navigating.

- Compared with the effect of any single factor, the effect of the cargo, ship, and environment is more significant, leading to a transition of the total risk performance of Handy bauxite carriers. The process between subsystems is an important part of the shipping process risk control of Handy bauxite carriers. In order to ensure safety of shipping, it is essential to respond to risks timely and effectively based on accurate multi-source risk data, such as ship maneuvering, environmental information, and dynamic cargo information from the water ingress alarm system and the radar fluid-level meter fixed in the cargo hold. Therefore, it is essential to make full use of the Internet of things and artificial intelligence to develop intelligent risk monitoring sensors and forecasting equipment for shipping processes on Handy bauxite carriers.

- Based on the accurate historical data of principal factors and a more detailed classification of risk performance, short-term process risk reasoning with high-quality can be realized. Detailed shipping data of bauxite and Handy carriers include, but are not limited to, information from the logbook, which can be used in parameter learning to obtain more accurate parameters of the HMM. This will allow real-time monitoring and pre-warnings of cargo liquefaction and ship stability for the navigator onboard and ship manager ashore.

- In total, 91% of the accidents on international routes caused by the liquefaction of solid bulk cargo involve Handy carriers [47], which are usually between 80 m and 190 m in length. When a Handy carrier navigates at sea with a wavelength of 90–150 m and the speed is close to the wave speed, the ship ends up hogging and sagging, threatening the structural strength of the ship. The stability of the ship is, thus, greatly reduced.

- If possible, Handy carriers with bauxite should avoid sailing in heavy seas. In the stormy season, some special voyages of bauxite shipping could be conducted by larger bulk carriers or special ore carriers, which would have strong stability even when bauxite is liquefied. Appropriate working conditions of the main engine and ballast water can reduce the vibration intensity in the cargo hold, thereby reducing the risk of early cargo liquefaction, and obtaining earlier pre-warnings and for a better response time.

6. Conclusions

The paper constructed a hidden Markov model of risk performance reasoning for bauxite shipping. Based on the HMM parameters, a transfer matrix between observation variables and the hidden risk status was obtained. The relationship between risk performance and principal factors was determined, and quantification of the total risk performance was carried out. The model and algorithm of risk performance reasoning was verified using cases of bauxite shipping processes. The effectiveness evaluation indexes of the total risk performance, the accuracy evaluation indexes of the factor risk performance, and the sensitivity evaluation index of the reasoning model were used to measure the performance of the HMM-based risk reasoning approach.

Some conclusions are proposed. Firstly, the approach to risk performance reasoning with HMM can effectively forecast the risk performance of bauxite shipping processes for Handy carriers. Furthermore, increasing the amount of training data and identifying key risk factors can help improve reasoning performance. The risk performance of cargo factors is critical to the overall risk state of Handy bauxite carriers. Compared with the influence of cargo factors, the coupling effect of multiple factors has a great influence, which leads to a leap in the overall risk scale onboard the ship. A benchmark model of risk assessment for the bauxite shipping process was built to monitor the risk threat due to the coupling effect of the cargo, environment, and ship; this is especially applicable to Handy carriers which do not have a special structural design for cargoes that may liquefy.

As with the global bauxite industry undertaking research into the behavior and characteristics of bauxite cargoes during ocean transportation, the crews and managers of Handy bauxite carriers can be armed with the knowledge of process risk and risk control; therefore, the risk performance analyzed in this paper is essential in relation to the bauxite shipping process, whose associated risk remains to be officially recognized by the IMO. The output of this paper can support the captain with quantitative risk decision-making, as opposed to previously used empirical decision-making, thereby laying a foundation for risk pre-warnings and the process safety of bauxite cargo.

Author Contributions: Conceptualization, S.H.; investigation, J.W.; methodology, J.W. and S.H.; software, J.W. and Y.Z.; supervision, Y.J. and J.F.; validation, Y.Z.; visualization, J.W.; writing—original draft, J.W.; writing—review and editing, Y.J., S.H., and J.F. All authors have read and agreed to the published version of the manuscript.

Funding: This work was supported by the National Science Foundation of China (NSFC) under grant No. 11671416 and 51909156, and the China Postdoctoral Science Foundation, grant number 2016M591651.

Acknowledgments: The authors acknowledge the technical support and funding support given by Shibin Zhang from Shanghai Normal University and Changhai Huang from Shanghai Maritime University. The authors further acknowledge M.V. Yuming for providing voyage data related to bauxite.

Conflicts of Interest: The authors declare no conflict of interest.

References

1. China's Imported Bauxite Was 82.62 Million Tons in 2018. Available online: http://www.mofcom.gov.cn/article/i/jyjl/k/201902/20190202834585.shtml (accessed on 15 February 2019).
2. Global Bauxite Working Group. *IMO-CCC4/INF.10. Global Bauxite Working Group Report on Research into the Bauxite during Shipping*; International Maritime Organization: London, UK, 2017.
3. International Association of Dry Cargo Shipowners. *Bulk Carrier Casualty Report (2003, the Previous Ten Years (1994–2003) and the Trends)*; International Association of Dry Cargo Shipowners: London, UK, 2004.
4. International Association of Dry Cargo Shipowners. Bulk Carrier Casualty Report (2005 to 2015). Available online: https://www.intercargo.org/bulk-carrier-casualty-report-2005-2015/ (accessed on 1 March 2016).

5. International Association of Dry Cargo Shipowners. Bulk Carrier Casualty Report (Years 2008 to 2017 and the Trends). Available online: https://www.intercargo.org/bulk-carrier-casualty-report-2017/ (accessed on 3 May 2018).

6. International Association of Dry Cargo Shipowners. Cargo Liquefaction Continues to Be a Major Risk for Dry Bulk Shipping. *Media Release*. 31 January 2019. Available online: https://www.intercargo.org/cargo-liquefaction-continues-to-be-a-major-risk-for-dry-bulk-shipping/ (accessed on 31 January 2019).

7. Insurance Companies Appeal Shipowners to Pay Attention to Bauxite Transportation. Available online: http://www.eworldship.com/html/2015/ship_finance_0109/97262.html (accessed on 9 January 2015).

8. Li, J.M.; Qi, J.; Zheng, Z.Y. Safety evolution mechanism for the maritime dangerous chemical transportation system under uncertainty conditions. *J. Harbin Eng. Univ.* **2014**, *35*, 707–712.

9. Ma, X.X.; Li, W.H.; Zhang, J.; Qiao, W.L.; Zhang, Y.D.; An, J. Accident occurrence regularity analysis of shipping ore concentrate powder which may liquify and countermeasures. *Navig. China* **2014**, *2*, 43–47.

10. Shen, J. Nickel Ore and its safe shipment. *Adv. Mater. Res.* **2012**, *396–398*, 2183–2187. [CrossRef]

11. Wang, H.L.; Koseki, J.; Sato, T.; Chiaro, G.; Tian, J.T. Effect of saturation on liquefaction resistance of iron ore fines and two sandy soils. *Soils Found.* **2016**, *56*, 732–744. [CrossRef]

12. Popek, M. Investigation of influence of organic polymer on TML value of the mineral concentrates. In *Proceedings of ESREL*; Kołowrocki, K., Ed.; Taylor and Francis Group: London, UK, 2005; pp. 1593–1596.

13. Popek, M. The influence of organic polymer on parameters determining ability to liquefaction of mineral concentrates. *Int. J. Mar. Navig. Saf. Sea Transp.* **2010**, *4*, 435–440.

14. Altun, O.; Göller, Z. The effect of filter aid reagent on CBI copper concentrate moisture. In Proceedings of the 23rd International Mining Congress and Exhibition of Turkey, Antalya, Turkey, 16–19 April 2013.

15. Wu, J.J.; Liu, Y.X.; Hu, S.P.; Zhao, Y.H. Hidden Markov model for risk estimation of ship carrying liquefiable cargoes. *China Saf. Sci. J.* **2017**, *27*, 73–78.

16. Munro, M.C.; Mohajerani, A. Moisture content limits of iron ore fines to prevent liquefaction during transport: Review and experimental study. *Int. J. Miner. Process.* **2016**, *148*, 137–146. [CrossRef]

17. Munro, M.C.; Mohajerani, A. Liquefaction incidents of mineral cargoes on board bulk carriers. *Adv. Mater. Sci. Eng.* **2016**, *2016*, 5219474. [CrossRef]

18. Spandonidis, C.C. Modelling and Numerical/Experimental Investigation of Granular Cargo Shift in Maritime Transportation. Ph.D. Thesis, National Technical University of Athens, Athens, Greece, 2016.

19. Montewka, J.; Goerlandt, F.; Kujala, P.; Lensu, M. Towards probabilistic models for the prediction of a ship performance in dynamic ice. *Cold Reg. Sci. Technol.* **2015**, *112*, 14–28. [CrossRef]

20. Wang, Y.G.; Tan, J.H. Research progress on ship stability and capsizing in random waves. *J. Ship Mech.* **2010**, *14*, 191–201.

21. Ren, Y.L.; Mou, J.M.; Li, Y.J.; Yi, K. Monte Carlo simulation for the grounding probability of ship maneuvering in approach channels. *J. Ship Mech.* **2014**, *18*, 532–539.

22. Zhao, Y.L.; Meng, S.X. Safety transportation of easily fluidized cargoes. *J. Dalian Marit. Univ.* **2012**, *38*, 15–18.

23. Holmes, R.; Williams, K.; Honeyands, T.; Orense, R.; Roberts, A.; Pender, M.; McCallum, D.; Krull, T. Bulk commodity characterisation for transportable moisture limit determination. In Proceedings of the International Mineral Processing Congress, Québec City, QC, Canada, 11–15 September 2016.

24. Ding, S. Numerical study on vibration characteristics of a cargo ship. *Ship Sci. Technol.* **2017**, *39*, 19–21.

25. Munro, M.C.; Mohajerani, A. Variation of the geotechnical properties of Iron Ore Fines under cyclic loading. *Ocean Eng.* **2016**, *126*, 411–431. [CrossRef]

26. Drzewieniecka, B. Safety aspect of handling and carriage of solid bulk cargoes by sea. *Sci. J. Marit. Univ. Szczec.* **2014**, *39*, 63–66.

27. Ju, L.; Vassalos, D.; Boulougouris, E. Numerical assessment of cargo liquefaction potential. *Ocean Eng.* **2016**, *120*, 383–388. [CrossRef]

28. Andrei, C.; Pazara, R.H. The impact of bulk cargoes liquefaction on ship's intact stability. *UPB Sci. Bull.* **2013**, *75*, 12–16.

29. Munro, M.C.; Mohajerani, A. Laboratory scale reproduction and analysis of the behaviour of iron ore fines under cyclic loading to investigate liquefaction during marine transportation. *Mar. Struct.* **2018**, *59*, 482–509. [CrossRef]

30. Daoud, S.; Ennour, S.; Said, I.; Bouassida, M. Quasi-static numerical modelling of ore cargo during shipping. In Proceedings of the XVI European Conference on Soil Mechanics and Geotechnical Engineering, Edinburgh, Scotland, UK, 13–17 September 2015.

31. Daoud, S.; Ennour, S.; Said, I.; Bouassida, M. Numerical analysis of cargo liquefaction mechanism under the swell motion. *Mar. Struct.* **2018**, *57*, 52–71.

32. Liu, D.G.; Zheng, Z.Y.; Wu, Z.L. Risk analysis system of underway ships in heavy sea. *J. Traffic Transp. Eng.* **2004**, *4*, 100–102.

33. IMO. *CCC.1/Circ.2. Carriage of Bauxite that May Liquefy*; International Maritime Organization: London, UK, 2015.

34. IMO. *CCC.1/Circ.2/Rev.1. Carriage of Bauxite that May Liquefy*; International Maritime Organization: London, UK, 2017.

35. Wu, J.; Hu, S.; Jin, Y.; Fei, J.; Fu, S. Performance simulation of the transportation process risk of bauxite carriers based on the Markov chain and cloud model. *J. Mar. Sci. Eng.* **2019**, *7*, 108. [CrossRef]

36. Chen, S.; Chung, G.; Kim, B.S.; Kim, T.W. Modified analogue forecasting in the hidden Markov framework for meteorological droughts. *Sci. China Technol. Sci.* **2019**, *1*, 151–162. [CrossRef]

37. Joshi, J.C.; Kumar, T.; Srivastava, S.; Sachdeva, D. Optimisation of Hidden Markov Model using Baum–Welch algorithm for prediction of maximum and minimum temperature over Indian Himalaya. *J. Earth Syst. Sci.* **2017**, *126*, 3. [CrossRef]

38. Fabbri, T.; Vicen-Bueno, R. Weather-Routing system based on METOC Navigation Risk assessment. *J. Mar. Sci. Eng.* **2019**, *7*, 127. [CrossRef]

39. Wen, Z.C.; Chen, Z.G. Network security situation prediction method based on hidden Markov model. *J. Cent. South Univ. Sci. Technol.* **2015**, *46*, 3689–3695.

40. Ghahramani, Z.; Jordan, M.I. Factorial hidden markov models. *Mach. Learn.* **1997**, *29*, 245–273. [CrossRef]

41. Hu, S.; Li, Z.; Xi, Y.; Gu, X.; Zhang, X. Path analysis of causal factors influencing marine traffic accident via structural equation numerical modeling. *J. Mar. Sci. Eng.* **2019**, *7*, 96. [CrossRef]

42. Ma, Z.X.; Ren, H.L. Research on the evaluation methods of formal safety assessment for ships. *Syst. Eng. Electron.* **2002**, *10*, 66–70.

43. Wen, Z.C.; Chen, Z.G.; Tang, J. Prediction of network security situation on the basis of time series analysis. *J. South China Univ. Technol. Nat. Sci. Ed.* **2016**, *5*, 137–143.

44. Zhang, X.N.; Lei, W.; Li, B. Bering fault detection and diagnosis method based on principal component analysis and hidden Markov model. *J. Xi'an Jiaotong Univ.* **2017**, *6*, 1–7.

45. Li, Z. Passenger Ro-Ro ship type evaluation for Taiwan straits. *Navig. China* **2010**, *33*, 99–102.

46. The Bahamas Maritime Authority. *"M.V Bulk Jupiter"—Report of the Marine Safety Investigation into the Loss of a Bulk Carrier in the South China Sea on January 2nd 2015*; Bahamas Maritime Authority: London, UK, 2015.

47. Munro, M.C.; Mohajerani, A. Bulk cargo liquefaction incidents during marine transportation and possible Causes. *Ocean Eng.* **2017**, *141*, 125–142. [CrossRef]

Article

A Novel Risk Assessment and Analysis Method for Correlation in a Complex System Based on Multi-Dimensional Theory

Zeyong Jiang, Tingdi Zhao, Shihai Wang * and Fuchun Ren

School of Reliability and Systems Engineering, Beihang University, Beijing 100191, China;
jiangzeyong@buaa.edu.cn (Z.J.); ztd@buaa.edu.cn (T.Z.); renfuchun@126.com (F.R.)
* Correspondence: wangshihai@buaa.edu.cn

Received: 3 April 2020; Accepted: 20 April 2020; Published: 25 April 2020

Abstract: With the rapid development of high integrations in large complex systems, such as aircraft, satellite, and railway systems, due to the increasingly complex coupling relationship between components within the system, local disturbances or faults may cause global effects on the system by fault propagation. Therefore, there are new challenges in safety analysis and risk assessment for complex systems. Aiming at analyzing and evaluating the inherent risks of the complex system with coupling correlation characteristics objectively, this paper proposes a novel risk assessment and analysis method for correlation in complex system based on multi-dimensional theory. Firstly, the formal description and coupling degree analysis method of the hierarchical structure of complex systems is established. Moreover, considering the three safety risk factors of fault propagation probability, potential severity, and fault propagation time, a multi-dimensional safety risk theory is proposed, in order to evaluate the risk of each element within the system effecting on the overall system. Furthermore, critical safety elements are identified based on Pareto rules, As Low As Reasonably Practicable (ALARP) principles, and safety risk entropy to support the preventive measures. Finally, an application of an avionics system is provided to demonstrate the effectiveness of the proposed method.

Keywords: safety; coupling correlation; risk assessment; multi-dimensional theory

1. Introduction

In recent years, due to the complex correlation of components in complex systems, local faults may have a great effect on the overall system by fault propagation [1–3]. Therefore, the safety and risk analysis of such complex systems has attracted more and more attention. Safety analysis and risk assessment aims to eliminate and control various hazards through the design system and take preventive measures to prevent accidents that will cause personal injury, equipment damage, and task failure during system operation. With the development of science and technology, a series of analysis methods for evaluating system failures and risk events have been developed, especially in high-risk fields such as aerospace, chemical, nuclear, and other industrial fields [4,5]. However, there are still a number of safety problems in these methods caused by the coupling and correlation characteristics in complex systems.

Traditional safety modeling and analysis methods are mainly based on the logical process of induction and deduction to carry out system safety analysis. From the local characteristics of the system or the direct relationship between internal components, these methods are used to find the root cause of safety problems and carry out safety work such as analysis, verification, assurance, etc. Typical analysis methods are Fault Tree Analysis (FTA) [6], Event Tree Analysis (ETA) [7,8], Failure Mode and Effects Analysis (FMEA) [9–11], Hazard and Operability Analysis (HAZOP), Probability Risk

Assessment (PRA) [12,13], etc. These methods have applications in nuclear power, chemical, and even aerospace. Zhou, X. [14] proposed a modified FMEA based on Dempster–Shafer evidence theory to analyze safety of aircraft turbine rotor blades. Rhee, Seung J. [15] used a Monte Carlo simulation and cost-based FMEA to account for the uncertainties in: detection time, fixing time, occurrence, delay time, down time, and model complex scenarios. Hyeon-ae Jang [16] proposed a time-dependent probabilistic approach of FMEA to evaluate safety of automotive-manufacturing. Liu, Yang [17] proposed an FTA-based method for risk decision-making in emergency response and applied it in H1N1 infectious diseases. Cheraghi, M. [18] proposed a fuzzy multi-attribute HAZOP technique and Analytic Hierarchy Process (AHP) to determine the weight of risk factors and to prioritize the hazards.

Moreover, with the development of the accident theory, a large number of modern methods for safety analysis, such as Markov process, Analysis and Design Language (AADL), Petri nets, Bayesian networks, etc., have also been invented. Feng, Q. [19] proposed the staged Bayesian failure model for girth welds of a pipeline, using the tree-type accident theory and Bayesian survival analysis method. Zhao, C. [20] applied the continuous-time Markov chains to analyze reliability of the reconfigurable integrated modular avionics. Singh, P. [21] applied Petri nets to estimate performability to ensure system dependability requirements and did the performance analysis of safety critical and control systems that helps to estimate the risk. Baouya, A. [22] presented AADL based on model-driven specification and probabilistic model checking to automatically analyze safety-based availability before synthesizing the embedded software product. John McDermid's team at the University of York in the United Kingdom proposed the theory and analysis techniques of safety case [23–26] to confirm that the system reaches an acceptable level of safety by establishing a correlation between safety requirements and safety evidence. The Functional Resonance Accident Model (FRAM) [27,28] was proposed by Erik Hollnagel, based on the principle of stochastic resonance in the system. However, the above safety analysis methods mainly focused on a qualitative description and the study of coupling mechanisms, and lack the quantitative analysis and evaluations of coupling and correlation relationships between components in the system.

Internationally, aviation criteria ARP 4754 (A) [29,30] recommended by the American Society of Automotive Engineers defines safety as a state where the risk is lower than the border risk. The domestic GJB 900A defines safety as the ability of a product not to cause personal injury or death, system damage, major property damage, or damage human health and the environment. For the measurement of risk, risk model of probability and severity is the most widely used, such as the civil aviation standard ARP 4761 [31], the US military standard MIL-STD-882E [32], the national military standard GJB 900A, etc., FTA, FMEA, ETA, and other reliability and safety analysis methods all use the models to evaluate risks. However, with the further understanding of the concept of risk, people have more research and cognition of the elements involved in risk. Mazzuchi T A. [33] developed a relationship for the probability of wire failure as a function of influencing factors in an aircraft environment in order to analyze wire failure in aircraft. Cour-Harbo A L. [34] presented a method for quantifying the probability of fatalities resulting from an uncontrolled descent of an unmanned aircraft conducting a beyond visual line-of-sight (BVLOS) flight so as to solve the major challenges to make a realistic and effective risk assessment of conducting operation of BVLOS. Li L. [35] proposed a new risk assessment method based on the cloud model, aiming to make an effective risk assessment method for subway operation by considering five aspects. Fayaz, M. [36] proposed an integrated risk index model based on hierarchical fuzzy logic for underground risk assessment to avoid occurrence of accidents due to underground facilities. Duan, Y. [37] presented a novel network security risk assessment approach by combining subjective and objective weights under uncertainty to effectively evaluate computer network security. Most of the above risk assessment methods focus on the analysis of accident probability and severity, and lack of multi-dimensional safety risk assessment methods by taking time factor related fault into consideration.

In view of the above considerations, this paper proposes a novel risk assessment and analysis method for correlation in complex system based on multi-dimensional theory, aiming at analyzing and evaluating the risk of complex system considering the coupling correlation, so as to identify the critical risk elements. Firstly, the formal description and coupling correlation analysis method of the hierarchical structure of complex systems based on the typical task-function-resource model is proposed, aiming to achieve a formal description of the coupling correlation between components within complex systems, providing the foundation for and analysis and evaluation of risk. Moreover, considering the three safety risk factors of propagation probability, potential severity, and propagation time, a multi-dimensional safety risk theory is proposed, in order to evaluate the risk of each element in the system effecting on the overall system from multiple perspectives. Furthermore, critical safety elements are identified based on Pareto rules, ALARP principles, and safety risk entropy to support the preventive measures.

The remainder of the paper is organized as follows. Section 2 describes the hierarchical model of complex systems and coupling correlation between elements in system. In Section 3, multi-dimensional safety risk theory and assessment are proposed. Section 4 introduces an application of avionics system. Section 5 presents the conclusions.

2. Coupling Correlation of Complex System

In a general sense, the adjective "complex" describes a system or component that by design or function or both is difficult to understand and verify [38]. Complex system is any system featuring a large number of interacting components that is often difficult to understand, and hard to solve [39,40]. Compared with simple systems, complex systems are usually characterized by more components and a high degree of coupling [41,42]. In the real world, there are a large number of systems that show the characteristics of complexity, such as ecosystem, social organization system, complex social technology system, complex electromechanical system, and complex equipment system [43–45]. The complex system concerned in this paper is mainly located in the complex engineering technology system, that is, a kind of complex system with engineering technology characteristics.

Coupling correlations refer to all kinds of association relationships between various elements in the system due to task and function requirements, such as resource reuse, information transfer, data sharing, etc. The strength of the coupling correlation can be quantified by the degree of coupling. For complex systems, the internal coupling correlations are more complicated. The complex coupling correlations increase the risk of fault propagation in the system. The establishment of a system model based on the coupling correlation is the basis for analyzing and quantifying the risk of system for fault propagation.

2.1. Hierarchical Model and Description of Complex Systems

2.1.1. Hierarchical Model

Generally, a system is built on the background of specific task requirements, that is, the use case scenarios of the system are planned in advance through requirements analysis. These planned use case scenarios can be defined as the task view or task layer of the system. Then, based on the system task planning, the necessary functional decomposition is needed, namely, what basic functions need to be established in order to achieve a specific task. Therefore, this paper defines such decomposed functions as the functional view or function layer of the system. However, the tasks and functions are in the design of the system logic layer. The final implementation still needs the support of the physical layer such as typical computing, storages, communication resources, etc. In other words, the configuration and mapping relationships from logic layer to the physical layer in the system require to be clarified and completed. This paper defines these general physical resources as the resource view or resource layer of the system.

In summary, when analyzing from the perspective of hierarchical decomposition, a hierarchical system model based on the task-function-resource layer can be established [46–48]. Then, the coupling relationships between the elements in the task-function-resource layer and between the layers can be considered. Based on the topology modeling theory, a topology model with elements as nodes and correlation relationships as connections can be formed. Finally, combined with hierarchical decomposition and coupling analysis, a complex system hierarchy model based on task-function-resource architecture is synthesized, as shown in Figure 1 and different colors and shapes are applied to present the elements in different layers for distinguishing. It is based on the assumption that the number of tasks, functions, and resources is unchanged during the time and the correlation relationships in systems are constant during the time.

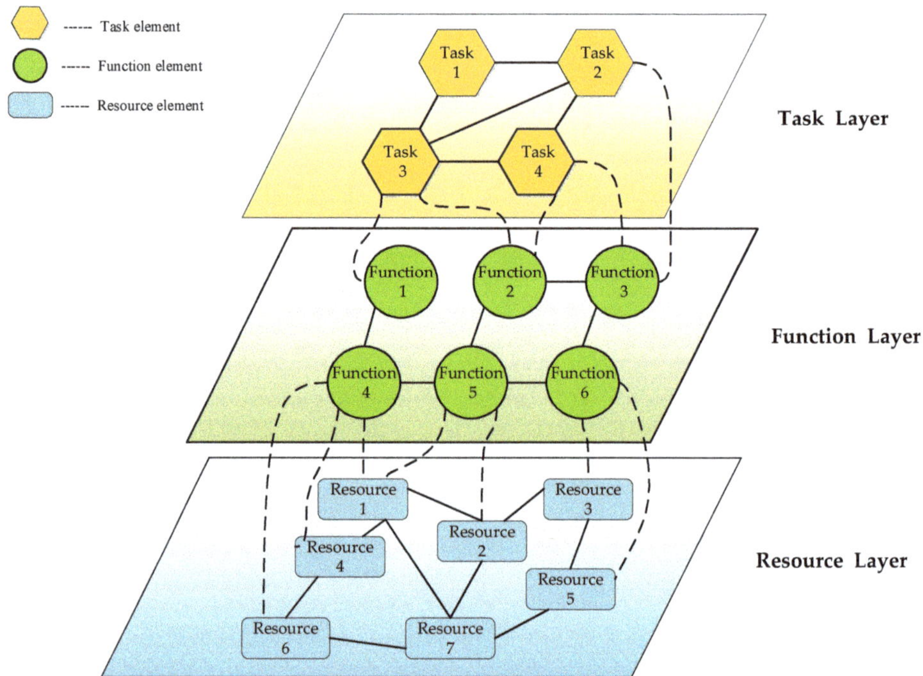

Figure 1. Typical task-function-resource architecture of complex systems. The straight lines represent the coupling correlation of elements in the same layer, and dashed lines represent coupling correlation across layers of elements in the different layers. Lines (connections) are bi-directional.

It is assumed that the destructive event of the system only originates from the fault of the element of resource layer, and the element of function layer exists as the use of the element of resource layer and the role of the caller. Therefore, in this paper, based on the fault propagation problem introduced by resource layer, the fault of the resource element is the fault trigger point, and the function element provides the propagation medium.

2.1.2. Formal Description of Hierarchical Model

For the task-function-resource hierarchy architecture of the system, from the perspective of the element set, the system's task element set, function element set, and resource element set can be defined separately. The task element t_i is a task unit established by the system requirement analysis. It is supported by a series of basic function elements. The task element set T can be expressed as a set of several task elements: $T = \{t_1, t_2, \ldots, t_k\}$. Similarly, the function element f_i is the basic function unit that supports the task implementation in the system. It is supported by a series of basic resource elements. The function element set F can be expressed as a set of several functional elements: $F = \{f_1, f_2, \ldots, f_m\}$. The resource element r_i is a physical or logical unit that supports the realization of function in the system. The system resource element set R can be represented as a set of several resource elements: $R = \{r_1, r_2, \ldots, r_n\}$.

Through the analysis of the system hierarchy architecture, in order to describe specific relational information, adjacency matrix can be used for the most direct formal record, that is, the mapping correlation matrix between task-function elements can be expressed as shown in Matrix (1).

$$M^{tf} = (M_{ij}^{tf})_{k \times m} = \begin{array}{c} t_1 \\ t_2 \\ \vdots \\ t_k \end{array} \begin{array}{cccc} f_1 & f_2 & \cdots & f_m \end{array} \atop \left[\begin{array}{cccc} M_{11}^{tf} & M_{12}^{tf} & \cdots & M_{1m}^{tf} \\ M_{21}^{tf} & M_{22}^{tf} & & M_{2m}^{tf} \\ \vdots & \vdots & \ddots & \vdots \\ M_{k1}^{tf} & M_{k2}^{tf} & \cdots & M_{km}^{tf} \end{array} \right] \tag{1}$$

where $M_{ij}^{tf} = 1$ means there is a direct correlation between task element t_i and function element f_j; $M_{ij}^{tf} = 0$ means no direct correlation.

Similarly, the function-resource element mapping correlation matrix can be expressed shown as Matrix (2).

$$M^{fr} = (M_{ij}^{fr})_{m \times n} = \begin{array}{c} f_1 \\ f_2 \\ \vdots \\ f_m \end{array} \begin{array}{cccc} r_1 & r_2 & \cdots & r_n \end{array} \atop \left[\begin{array}{cccc} M_{11}^{fr} & M_{12}^{fr} & \cdots & M_{1n}^{fr} \\ M_{21}^{fr} & M_{22}^{fr} & & M_{2n}^{fr} \\ \vdots & \vdots & \ddots & \vdots \\ M_{m1}^{fr} & M_{m2}^{fr} & \cdots & M_{mn}^{fr} \end{array} \right] \tag{2}$$

where $M_{ij}^{fr} = 1$ means there is a direct correlation between function element f_i and resource element r_j; $M_{ij}^{fr} = 0$ means no direct correlation.

If requiring to further record the cross-layer correlation relationship between task and resource elements, we can define and obtain it by matrix operation $M^{tr} = M^{tf} \times M^{fr}$ shown in Matrix (3). However, in general, the practical significance of this cross-layer correlation is not obvious. It is the focus on the system design to clarify the software-hardware configuration mapping relationships from functions to resources. Therefore, this paper will focus on correlations from functions to resources.

$$M^{tr} = M^{tf} \times M^{fr} = \begin{array}{c} t_1 \\ \vdots \\ t_n \end{array} \begin{array}{ccc} r_1 & \cdots & r_n \end{array} \atop \left[\begin{array}{ccc} M_{11}^{tf} \times M_{11}^{fr} + \ldots M_{1m}^{tf} \times M_{m1}^{fr} & \cdots & M_{11}^{tf} \times M_{11}^{fr} + \ldots M_{1m}^{tf} \times M_{mn}^{fr} \\ \vdots & \ddots & \vdots \\ M_{k1}^{tf} \times M_{k1}^{fr} + \ldots M_{km}^{tf} \times M_{mn}^{fr} & \cdots & M_{k1}^{tf} \times M_{k2}^{fr} + \ldots M_{km}^{tf} \times M_{mn}^{fr} \end{array} \right] \tag{3}$$

where $M_{ij}^{tr} \geq 1$ means there is a direct correlation between task element t_i and resource element r_j. $M_{ij}^{tr} = 0$ means no direct correlation.

2.2. Analysis of Coupling Degree

As the physical layer within the system, the form of coupling between the resource layer is also the most obvious: on the one hand, this coupling may result from the functional/logical coupling generated by each resource element serving the same function; on the other hand, it may also cause direct material or information transfer between resource elements, thus introducing specific coupling relationships. Both of the above two coupling forms can be defined as direct coupling. In contrast, a more complex form of indirect association between groups of coupled resource elements is generated due to the addition of the resource-sharing form. This form of coupling can be defined as indirect/cascading coupling. In order to quantitatively describe the direct and indirect coupling relationships within the system hierarchy, this study takes the resource layer as an example to define and distinguish the two coupling concepts.

1. Direct coupling degree matrix

The direct coupling degree is used to characterize the direct coupling relationship between elements. It represents the situation where there are direct information interactions, material exchanges or being occupied by the same other layer elements in the layer. The direct coupling degree matrix C^d is represented in Matrix (4):

$$C^d = \left(C^d_{ij}\right)_{n\times n} = \begin{array}{c} \\ r_1 \\ r_2 \\ \vdots \\ r_n \end{array} \begin{array}{cccc} r_1 & r_2 & \cdots & r_n \\ \left[\begin{array}{cccc} C^d_{11} & C^d_{12} & \cdots & C^d_{1n} \\ C^d_{21} & C^d_{22} & & C^d_{2n} \\ \vdots & & \ddots & \vdots \\ C^d_{n1} & C^d_{n2} & \cdots & C^d_{nn} \end{array}\right] \end{array} \tag{4}$$

where $C^d_{ij} = 0$ means no direct correlation; $C^d_{ij} = 1$ means correlation degree between resource element r_i and resource element r_j is 1, that, is, the fault propagation from resource element r_i to resource element r_j only needs 1 step.

2. Indirect coupling degree matrix

According to the fault propagation theory and cascading failure theory, the fault of a single element will not only affect the element itself, but also cause a cascading effect by the correlation between elements, causing the fault propagation and diffusion, and the more serious situation may affect the normal operation of the whole system. Thus, simply establishing the concept of direct coupling degree is insufficient to assess the potential risk introduced by multiple coupling correlation of elements. In contrast, the indirect coupling degree is more efficient to reflect the degree of such risk.

The indirect coupling degree matrix C^c_{ij} characterizes the indirect coupling relationships between elements, which is an extension of the direct coupling degree. It can be represented by the indirect coupling degree matrix C^c as presented in Matrix (5).

$$C^c = \left(C^c_{ij}\right)_{n\times n} = \begin{array}{c} \\ r_1 \\ r_2 \\ \vdots \\ r_n \end{array} \begin{array}{cccc} r_1 & r_2 & \cdots & r_n \\ \left[\begin{array}{cccc} C^c_{11} & C^c_{12} & \cdots & C^c_{1n} \\ C^c_{21} & C^c_{22} & & C^c_{2n} \\ \vdots & & \ddots & \vdots \\ C^c_{n1} & C^c_{n2} & \cdots & C^c_{nn} \end{array}\right] \end{array} \tag{5}$$

Based on the direct coupling degree matrix C^d, the shortest fault propagation path order is calculated based on the Floyd algorithm, and that means the indirect coupling degree matrix C^c is generated. Among them, the element C^c_{ij} in the matrix is a natural number. When $C^c_{ij} = 0$, it means no indirect coupling relationship. When $C^c_{ij} = n$, it means that coupling correlation degree between resource element r_i and resource element r_j is n, indicating that fault propagation from resource element r_i to resource element r_j needs n steps.

The basic process of the Floyd algorithm is to start from the direct coupling matrix C^d and recursively update n times. Each update process introduces a new transition node to compare whether the path optimization can be achieved, until all nodes are introduced. Meanwhile, by using Floyd algorithm, the shortest path matrix C^r is obtained, where C^r_{ij} represents the next resource element that fault propagation from resource element r_i to resource element r_j should go through. Then, the order of the resource elements which the shortest path of fault propagation from resource element r_i to resource element r_j should go through can be deduced in turn.

2.3. Related Factors of Risk

2.3.1. Potential Severity

Further, when a risk quantification is required in view of coupling correlations, a potential severity matrix S_p can be established as Matrix (6):

$$
S_p = \left(S_p\right)_{n \times n} =
\begin{array}{c}
\\ r_1 \\ r_2 \\ \vdots \\ r_n
\end{array}
\begin{array}{c}
r_1 \quad r_2 \quad \cdots \quad r_n \\
\left[
\begin{array}{cccc}
S_{11} & S_{12} & \cdots & S_{1n} \\
S_{21} & S_{22} & & S_{2n} \\
\vdots & \vdots & \ddots & \vdots \\
S_{n1} & S_{n2} & \cdots & S_{nn}
\end{array}
\right]
\end{array}
\tag{6}
$$

The potential severity between resource elements will decrease non-linearly with the indirect coupling degree (such as the impact of radio waves, noise, etc.), that is, due to the natural elasticity and robustness of the system, the more propagation steps a potential fault needs, the lower its effect will be. Therefore, a function relationship between the potential severity and the indirect coupling degree is required to be established. According to the characteristics of the membership relationship between the two factors in the shape of the graph. The typical mapping relationship function is divided into normal type, Γ type, and Cauchy type [49–51], and each type is divided into smaller-type, middle-type, and larger-type [52,53]. Because the degree of propagation effect decreases nonlinearly with the coupling degree, a typical smaller-type of Cauchy type membership function [54,55] $S_p(C^c)$ is used for fitting in this paper shown in Equation (7).

$$
S_p(C^c) =
\begin{cases}
\dfrac{1}{1+a(C^c - c)^2}, & C^c > c \\
1, & C^c \le c
\end{cases}
\tag{7}
$$

where C^c represents the coupling degree (positive integer) which can be obtained from the indirect coupling degree matrix. S_p is the potential severity; a and c are constant, and need to be further quantified.

Moreover, the factor of safety critical degree $SCG = \{g_1, g_2, \ldots, g_n\}$ of resource elements requires to be considered. In other words, there is difference in the fault effect strength in different resource elements. Therefore, the SCG factor needs to be added to the potential severity matrix, $S = S_p \times SCG$ forming updated potential severity matrix S.

2.3.2. Propagation Probability and Propagation Time

Ideally, the original data should be determined by experimental statistics. However, in the case of insufficient experimental data, the expected data can be obtained by simulation complex system or modified by referring to expert experience. For example, for the direct propagation probability and direct propagation time, from the perspective of related faults, based on the analysis of the fault effect mechanism between elements, fault correlation effect (simulation) test work can be carried out. Based on the test data, the frequency and average time of fault propagation are calculated and counted as the expected values of the direct propagation probability matrix C_d^P and the direct propagation time matrix C_d^T. In the paper, fault injection [56,57] is applied in the simulation system for a large number of times (usually 10,000) to record and obtain average propagation probability and propagation time [58,59]. In general, if the sample size is large enough, then the average value can be regarded as the actual value [60–62].

If fault propagation from resource element r_i to resource element r_j needs n steps (can be obtain from the shortest path matrix), the probability of each propagation step is $p_1, p_2, \ldots, p_n$ (can be obtained from the direct propagation probability matrix C_d^P). Indirect propagation probability $C_{c(ij)}^P$ that fault in element r_i to element r_j can be calculated by Equation (8) and then indirect propagation probability matrix C_c^P is formed.

$$C_{c(ij)}^P = \prod_{m=1}^{n} p_m \tag{8}$$

Similarly, if the fault propagation from resource element r_i to resource element r_j needs n steps (it can be obtained from the shortest path matrix), and the time of each propagation step is $t_1, t_2, \dots,$ t_n (it can be obtain from the direct propagation time matrix C_d^T), indirect propagation time $C_{c(ij)}^T$ that fault propagation takes from element r_i to element r_j can be calculated by Equation (9). Then indirect propagation time matrix C_c^T is formed.

$$C_{c(ij)}^T = \sum_{m=1}^{n} p_m \tag{9}$$

3. Multi-Dimensional Safety Risk Theory

3.1. Multi-Dimensional Safety Risk Model

Generally, the safety risk of a system is measured in two dimensions, which is to quantify the safety risk from two dimensions: the probability of a dangerous event and the severity of the potential effect as shown in Equation (10). However, it is incomprehensive to fully characterize the safety risk characteristics of the system by analyzing and evaluating safety risks from only two dimensions. Therefore, this paper takes another dimension (propagation time) into consideration and proposes a new theory to quantify safety risk from three dimensions: probability, severity and time, and compare the effect weight and correlation of each element, in order to analyze and evaluate the safety risk comprehensively. Combined with the risk concept of Terje Aven [63], the multi-dimensional safety risk model can be formalized as presented in Equation (11).

$$R = f(P, S) \tag{10}$$

where P is the probability of a dangerous event, S is the severity of the potential effect.

$$R = f(P, S, T) \tag{11}$$

where P is the fault propagation probability, S is the potential severity, and T is the fault propagation time.

3.2. Calculation of Multi-Dimensional Safety Risk Model

Traditional risk assessment often adopts qualitative/semi-quantitative methods. The basic rule is to classify risk factors into different levels qualitatively based on experience, and then refer to the risk assessment model for semi-quantitative risk assessment. The core reason for using the qualitative/semi-quantitative risk assessment method is that the risk factors have different dimensional units, and the resulting risk values can be considered as the normalized result after empirical classification. In GJB 900A [64], the probability and severity are classified into five levels and four levels, respectively, and based on expert scoring method [65,66], different probability levels and severity levels corresponding to different risk values are shown in Table 1.

Table 1. Risk index matrix based on GJB 900A.

Probability Level	Severity Level			
	1	**2**	**3**	**4**
1	1	3	7	13
2	2	5	9	16
3	4	6	11	18
4	8	10	14	19
5	12	15	17	20

Therefore, risk value in Table 1 mapping to the two-dimensional space, then Euclidean distance between the risk assessment point $R(P, S)$ as shown in Figure 2 and the space origin is introduced to calculate risk evaluation values as shown in Equation (12). a, b is the preference correction factors.

$$R = \sqrt[2]{(a * P)^2 + (b * S)^2} \tag{12}$$

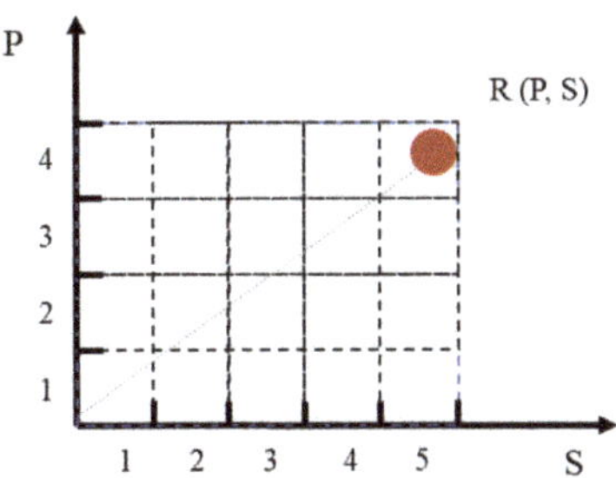

Figure 2. Traditional risk model space of risk factors.

According to Table 1, $R = f(P,S)$, $1 = f(1, 1)$, $2 = f(2, 1)$, $3 = f(4, 1) \ldots$, and based on Equation (12), 'regress' function in MATLAB is applied to implement multiple linear regression fitting, obtaining $a = 2.2$, $b = 3.3$.

Similarly, based on the multi-dimensional safety risk theory, this paper uses a five-level risk factor level method based on expert experience [67]. In other words, the degree from light to heavy is level 1 to level 5. Similarly, based on GJB 900A and expert scoring method, different risk values corresponding to different propagation probability, severity and propagation time are obtained as shown in Table 2. Therefore, the actual parameter values of the safety risk factor propagation probability P, potential severity S, and propagation time T can be quantified into risk factor level.

Table 2. Risk index of multi-dimensional safety risk model.

Risk	Probability Level	Severity Level	Propagation Time
1	1	1	1
1	2	1	1
1	1	1	2
1	1	2	1
1	3	1	1
2	2	1	2
...	...	...	...
13	3	3	3
...	...	...	...
25	5	5	5

Based on multi-dimensional safety risk model, the risk factors *P*, *S*, *T* are mapped to the three-dimensional space shown as Figure 3. The improved Euclidean distance between the risk assessment point $R(p, s, t)$ and the space origin is introduced to calculate risk evaluation values as shown in Equation (13).

$$R = \sqrt[2]{[a * f_1(p)]^2 + [b * f_2(s)]^2 + [c * f_3(t)]^2} \tag{13}$$

where $f_1(p)$, $f_2(s)$, $f_3(t)$ are risk factor levels, which actual parameter values of risk factors *P*, *S*, *T* are classified into, respectively; *a*, *b*, *c* is the preference correction factors. Based on Table 2 and Equation (13), 'regress' function in MATLAB is applied to implement multiple linear regression fitting, obtaining $a = 2.2$, $b = 3.3$, $c = 2.7$.

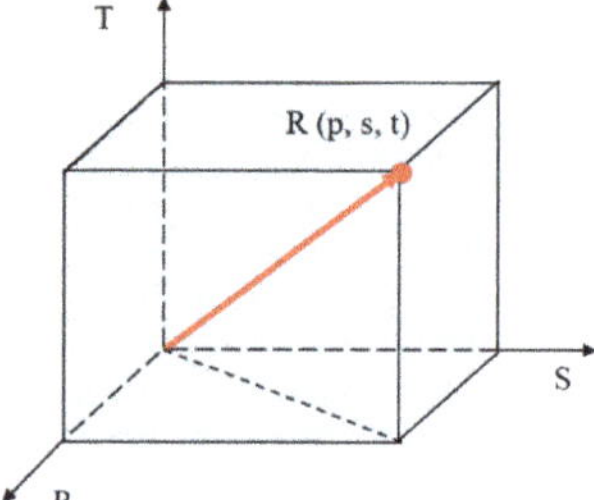

Figure 3. The three-dimensional model space of risk factors.

In addition, total safety risk value R_N of the system and safety risk ratio η_i of element *i* is calculated as shown in Equations (14) and (15).

$$R_N = \sum_{i=1}^{n} R_i \tag{14}$$

$$\eta_i = \frac{R_i}{R_N} \times 100\% \tag{15}$$

3.3. Evaluation of Multi-Dimensional Safety Risk Model

1. Pareto rule

The safety risk ratio characterizes the extent to which each element in the system contributes to the total safety risk value of the system, and from this, the critical safety factor in the system can be intuitively identified. According to Pareto rule [68,69], when distinguishing safety-critical links, it can be considered that 80% of accidents are originated from 20% of dangerous sources. Therefore, the value of the safety risk ratio η_i is sorted in descending order, and the first 20% of the values of η_i are defined as safety-critical elements, and the last 80% are defined as general safety elements.

2. ALARP principle

As a project risk criterion generally adopted by domestic and foreign institutions, the principle of ALARP (As Low As Reasonably Practicable) [70,71] sets two risk "boundaries" based on the value of safety risk and related experience: intolerable boundary and negligible boundary [70], meanwhile forming three risk region and level: serious risk region, ALARP region and negligible region, and the top extreme of the principle is "accident", and the bottom extreme is "safety". ALARP rule is shown as Figure 4. The values of the regions and boundaries of ALARP principle are all relative, and there is no standard of definition [72,73]. In practice, expert evaluation method considering potential severity, propagation probability, and propagation time can be applied to determine final values of the boundaries [70,74,75]. Meanwhile, alternative values of the boundaries are also obtained. Finally, compared and analyzed results of final values and alternative values of boundaries, final results can be determined. ALARP region means risk value in this region is reasonably acceptable. Therefore, according to the ALARP principle, this paper classifies risk value of each element into different regions, in order to make further research to propose preventive measurements, so as to reduce the risk level and improve system safety.

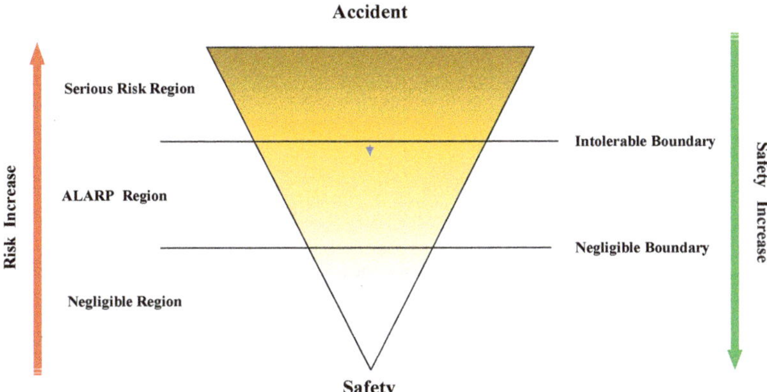

Figure 4. As Low As Reasonably Practicable (ALARP) model.

3. Safety risk entropy

The essence of entropy [76,77] is considered as a measure of the degree of disorder in the system. Currently, there are three typical definitions: Clausius entropy, Boltzmann entropy, Shannon entropy. Therefore, in this paper, safety risk entropy is defined as the measure of all random factors in system safety risk. Through the previous system's safety risk analysis, it was found that the randomness mainly derives from the probabilistic characteristics of each step of the fault propagation process. Therefore, according to the definition of Shannon entropy, it is assumed that the fault propagation from resource element r_i to resource element r_j requires n steps, and the probability of fault propagation for each step is $p_1, p_2, \ldots, p_n$ (based on the direct propagation matrix C_d^P). Then, based on Shannon entropy, H_{ij} means the effect of safety risk entropy that from resource element resource r_i to the resource element r_j, as shown in Equation (16). In other words, H_{ij} represents the uncertainty risk of fault propagation from resource element r_i to resource element r_j. Moreover, total safety risk entropy H_i of resource element r_i, which effects the overall system calculated in Equation (17). The higher the safety risk entropy value is, the greater the uncertainty risk caused by the fault in this element effects on the system is.

$$H_{ij} = -\sum_{m=1}^{n} p_m \ln p_m \tag{16}$$

$$H_i = \sum_{j=1}^{n} H_{ij} \tag{17}$$

According to comprehensive analysis on results of Pareto rule, ALARP principle and safety risk entropy, aimed at the serious risk region and critical risk factors, the coupling correlations are further researched to propose preventive measurements, so as to reduce the risk level and improve system safety.

4. Case Study and Discussion

4.1. Coupling Correlation of Complex System

4.1.1. Hierarchical Model and Description

1. Hierarchical model

Integrated modular avionics (IMA) [78,79] is a shared set of flexible, reusable, and interoperable hardware and software resources. When integrated, these resources can form a platform that provides service, designed and verified to a defined set of safety and performance requirements, to host applications performing aircraft functions [80]. Based on ASAAC criterion [81], IMA system is managed by a three-layer model: Aircraft Level (AL), Integration Area Level (IAL) and Resource Element Level (REL). This three-level hierarchy of IMA is typical task-function-resource model.

On the basis of the initial design plan of a certain aircraft, this integrated modular avionics (IMA) system contains three functions: navigation, communication, and integrated management; nine system resources: GPM (Graphics Processing Module), GPM, DPM (Data Processing Module), DPM, SPM (Signal Processing Module), SPM, PCM (Power Conversion Module), PCM, NSM (Network Support Module). The IMA system task-function-resource mapping relationship and details of function-resource mapping relationship are shown in Figure 5 and Table 3.

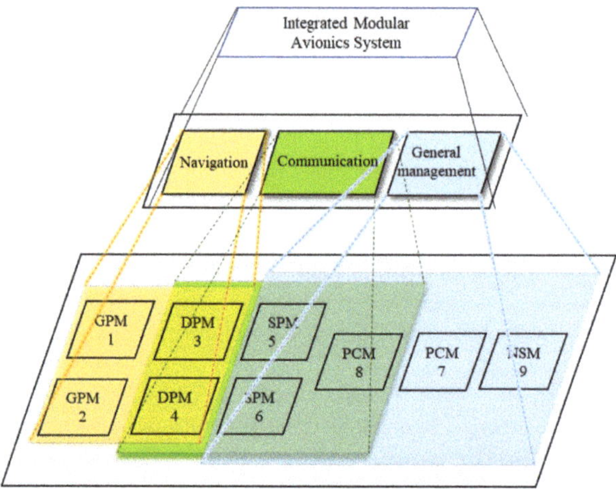

Figure 5. Task-function-resource model of the Integrated modular avionics (IMA) system. GPM, Graphics Processing Module; DPM, Data Processing Module; SPM, Signal Processing Module; PCM, Power Conversion Module; NSM, Network Support Module.

Table 3. Details of function-resource mapping relationship of the IMA system.

Function Layer	Resource Layer/Resource Serial Number				
	GPM	DPM	SPM	PCM	NSM
navigation	1, 2	3, 4	/	/	/
communication	/	3, 4	5, 6	8	/
integrated management	/		5, 6	7, 8	9

It is generally considered that in the IMA system, the top-level system functional entities are unique, and it can be considered that there is only one element in the IMA system task set $T = \{t_1\}$, which is aimed to complete the management of the entire IMA system to support the operation of the system. Therefore, it can be ignored. Then, function element set $F = \{f_1, f_2, f_3\}$; resource element set $R = \{r_1, r_2, \ldots, r_9\}$.

The safety critical grade SCG of the resource elements is divided into 3 levels (larger numbers indicate higher SCG), and based on experience set $SCG_{1\times9} = \{1, 2, 3, 3, 2, 1, 2, 2, 3\}$.

2. Coupling degree Matrix

Function-resource element mapping coupling matrix M^{fr} is presented in Matrix (18).

$$M^{fr} = (M^{fr}_{ij})_{3\times9} = \begin{bmatrix} 1 & 1 & 1 & 1 & 0 & 0 & 0 & 0 & 0 \\ 0 & 0 & 1 & 1 & 1 & 1 & 0 & 1 & 0 \\ 0 & 0 & 0 & 0 & 1 & 1 & 1 & 1 & 1 \end{bmatrix} \tag{18}$$

4.1.2. Coupling Degree and Related Factors

1. Direct coupling degree matrix

Direct coupling relationship caused by different resource elements serving same function can be presented by direct coupling degree matrix C^d shown in Matrix (19).

$$C^d = (C^d_{ij})_{9\times9} = \begin{bmatrix} 0 & 1 & 1 & 1 & 0 & 0 & 0 & 0 & 0 \\ 1 & 0 & 1 & 1 & 0 & 0 & 0 & 0 & 0 \\ 1 & 1 & 0 & 1 & 1 & 1 & 0 & 1 & 0 \\ 1 & 1 & 1 & 0 & 1 & 1 & 0 & 1 & 0 \\ 0 & 0 & 1 & 1 & 0 & 1 & 1 & 1 & 1 \\ 0 & 0 & 1 & 1 & 1 & 0 & 1 & 1 & 1 \\ 0 & 0 & 0 & 0 & 1 & 1 & 0 & 1 & 1 \\ 0 & 0 & 1 & 1 & 1 & 1 & 1 & 0 & 1 \\ 0 & 0 & 0 & 0 & 1 & 1 & 1 & 1 & 0 \end{bmatrix} \tag{19}$$

2. Indirect coupling degree

Based on Floyd algorithm, fault propagation path order is calculated referring to direct coupling matrix C^d, and then, the indirect coupling degree matrix C^c and the shortest path matrix C^r of fault propagation are generated as shown in Matrix (20) and Matrix (21), respectively.

C^r_{ij} represents the next element that the fault propagation from element i to element j should pass. For instance, fault propagation from element 1 to element 9, according to $C^r_{19} = 3$, it can be inferred that fault in element 1 will propagate to element 3 first. Then, according to $C^r_{39} = 5$, it can be inferred that fault will propagate from element 3 to element 5 second. Finally, according to $C^r_{59} = 9$, it can be seen that fault will propagate from element 5 to element 9. Therefore, the fault propagation path from element 1 to element 9 is formed: $1 \rightarrow 3 \rightarrow 5 \rightarrow 9$. So, other fault propagation paths can be deduced by analogy.

$$C^c = (C^c_{ij})_{9\times9} = \begin{bmatrix} 0 & 1 & 1 & 1 & 2 & 2 & 3 & 2 & 3 \\ 1 & 0 & 1 & 1 & 2 & 2 & 3 & 2 & 3 \\ 1 & 1 & 0 & 1 & 1 & 1 & 2 & 1 & 2 \\ 1 & 1 & 1 & 0 & 1 & 1 & 2 & 1 & 2 \\ 2 & 2 & 1 & 1 & 0 & 1 & 1 & 1 & 1 \\ 2 & 2 & 1 & 1 & 1 & 0 & 1 & 1 & 1 \\ 3 & 3 & 2 & 2 & 1 & 1 & 0 & 1 & 1 \\ 2 & 2 & 1 & 1 & 1 & 1 & 1 & 0 & 1 \\ 3 & 3 & 2 & 2 & 1 & 1 & 1 & 1 & 0 \end{bmatrix} \tag{20}$$

$$
C^r = \left(C^r_{ij}\right)_{9\times9} =
\begin{bmatrix}
1 & 2 & 3 & 4 & 3 & 3 & 3 & 3 & 3 \\
1 & 2 & 3 & 4 & 3 & 3 & 3 & 3 & 3 \\
1 & 2 & 3 & 4 & 5 & 6 & 5 & 8 & 5 \\
1 & 2 & 3 & 4 & 5 & 6 & 5 & 8 & 5 \\
3 & 3 & 3 & 4 & 5 & 6 & 7 & 8 & 9 \\
3 & 3 & 3 & 4 & 5 & 6 & 7 & 8 & 9 \\
5 & 5 & 5 & 5 & 5 & 6 & 7 & 8 & 9 \\
3 & 3 & 3 & 4 & 5 & 6 & 7 & 8 & 9 \\
5 & 5 & 5 & 5 & 5 & 6 & 7 & 8 & 9
\end{bmatrix}
\tag{21}
$$

3. Potential severity

The potential severity effect is greatest when there is a direct coupling correlation relationship between the elements (the coupling degree is 1), so the corresponding potential severity value is set to 1. In addition, when the coupling degree is 5 or above, the degree of effect is the smallest, and the corresponding potential severity value is set to 0.1. Based on Equation (7), $S_p(1) = \frac{1}{1+a(1-c)^2} = 1$, $S_p(5) = \frac{1}{1+a(5-c)^2} = 0.1$, as shown in equation set (22), and then solving equation set (22), obtaining $c = 1$, $a = 0.56$. Then, based on Equation (7), obtaining $S_p(2) = 0.64, S_p(3) = 0.31, S_p(4) = 0.17$. Therefore, potential severity matrix S_p is shown as Matrix (23).

$$
\begin{cases}
\dfrac{1}{1+a(1-c)^2} = 1 \\[2mm]
\dfrac{1}{1+a(5-c)^2} = 0.1
\end{cases}
\tag{22}
$$

$$
S = \left(S_p\right)_{9\times9} =
\begin{bmatrix}
0.00 & 1.00 & 1.00 & 1.00 & 0.64 & 0.64 & 0.31 & 0.64 & 0.31 \\
1.00 & 0.00 & 1.00 & 1.00 & 0.64 & 0.64 & 0.31 & 0.64 & 0.31 \\
1.00 & 1.00 & 0.00 & 1.00 & 1.00 & 1.00 & 0.64 & 1.00 & 0.64 \\
1.00 & 1.00 & 1.00 & 0.00 & 1.00 & 1.00 & 0.64 & 1.00 & 0.64 \\
0.64 & 0.64 & 1.00 & 1.00 & 0.00 & 1.00 & 1.00 & 1.00 & 1.00 \\
0.64 & 0.64 & 1.00 & 1.00 & 1.00 & 0.00 & 1.00 & 1.00 & 1.00 \\
0.31 & 0.31 & 0.64 & 0.64 & 1.00 & 1.00 & 0.00 & 1.00 & 1.00 \\
0.64 & 0.64 & 1.00 & 1.00 & 1.00 & 1.00 & 1.00 & 0.00 & 1.00 \\
0.31 & 0.31 & 0.64 & 0.64 & 1.00 & 1.00 & 1.00 & 1.00 & 0.00
\end{bmatrix}
\tag{23}
$$

In addition, considering the safety critical grade of the resource elements $SCG_{1\times9} = \{1, 2, 3, 3, 2, 1, 2, 2, 3\}$, and based on Equation (7), the final potential severity matrix S is presented as Matrix (24).

$$
S = S_{9\times9} =
\begin{bmatrix}
0.00 & 2.00 & 3.00 & 3.00 & 1.28 & 0.64 & 0.62 & 1.28 & 0.93 \\
1.00 & 0.00 & 3.00 & 3.00 & 1.28 & 0.64 & 0.62 & 1.28 & 0.93 \\
1.00 & 2.00 & 0.00 & 3.00 & 2.00 & 1.00 & 1.28 & 2.00 & 1.92 \\
1.00 & 2.00 & 3.00 & 0.00 & 2.00 & 1.00 & 1.28 & 2.00 & 1.92 \\
0.64 & 1.28 & 3.00 & 3.00 & 0.00 & 1.00 & 2.00 & 2.00 & 3.00 \\
0.64 & 1.28 & 3.00 & 3.00 & 2.00 & 0.00 & 2.00 & 2.00 & 3.00 \\
0.31 & 0.62 & 1.92 & 1.92 & 2.00 & 1.00 & 0.00 & 2.00 & 3.00 \\
0.64 & 1.28 & 3.00 & 3.00 & 2.00 & 1.00 & 2.00 & 0.00 & 3.00 \\
0.31 & 0.62 & 1.92 & 1.92 & 2.00 & 1.00 & 2.00 & 2.00 & 0.00
\end{bmatrix}
\tag{24}
$$

4. Propagation probability

Fault injection is applied in simulation of the IMA system for 10,000 times, and the direct propagation probability matrix C_d^P and the direct propagation time matrix C_d^T are obtained as shown in Matrix (25) and Matrix (26).

$$C_d^p = (C_d^p)_{9\times9} = \begin{bmatrix} 0.0 & 1.0 & 0.8 & 0.9 & 0.0 & 0.0 & 0.0 & 0.0 & 0.0 \\ 1.0 & 0.0 & 0.8 & 1.0 & 0.0 & 0.0 & 0.0 & 0.0 & 0.0 \\ 0.8 & 0.8 & 0.0 & 1.0 & 0.8 & 0.7 & 0.0 & 0.9 & 0.0 \\ 0.9 & 1.0 & 1.0 & 0.0 & 0.8 & 0.8 & 0.0 & 0.7 & 0.0 \\ 0.0 & 0.0 & 0.8 & 0.8 & 0.0 & 1.0 & 0.9 & 0.7 & 0.9 \\ 0.0 & 0.0 & 0.7 & 0.8 & 1.0 & 0.0 & 0.7 & 0.8 & 0.9 \\ 0.0 & 0.0 & 0.0 & 0.0 & 0.9 & 0.7 & 0.0 & 0.8 & 1.0 \\ 0.0 & 0.0 & 0.9 & 0.7 & 0.7 & 0.8 & 0.8 & 0.0 & 1.0 \\ 0.0 & 0.0 & 0.0 & 0.0 & 0.9 & 0.9 & 1.0 & 1.0 & 0.0 \end{bmatrix} \tag{25}$$

$$C_d^T = (C_d^T)_{9\times9} = \begin{bmatrix} 0.0 & 0.5 & 0.7 & 0.6 & 0.0 & 0.0 & 0.0 & 0.0 & 0.0 \\ 0.5 & 0.0 & 0.8 & 0.4 & 0.0 & 0.0 & 0.0 & 0.0 & 0.0 \\ 0.7 & 0.8 & 0.0 & 0.6 & 0.7 & 0.7 & 0.0 & 0.9 & 0.0 \\ 0.6 & 0.4 & 0.6 & 0.0 & 0.8 & 0.7 & 0.0 & 0.8 & 0.0 \\ 0.0 & 0.0 & 0.7 & 0.8 & 0.0 & 0.5 & 0.6 & 0.7 & 0.4 \\ 0.0 & 0.0 & 0.7 & 0.7 & 0.5 & 0.0 & 0.7 & 0.5 & 0.6 \\ 0.0 & 0.0 & 0.0 & 0.0 & 0.6 & 0.7 & 0.0 & 0.6 & 0.5 \\ 0.0 & 0.0 & 0.9 & 0.8 & 0.7 & 0.5 & 0.6 & 0.0 & 0.8 \\ 0.0 & 0.0 & 0.0 & 0.0 & 0.4 & 0.6 & 0.5 & 0.8 & 0.0 \end{bmatrix} \tag{26}$$

where the unit in C_d^P and C_d^T is percentage (%) and second (s), respectively.

Based on direct propagation probability matrix C_d^P, referring to propagation path, indirect propagation probability is calculated by Equation (8). For example, fault propagation path from element 1 to element 9 is $1 \to 3 \to 5 \to 9$. The propagation probability from element 1 to element 3, and element 3 to element 5, and element 5 to element 9 is 0.8, 0.8, 0.9, respectively, based on $C_{19}^P = 0.8$, $C_{39}^P = 0.8$, $C_{59}^P = 0.9$ in the direct propagation probability C_d^P. Then, fault propagation probability from element 1 to element 9 is $0.8 \times 0.8 \times 0.9 = 0.576$. Similarly, whole indirect propagation probability matrix C_c^P is formed as shown in Matrix (27).

$$C_c^p = (C_c^p)_{9\times9} = \begin{bmatrix} 0.00 & 1.00 & 0.80 & 0.90 & 0.64 & 0.56 & 0.576 & 0.72 & 0.576 \\ 1.00 & 0.00 & 0.80 & 1.00 & 0.64 & 0.56 & 0.576 & 0.72 & 0.576 \\ 0.80 & 0.80 & 0.00 & 1.00 & 0.80 & 0.70 & 0.72 & 0.90 & 0.72 \\ 0.90 & 1.00 & 1.00 & 0.00 & 0.80 & 0.80 & 0.72 & 0.70 & 0.72 \\ 0.64 & 0.64 & 0.80 & 0.80 & 0.00 & 1.00 & 0.90 & 0.70 & 0.90 \\ 0.56 & 0.56 & 0.70 & 0.80 & 1.00 & 0.00 & 0.70 & 0.80 & 0.90 \\ 0.576 & 0.576 & 0.72 & 0.72 & 0.90 & 0.70 & 0.00 & 0.80 & 1.00 \\ 0.72 & 0.72 & 0.90 & 0.70 & 0.70 & 0.80 & 0.80 & 0.00 & 1.00 \\ 0.576 & 0.576 & 0.72 & 0.72 & 0.90 & 0.90 & 1.00 & 1.00 & 0.00 \end{bmatrix} \tag{27}$$

5. Propagation time

Similarly, based on direct propagation time C_d^T, referring to propagation path, indirect propagation time is calculated by Equation (9). For instance, fault propagation path from element 1 to element 9 is $1 \to 3 \to 5 \to 9$. The fault propagation time from element 1 to element 3, and element 3 to element 5, and element 5 to element 9 is 0.7, 0.7, 0.4, respectively, based on $C_{19}^T = 0.7(s)$, $C_{39}^T = 0.7(s)$, $C_{59}^T = 0.4(s)$

in the direct propagation time C_d^T. Then propagation time from element 1 to element 9 is $0.7 + 0.7 + 0.4$ = 1.8(s). Similarly, the whole indirect propagation time matrix C_c^T is formed as shown in Matrix (28).

$$C_c^T = (C_c^T)_{9\times9} = \begin{bmatrix} 0.0 & 0.5 & 0.7 & 0.6 & 1.4 & 1.4 & 2.0 & 1.6 & 1.8 \\ 0.5 & 0.0 & 0.8 & 0.4 & 1.5 & 1.5 & 2.1 & 1.7 & 1.9 \\ 0.7 & 0.8 & 0.0 & 0.6 & 0.7 & 0.7 & 1.3 & 0.9 & 1.1 \\ 0.6 & 0.4 & 0.6 & 0.0 & 0.8 & 0.7 & 1.4 & 0.8 & 1.2 \\ 1.4 & 1.5 & 0.7 & 0.8 & 0.0 & 0.5 & 0.6 & 0.7 & 0.4 \\ 1.4 & 1.5 & 0.7 & 0.7 & 0.5 & 0.0 & 0.7 & 0.5 & 0.6 \\ 2.0 & 2.1 & 1.3 & 1.4 & 0.6 & 0.7 & 0.0 & 0.6 & 0.5 \\ 1.6 & 1.7 & 0.9 & 0.8 & 0.7 & 0.5 & 0.6 & 0.0 & 0.8 \\ 1.8 & 1.9 & 1.1 & 1.2 & 0.4 & 0.6 & 0.5 & 0.8 & 0.0 \end{bmatrix} \tag{28}$$

4.2. Risk Assessment

1. Classification of risk factors

Considering numerical ranges within the indirect propagation probability matrix C_c^P, the potential severity matrix S, and the indirect propagation time matrix C_c^T, qualitative risk factor level rules are given (from light to heavy, respectively 1 to 5), as shown in Table 4 based on experience.

Table 4. Classification level of risk factor.

Risk Factor Level	C_c^P	S	C_c^T/s
1	$(0, 0.3)$	$(0, 0.6)$	$(2, +\infty)$
2	$(0.3, 0.5)$	$(0.6, 1.2)$	$(1.5, 2)$
3	$(0.5, 0.7)$	$(1.2, 1.8)$	$(1, 1.5)$
4	$(0.7, 0.9)$	$(1.8, 2.4)$	$(0.5, 1)$
5	$(0.9, 1)$	$(2.4, +\infty)$	$(0, 0.5)$

2. Calculation of multi-dimensional safety risk

Based on Table 2, the safety risk factors P, S, T are converted into uniform safety risk level. Equation (13) is used to calculate multi-dimensional safety risk value, as shown in Matrix (29). Total risk value that element i effect on the overall system is calculated as shown in Matrix (30).

$$P = R_{9\times9} = \begin{bmatrix} 0.00 & 21.85 & 21.59 & 21.59 & 14.39 & 12.36 & 10.78 & 14.30 & 10.78 \\ 18.62 & 0.00 & 21.59 & 23.99 & 14.39 & 12.36 & 9.72 & 14.30 & 10.78 \\ 15.42 & 19.19 & 0.00 & 22.58 & 19.19 & 14.27 & 15.53 & 19.19 & 17.81 \\ 15.42 & 21.85 & 22.58 & 0.00 & 19.19 & 15.42 & 15.53 & 18.29 & 17.81 \\ 12.36 & 14.39 & 21.59 & 21.59 & 0.00 & 18.62 & 19.19 & 18.29 & 23.06 \\ 12.36 & 14.39 & 20.80 & 21.59 & 21.85 & 0.0 & 18.29 & 20.83 & 21.59 \\ 9.14 & 9.72 & 17.81 & 17.81 & 19.19 & 14.27 & 0.0 & 19.19 & 23.9 \\ 12.25 & 14.30 & 21.59 & 20.80 & 18.29 & 17.41 & 19.19 & 0.0 & 22.58 \\ 9.14 & 10.78 & 17.81 & 17.81 & 20.83 & 15.42 & 21.85 & 20.29 & 0.0 \end{bmatrix} \tag{29}$$

$$R_i = [127.66, 125.76, 142.18, 143.08, 149.11, 151.71, 134.13, 144.42, 136.95] \tag{30}$$

System total safety risk assessment value $R_N = 1255$, and based on Equation (15), safety risk ratio is shown in Matrix (31).

$$\eta_i = [10.17, 10.02, 11.33, 11.40, 11.88, 12.09, 10.69, 11.51, 10.91] \tag{31}$$

3. Results of risk

- Pareto rule

The results of R_i and η_i are sorted in descending order as shown in Table 5. Then Pareto chart is presented as Figures 6 and 7.

Table 5. Results of Pareto rule.

Rank	Element Number	R_i	η_i	Accumulated Value
1	6	151.71	12.09	12.09%
2	5	149.11	11.88	23.97%
3	8	146.42	11.51	35.48%
4	4	146.08	11.40	46.88%
5	3	143.18	11.33	58.21%
6	9	133.95	10.91	69.12%
7	7	131.13	10.69	79.81%
8	1	127.66	10.17	89.98%
9	2	125.76	10.02	100.00%

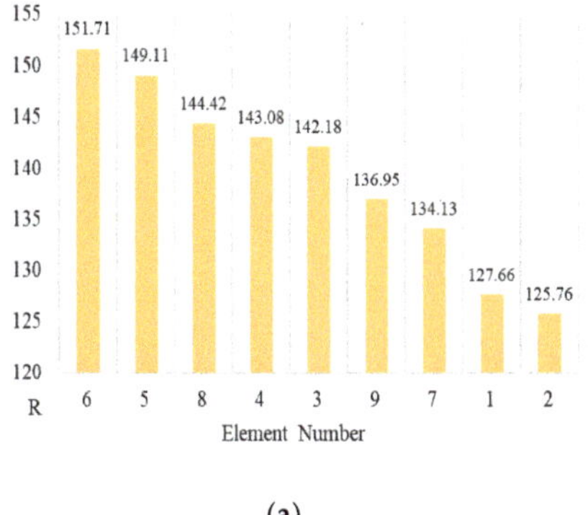
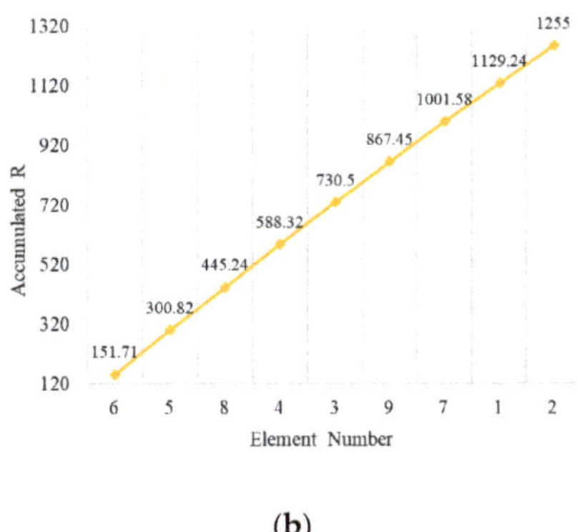

(a)

(b)

Figure 6. Chart of Pareto rule of the risk value R. (**a**) the risk value R of the single element in descending order; (**b**) the accumulated risk value R.

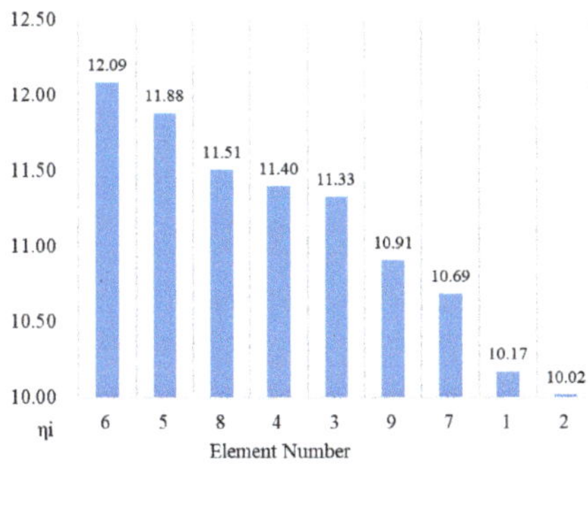
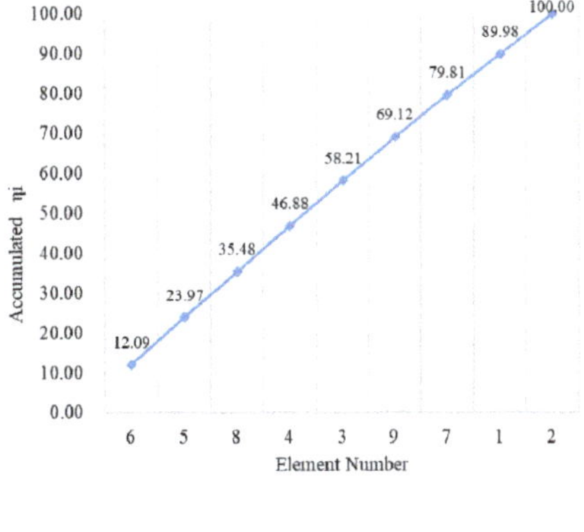

(a)

(b)

Figure 7. Chart of Pareto rule of η_i. (**a**) the safety risk ratio η_i of the single element in descending order; (**b**) the accumulated the safety risk ratio η_i.

- ALARP principle

Under the ALARP principle, two risk "boundaries" are set based on expert experience: the intolerable boundary and the negligible boundary are 145 and 130, respectively, and the alternative values of the boundaries are 140 and 130, respectively. On the condition of alternative values, element 3, 4, 5, 6, 8 are all in the serious risk region. According to Pareto rule, when distinguishing safety-critical links, it can be considered that 80% of accidents are originated from 20% of dangerous sources, but the accumulated risk value of the elements in the serious risk region is well over 20%, which is a violation of the Pareto rule. Therefore, value of the boundaries 145, 130 are determined. The risk level of each element is presented in Figure 8.

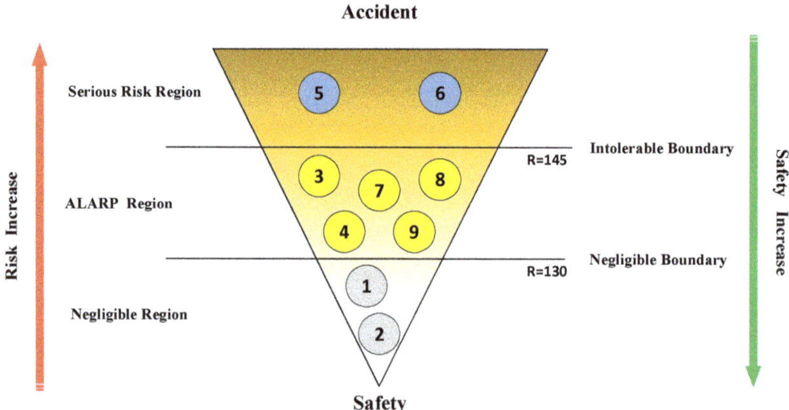

Figure 8. Results of ALARP model chart.

- Safety risk entropy

Based on Equation (16), entropy matrix H is presented as Matrix (32).

$$H = H_{9\times9} = \begin{bmatrix} 0.00 & 0.00 & 0.17 & 0.09 & 0.35 & 0.42 & 0.45 & 0.27 & 0.45 \\ 1.00 & 0.00 & 0.17 & 0.00 & 0.35 & 0.42 & 0.45 & 0.27 & 0.45 \\ 0.17 & 0.17 & 0.00 & 0.00 & 0.17 & 0.24 & 0.27 & 0.09 & 0.27 \\ 0.09 & 0.00 & 0.00 & 0.00 & 0.17 & 0.17 & 0.27 & 0.24 & 0.27 \\ 0.35 & 0.35 & 0.17 & 0.17 & 0.00 & 0.00 & 0.09 & 0.24 & 0.09 \\ 0.42 & 0.42 & 0.24 & 0.17 & 0.00 & 0.00 & 0.24 & 0.17 & 0.09 \\ 0.45 & 0.45 & 0.27 & 0.27 & 0.09 & 0.24 & 0.00 & 0.17 & 0.00 \\ 0.27 & 0.27 & 0.09 & 0.24 & 0.24 & 0.17 & 0.17 & 0.00 & 0.00 \\ 0.45 & 0.45 & 0.27 & 0.27 & 0.09 & 0.09 & 0.00 & 0.00 & 0.00 \end{bmatrix} \tag{32}$$

Based on Equation (17), total safety risk entropy H_i of element i which effects on the overall system is calculated as shown in Matrix (33). The safety risk entropy of the single element in descending order and the accumulated the safety risk entropy are presented in Figure 9.

$$H_i = [2.24, 2.14, 1.43, 1.25, 1.51, 1.81, 1.97, 1.49, 1.64] \tag{33}$$

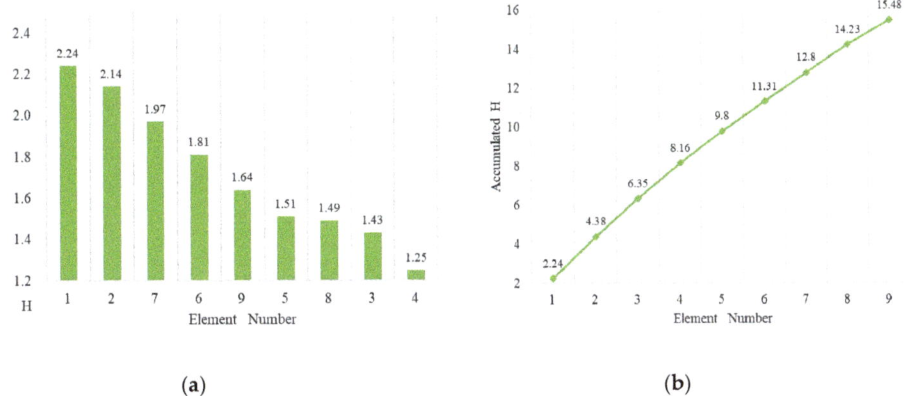

(a)　　　　　　　　　　　　　　(b)

Figure 9. Chart of results of safety risk entropy *H*. (**a**) the safety risk entropy of the single element in descending order; (**b**) the accumulated the safety risk entropy.

4.3. Discussion

1. According to the sorting results in Figure 6, Figure 7, and Pareto Rule, elements 5 and 6 can be defined as the safety-critical elements; and elements 1, 2, 3, 4, 7, 8, and 9 are defined as safety-general elements.
2. Based on Figure 8, element 5 and 6 are in the serious risk region; elements 3, 4, 7, 8, and 9 are in the ALARP region; and elements 1 and 2 are in the negligible region.
3. From Figure 9, elements 1 and 2 have higher uncertainty; elements 6, 7, and 9 have moderate uncertainty; and elements 3, 4, 5, 8 are with lower uncertainty.
4. In summary, elements 5 and 6 are the safety-critical elements and located in a serious risk region, which has a serious effect on the overall system. Simultaneously, it has a certain degree of uncertainty. Therefore, corresponding measures must be taken to ensure the safety of elements 5 and 6 in order to decrease the system risk. In other words, more attention must be paid to Signal Processing Module in this avionics system. Elements 3, 4, 7, 8, and 9 are in the ALARP region and have lower uncertainty. This means that the risk caused by these elements in the region are acceptable. As a consequence, Data Processing Module, Power Conversion Module, and Network Support Module should be given due attention if the conditions permit. Although elements 1 and 2 have higher uncertainty, they are located in the negligible region. Consequently, Graphics Processing Module can be ignored under limited conditions. If the conditions permit, in view of the higher uncertainty of elements 1 and 2 (Graphics Processing Module), by increasing the reliability of elements 1 and 2 and ensuring the reliability of the element's correlation with other elements, such as ensuring the reliability of the data transmission channel between elements 1 and 2 and other elements, and the reliability of the information transmission bus, etc. Based on these measures, the fault propagation from elements 1 and 2 to other elements can be reduced, so as to reduce risk to overall systems of high uncertainty of elements 1 and 2.

5. Conclusions

While aiming to address the insufficiency of traditional safety risk analysis and risk assessment technology to solve coupling problems between components in complex systems, this study proposed a novel risk assessment and analysis method for correlation in complex systems based on multi-dimensional theory. Firstly, a matrix-based hierarchical model for the complex system is presented and correlation relationships between elements in the system were established. Furthermore, based on correlation relationship, the multi-dimensional theory and model are proposed in order to evaluate risk more objectively. Moreover, based on the Pareto rule, ALARP principle, and safety risk entropy, the critical risk elements are identified, which provides a theoretical basis for putting forward preventive measures, so as to ensure and improve system safety. Compared with the current methods and technologies, the method proposed in this paper mainly reflects the advantages of two aspects. On the one hand, the hierarchical model is modeled in a matrix manner, and the association relationship of each element in the complex system is quickly and accurately analyzed, which reduces the skill requirements of analysts. On the other hand, it provides a feasible and multi-faceted analysis method for the risk assessment of systems in view of fault propagation, which is the core judgment criterion for identifying critical risk factors and of great significance for ensuring system safety.

Author Contributions: Conceptualization, Z.J. and T.Z.; methodology, Z.J., S.W., and F.R.; formal analysis, Z.J. and F.R.; writing—original draft preparation, Z.J.; writing—review and editing, Z.J. and S.W.; project administration, T.Z.; funding acquisition, T.Z. and S.W. All authors have read and agreed to the published version of the manuscript.

Funding: This research was funded by Foundation of No.61400020404.

Conflicts of Interest: The authors declare that there are no conflicts of interest.

References

1. Zhu, Z.; Feng, Y.; Lu, C.; Fei, C. Efficient Driving Plan and Validation of Aircraft NLG Emergency Extension System via Mixture of Reliability Models and Test Bench. *Appl. Sci.* **2019**, *9*, 3578. [CrossRef]
2. Song, S.; Ko, T.K.; Choi, Y.; Lee, S. A Novel Fault Diagnosis Method for High-Temperature Superconducting Field Coil of Superconducting Rotating Machine. *Appl. Sci.* **2019**, *10*, 223. [CrossRef]
3. Xiao, D.; Ding, J.; Li, X.; Huang, L. Gear Fault Diagnosis Based on Kurtosis Criterion VMD and SOM Neural Network. *Appl. Sci.* **2019**, *9*, 5424. [CrossRef]
4. Chu, J.; Zhao, T.; Jiao, J.; Chen, Z.; Ren, F. Reliability Modelling and Evaluation for LTD System Based on Load-Sharing Model. *Appl. Sci.* **2019**, *9*, 5528. [CrossRef]
5. Jiao, J.; Wei, M.; Yuan, Y.; Zhao, T. Risk Quantification and Analysis of Coupled Factors Based on the DEMATEL Model and a Bayesian Network. *Appl. Sci.* **2020**, *10*, 317. [CrossRef]
6. Vesely, W.E.; Goldberg, F.F.; Roberts, N.H.; Haasl, D.F. *Fault Tree Handbook*; Nuclear Regulatory Commission: Washington, DC, USA, 1981.
7. Papazoglou, I.A. Functional Block Diagrams and Automated Construction of Event Trees. *Reliab. Eng. Syst. Saf.* **1998**, *61*, 185–214. [CrossRef]
8. Andrews, J.D.; Dunnett, S.J. Event-Tree Analysis Using Binary Decision Diagrams. *IEEE Trans. Reliab.* **2000**, *49*, 230–238. [CrossRef]
9. Amzen, H.E. Failure Mode and Effect Analysis: A Powerful Engineering Tool for Component and System Optimization. In Proceedings of the Fifth Reliability and Maintainability Conference, New York, NY, USA, 18–20 July 1966; pp. 355–371.
10. Teng, S.; Ho, S.Y. Failure Mode and Effects Analysis: An Integrated Approach for Product Design and Process Control. *Int. J. Qual. Reliab. Manag.* **1996**, *13*, 8–26. [CrossRef]
11. Mahdiyar, A.; Jahed Armaghani, D.; Koopialipoor, M.; Hedayat, A.; Abdullah, A.; Yahya, K. Practical Risk Assessment of Ground Vibrations Resulting from Blasting, Using Gene Expression Programming and Monte Carlo Simulation Techniques. *Appl. Sci.* **2020**, *10*, 472. [CrossRef]
12. Schuëller, G.I. Impact of probability risk assessment on containment. *Nucl. Eng. Des.* **1984**, *80*, 203–216. [CrossRef]
13. Li, J. The Application of Probability Risk Assessment to the Safety Management in a Nuclear Power Plant. *J. Chin. Peoples Armed Police Force Acad.* **2010**, *26*, 43–46.
14. Zhou, X.; Tang, Y. Modeling and Fusing the Uncertainty of FMEA Experts Using an Entropy-Like Measure with an Application in Fault Evaluation of Aircraft Turbine Rotor Blades. *Entropy* **2018**, *20*, 864. [CrossRef]
15. Rhee, S.J.; Ishii, K. Using cost based FMEA to enhance reliability and serviceability. *Adv. Eng. Inform.* **2003**, *17*, 179–188. [CrossRef]
16. Jang, H.A.; Min, S. Time-Dependent Probabilistic Model for Hierarchical Structure in Failure Mode and Effect Analysis. *Appl. Sci.* **2019**, *9*, 4265. [CrossRef]
17. Liu, Y.; Yan, Z.P.; Yuan, Y.; Li, H. A FTA-based method for risk decision-making in emergency response. *Comput. Oper. Res.* **2014**, *42*, 49–57. [CrossRef]

18. Cheraghi, M.; Eslami Baladeh, A.; Khakzad, N. A fuzzy multi-attribute HAZOP technique (FMA-HAZOP): Application to gas wellhead facilities. *Saf. Sci.* **2019**, *114*, 12–22. [CrossRef]

19. Feng, Q.; Sha, S.; Dai, L. Bayesian Survival Analysis Model for Girth Weld Failure Prediction. *Appl. Sci.* **2019**, *9*, 1150. [CrossRef]

20. Zhao, C.; Wang, P.; Yan, F. Reliability Analysis of the Reconfigurable Integrated Modular Avionics Using the Continuous-Time Markov Chains. *Int. J. Aerosp. Eng.* **2018**, *2018*, 5213249. [CrossRef]

21. Singh, P.; Singh, L. Verification of safety critical and control systems of Nuclear Power Plants using Petri nets. *Ann. Nucl. Energy* **2019**, *132*, 584–592. [CrossRef]

22. Baouya, A.; Ait Mohamed, O.; Bennouar, D.; Ouchani, S. Safety analysis of train control system based on model-driven design methodology. *Comput. Ind.* **2019**, *105*, 1–16. [CrossRef]

23. Wilson, S.P.; Kelly, T.P.; McDermid, J. Safety Case Development: Current Practice, Future Prospects. In *Safety and Reliability of Software Based Systems*; Springer: London, UK, 1997; pp. 135–156.

24. Kelly, T.P. Arguing Safety—A Systematic Approach to Safety Case Management . Ph.D. Thesis, Department of Computer Science University of York, York, UK, September 1998.

25. Alexander, R.; Kelly, T.P.; Kurd, Z.; McDermid, J.A. Safety Cases for Advanced Control Software: Safety Case Patterns. Ph.D. Thesis, Department of Computer Science University of York, York, UK, 2007.

26. Iwu, F.; Galloway, A.; McDermid, J.; Toyn, I. Integrating Safety and Formal Analyses Using UML and PFS. *Reliab. Eng. Syst. Saf.* **2007**, *92*, 156–170. [CrossRef]

27. Hollnagel, E.; Goteman, O. The Functional Resonance Accident Model. *Proc. Cogn. Syst. Eng. Process Plant* **2004**, *2004*, 155–161.

28. Hollnagel, E. *FRAM: The Functional Resonance Analysis Method: Modelling Complex Socio-Technical Systems*; Ashgate Publishing, Ltd.: Farnham, UK, 2012.

29. SAE. *ARP 4754 Certification Considerations for Highly-Integrated or Complex Aircraft Systems*; SAE: Warrendale, PA, USA, 1996.

30. SAE. *ARP 4754A Guidelines for Development of Civil Aircraft and Systems*; SAE: Warrendale, PA, USA, 2010.

31. SAE. *ARP 4761 Guidelines and Methods for Conducting the Safety Assessment Process on Civil Airborne Systems and Equipment*; SAE: Warrendale, PA, USA, 1996.

32. *MIL-STD-882E Department of Defence Standard Practice: System Safety*; Defense Acquisition University: Fort Belvoir, VA, USA, 2011.

33. Mazzuchi, T.A.; Linzey, W.G.; Bruning, A. A paired comparison experiment for gathering expert judgment for an aircraft wiring risk assessment. *Reliab. Eng. Syst. Saf.* **2008**, *93*, 722–731. [CrossRef]

34. Cour-Harbo, A.L. Quantifying risk of ground impact fatalities of power line inspection BVLOS flight with small unmanned aircraft. In Proceedings of the 2017 International Conference on Unmanned Aircraft Systems, Miami, FL, USA, 13–16 June 2017.

35. Li, L.; Wu, Y.; Guo, G.; Shi, J. Research on risk assessment method of subway operation based on cloud model. In Proceedings of the 2016 35th Chinese Control Conference (CCC), Chengdu, China, 27–29 July 2016.

36. Fayaz, M.; Ullah, I.; Park, D.; Kim, K.; Kim, D. An Integrated Risk Index Model Based on Hierarchical Fuzzy Logic for Underground Risk Assessment. *Appl. Sci.* **2017**, *7*, 1037. [CrossRef]

37. Duan, Y.; Cai, Y.; Wang, Z.; Deng, X. A Novel Network Security Risk Assessment Approach by Combining Subjective and Objective Weights under Uncertainty. *Appl. Sci.* **2018**, *8*, 428. [CrossRef]

38. Weng, G.; Bhalla, U.S.; Iyengar, R. Complexity in biological signaling systems. *Science* **1999**, *284*, 92–96. [CrossRef]

39. Rind, D. Complexity and climate. *Science* **1999**, *284*, 105–107. [CrossRef]

40. Chan, S. *Complex Adaptive Systems, in: Research Seminar in Engineering Systems*; MIT Press: Cambridge, MA, USA, 2001; pp. 1–9.

41. Ladyman, J.; Lambert, J.; Wiesner, K. What is a complex system? *Eur. J. Philos. Sci.* **2013**, *3*, 33–67. [CrossRef]

42. Dietz, T.; Klamroth, K.; Kraus, K.; Ruzika, S.; Schäfer, L.E.; Schulze, B.; Stiglmayr, M.; Wiecek, M.M. Introducing multiobjective complex systems. *Eur. J. Oper. Res.* **2020**, *280*, 581–596. [CrossRef]

43. Mokshin, A.V.; Mokshin, V.V.; Sharnin, L.M. Adaptive genetic algorithms used to analyze behavior of complex system. *Commun. Nonlinear Sci.* **2019**, *71*, 174–186. [CrossRef]

44. Nair, A.; Reckien, D.; van Maarseveen, M.F.A.M. A generalised fuzzy cognitive mapping approach for modelling complex systems. *Appl. Soft Comput.* **2019**, *84*, 105754. [CrossRef]

45. Chaabane, S.; Trentesaux, D. Coping with disruptions in complex systems: A framework. *IFAC-PapersOnLine* **2019**, *52*, 2413–2418. [CrossRef]

46. Chen, L.; Jiao, J.; Wei, Q.; Zhao, T. An improved formal failure analysis approach for safety-critical system based on MBSA. *Eng. Fail. Anal.* **2017**, *82*, 713–725. [CrossRef]

47. Wei, Q.X. A Research of Formal Verification of System Safety based on Model Checking. Master's Thesis, Beihang University, Beijing, China, 2017.

48. Wang, H.; Zhao, T.; Ren, F.; Jiang, Z. Integrated modular avionics system safety analysis based on model checking. In Proceedings of the 2017 Annual Reliability and Maintainability Symposium (RAMS), Orlando, FL, USA, 23–26 January 2017.

49. Medasani, S.; Kim, J.; Krishnapuram, R. An overview of membership function generation techniques for pattern recognition. *Int. J. Approx. Reason.* **1998**, *19*, 391–417. [CrossRef]

50. Dombi, J. Membership function as an evaluation. *Fuzzy Sets Syst.* **1990**, *35*, 1–21. [CrossRef]

51. Rakus-Andersson, E. The new approach to the construction of parametric membership functions for fuzzy sets with unequal supports. *Procedia Comput. Sci.* **2017**, *112*, 2057–2065. [CrossRef]

52. Liu, K.; Li, J.; Yang, L. The measure and improvement of fuzzy decision in membership function determination method. *Shaanxi Inst. Technol.* **2005**, *21*, 68–71.

53. Yadav, H.B.; Yadav, D.K. Construction of Membership Function for Software Metrics. *Procedia Comput. Sci.* **2015**, *46*, 933–940. [CrossRef]

54. Wang, X.L. The Measure and Improvement of Fuzzy Decision in Membership Function Determination Method. Ph.D. Thesis, Nanjing University of Aeronautics and Astronautics, Nanjing, China, 2016.

55. Hasuike, T.; Katagiri, H.; Tsubaki, H. An Interactive Algorithm to Construct an Appropriate Nonlinear Membership Function Using Information Theory and Statistical Method. *Procedia Comput. Sci.* **2015**, *61*, 32–37. [CrossRef]

56. Zhang, R.P. Research on Fault Samples Selection and Fault Injection Method for Flight Control Systems. Master's Thesis, Nanjing University of Aeronautics and Astronautics, Nanjing, China, 2017.

57. Eslami, M.; Ghavami, B.; Raji, M.; Mahani, A. A survey on fault injection methods of digital integrated circuits. *Integration* **2020**, *71*, 154–163. [CrossRef]

58. Azimi, S.; Du, B.; Sterpone, L. Evaluation of transient errors in GPGPUs for safety critical applications: An effective simulation-based fault injection environment. *J. Syst. Architect.* **2017**, *75*, 95–106. [CrossRef]

59. Kirbiš, G.; Selčan, D.; Kramberger, I. Software Reliability Validation and Verification Using Fault Injection Techniques on a Fault Tolerant Processor. *IFAC-PapersOnLine* **2015**, *48*, 252–257. [CrossRef]

60. Zhang, R.; Xiao, L.; Li, J.; Cao, X.; Qi, C.; Li, J.; Wang, M. A fast fault injection platform of multiple SEUs for SRAM-based FPGAs. *Microelectron. Reliab.* **2018**, *82*, 147–152. [CrossRef]

61. Kim, M.C.; Seo, J.; Jung, W.; Choi, J.G.; Kang, H.G.; Lee, S.J. Evaluation of effectiveness of fault-tolerant techniques in a digital instrumentation and control system with a fault injection experiment. *Nucl. Eng. Technol.* **2019**, *51*, 692–701. [CrossRef]

62. Xu, G.X. Research of Software-Implemented Fault Injection and Reliability Evaluation Methods in Distributed Real-Time System. Doctoral Thesis, Chongqing University, Chongqing, China, 2011.

63. Steen, R.; Aven, T. A Risk Perspective Suitable for Resilience Engineering. *Saf. Sci.* **2011**, *49*, 292–297. [CrossRef]

64. COSTIND. *GJB 900A-2012, General Requirements for Materiel Safety Program*; COSTIND: Beijing, China, 2012.

65. Tang, K.H.D.; Md Dawal, S.Z.; Olugu, E.U. Integrating fuzzy expert system and scoring system for safety performance evaluation of offshore oil and gas platforms in Malaysia. *J. Loss Prevent. Proc. Ind.* **2018**, *56*, 32–45. [CrossRef]

66. Campagne, C.S.; Roche, P.; Gosselin, F.; Tschanz, L.; Tatoni, T. Expert-based ecosystem services capacity matrices: Dealing with scoring variability. *Ecol. Indic.* **2017**, *79*, 63–72. [CrossRef]

67. Hokstad, P.; Utne, I.B.; Vatn, J. *Risk and Interdependencies in Critical Infrastructures*; Springer: Lund, Sweden, 2012.

68. Tsai, S.B.; Xue, Y.Z. Models for forecasting growth trends in renewable energy. *Renew. Sustain. Energy Rev.* **2017**, *77*, 1069–1078. [CrossRef]

69. Jure, M.; Janez, K.; Tomaz, B. Methodology for Searching Representative Elements. *Appl. Sci.* **2019**, *9*, 3482–3497.

70. Melchers, R.E. On the ALARP Approach to Risk Management. *Reliab. Eng. Syst. Saf.* **2001**, *71*, 201–208. [CrossRef]

71. Martin, C.J.; Whitby, M. Application of ALARP to extremity doses for hospital workers. *J. Radiol. Prot.* **2003**, *23*, 405–421. [CrossRef] [PubMed]

72. Andrew Hopkins. Risk-management and rule-compliance: Decision-making in hazardous industries. *Saf. Sci.* **2011**, *49*, 110–120. [CrossRef]

73. Seminatore, A.A.; Ghelardoni, L.; Ceccarelli, A.; Falai, L.; Schultheis, M.; Malinowsky, B. ALARP (A Railway Automatic Track Warning System Based on Distributed Personal Mobile Terminals). *Procedia Soc. Behav. Sci.* **2012**, *48*, 2081–2090. [CrossRef]

74. Van Coile, R.; Jomaas, G.; Bisby, L. Defining ALARP for fire safety engineering design via the Life Quality Index. *Fire Saf. J.* **2019**, *107*, 1–14. [CrossRef]

75. Sørskår, L.I.K.; Selvik, J.T.; Abrahamsen, E.B. On the use of the vision zero principle and the ALARP principle for production loss in the oil and gas industry. *Reliab. Eng. Syst. Saf.* **2019**, *191*, 106541. [CrossRef]

76. An Institution of Civil Engineers. *Risk Analysis and Management for Projects (RAMP)*; ICE Publishing: London, UK, 2009.

77. Lorenc, A.; Kuźnar, M. An Intelligent System to Predict Risk and Costs of Cargo Thefts in Road Transport. *Int. J. Eng. Technol. Innov.* **2018**, *8*, 284–293.

78. Watkins, C.B. Integrated Modular Avionics: Managing the allocation of shared intersystem resources. In Proceedings of the 2006 IEEE/AIAA 25th Digital Avionics Systems Conference, Portland, OR, USA, 15–18 October 2006; pp. 1–12.

79. Prisaznuk, P.J. Integrated modular avionics. In Proceedings of the IEEE 1992 National Aerospace and Electronics Conference (NAECON 1992), Dayton, OH, USA, USA, 18–22 May 1992; pp. 39–45.

80. *Integrated Modular Avionics (IMA) Development Guidance and Certification Considerations*; Radio Technical Commission for Aeronautics DO-297; RTCA: Washington, DC, USA, 2005.

81. *STANAG 4626, Final Draft of Proposed Standards for Software*; ASAAC, NATO: Brussels, Belgium, 2004.

 *applied sciences*

Article

Reliability-Based Preventive Maintenance Strategy of Truck Unloading Systems

Awsan Mohammed *, Ahmed Ghaithan, Mashel Al-Saleh and Khalaf Al-Ofi

Construction Engineering and Management Department, College of Environmental Design, King Fahd University of Petroleum and Minerals, Dhahran 31261, Saudi Arabia; ahmedgh@kfupm.edu.sa (A.G.); g201214380@kfupm.edu.sa (M.A.-S.); kaluwfi@kfupm.edu.sa (K.A.-O.)
* Correspondence: awsan.mohammed@kfupm.edu.sa; Tel.: +966-13-8607575

Received: 11 September 2020; Accepted: 30 September 2020; Published: 5 October 2020

Abstract: The unloading of petroleum products is a complex and potentially dangerous operation since the unloading system contains complex interdependency components. Any failures in one of its components lead to a cut in the petroleum supply chain. Therefore, it is important to assess and evaluate the reliability of the unloading system in order to improve its availability. In this context, this paper presents the operation philosophy of the truck unloading system, failure modes of the components within the system, and a bottom-up approach to analyze the reliability of the system. In addition, it provides reliability data, such as failure rates, and mean time between failures of the system components. Furthermore, the reliability of the whole system was calculated and is presented for different time periods. The critical components, which are major contributors towards the system reliability, were identified. To enhance the system reliability, a reliability-based preventive maintenance strategy for the critical components was implemented. In addition, the preventive maintenance scheduling was identified based on the reliability plots of the unloading system. The best schedule for preventive maintenance of the system was determined based on the reliability function to be every 45 days for maintaining the system reliability above 0.9. Findings reveal that the reliability of the unloading system was significantly improved. For instance, the system reliability at one year improved by 80%, and this ratio increased dramatically as the time period increased.

Keywords: reliability; truck unloading system; petroleum equipment; preventive maintenance

---

1. Introduction

The supply chain of petroleum products faces major challenges, i.e., demand growth and the complexity of fluid transportation. The petroleum supply chain contains multiple stations extending from oil wells in exploration and production areas to the final destination. Each station has its own difficulties and challenges in the contribution towards the success of the safe and continuous supply. Undoubtedly, meeting a regional demand of petroleum products requires an uninterrupted safe operation. In the last station, the petroleum products pass through a process called Custody Transfer in which the ownership of the products transfers from one party to another. These massive quantities need to be measured accurately when their possession is being transferred to reserve the rights of each party. Nevertheless, fluid measurements are complex due to multiple factors which, when they deviate, may have a significance impact of the genuineness of the measurements. For instance, if the fluid volume is measured directly in a Custody Transfer process, then fluid pressure and temperature have to be maintained at certain levels to have an accurate volume measurement.

In bulk plants and distribution points, the petroleum products are transferred from truck tanks to storage tanks through the truck unloading system, in which the fluid passes through multiple mechanical components and instruments, among which is the flow meter where the quantity of the fluid is being measured in a process called truck unloading. The truck unloading system is a vital part of any supply chain and can be available in many applications such as airports, petroleum distribution plants and some other stations, where the high traffic requires large quantities of fuel. Apparently, any failure in these systems leads to a cut in the petroleum supply chain. Therefore, it is important to enhance the reliability of unloading systems by performing a deeper study of the reliability of their components to find out how each component contributes to the system reliability.

The truck unloading system is an integration of a large number of different mechanisms working as one unit to simultaneously unload the petroleum products and genuinely measure its quantity with high measurement accuracy and high reliability to have an uninterrupted operation. It contains a variety of interacting mechanical, piping, electrical, and instrumentation components. The sequence of operation starts when trucks arrive to unload the petroleum product to storage tanks. The unloading arms are then connected to the bottom of the truck and the vapor arm is connected to the top of the truck to prevent a vacuum, which might cause the truck to collapse. Through the effect of the gravitational force and the difference in elevation, the fluid will start flowing slowly from the truck tank into the unloading arms until it reaches the pump, and then fluid will be pumped in the desired predefined rate in the pre-set controller, which is considered the brain of the system. Passing through the system piping, the fluid eventually reaches the outlet with the same quantity measured by the flow meter.

Because petroleum products are dangerous and flammable, they require special attention from their extraction until they reach the final customer. The unloading system is an important part of the long petroleum supply chain, which plays a connecting role between different modes of transportation. The unloading system contains many linked, sophisticated and complex components. Any failures in one of its components may lead to a cut in the whole petroleum supply chain. Therefore, it is important to evaluate the reliability of the unloading system by performing a deep study of the reliability of its components to find out how each component contributes to the system reliability. In this context, this paper presents the operation philosophy of the truck unloading system, failure modes of the components within the system, and a bottom-up approach to analyze the reliability of the system.

Despite the importance of the truck unloading systems, and to the best of our knowledge, the reliability-based preventive maintenance analysis has not been carried out on the unloading systems. As will be discussed in the literature review section, most of the studies have generally considered the failure rates and reliability of pipeline transportation applications. In fact, most of these studies have been carried out for reliability analysis of the component levels, regardless of the system effect of the behavior of these components.

In this context, this paper assesses the current level of the overall truck unloading system reliability and the failure modes of the components of the unloading system. It assesses the truck unloading system reliability during the lifetime of the unloading system, investigates the effect of the preventive maintenance on the truck unloading system reliability. Then, it establishes a reliability-based preventive maintenance strategy for the truck unloading systems and identifies the major contributors to the system reliability for design improvement purposes.

The rest of this paper is organized as follows; the relevant studies are reviewed in Section 2, followed by a description of the materials and methods in Section 3. Section 4 presents the results and discussion. Finally, the paper is concluded and directions for further research are highlighted in Section 5.

2. Literature Review

In this paper, the main concern is the reliability of unloading systems, which includes a variety of pipeline components. Reliability of oil and gas pipelines and equipment has always been identified as a major problem in the industry. These systems work continuously for a long period of time. Thus, a minor failure of operation in any of their parts can have a major impact on the overall system performance. Safety and availability of pipeline components are the most important factors contributing to the overall system quality. Consequently, reliability is crucial for achieving the objectives of any industrial system. This section focuses on the reliability of pipeline sources and the reliability of applications similar to truck loading/unloading systems. In addition, it reviews the state-of-the-art papers related to reliability-based maintenance of similar applications.

During the design process of any systems, reliability is usually evaluated and assessed [1]. Several published papers have focused on the reliability of system components and units and their improvement [2–4]. The reliability of many systems has been evaluated [5–7]. It is mentioned that systems reliability is diminishable. Therefore, a minimum reliability threshold must be set to maintain (restore) a system and enhance its reliability. The repair process can be broken down into further unique subtasks and delays, and various maintenance measures, which include mean repair time (MTTR), the median repair time, mode, most likely repair time, and maintenance durations.

Nataša at al. [8] addressed the spare parts inventory in aircraft maintenance. The authors proposed a decision-making process for planning and controlling the non-repairable aircraft spare parts inventory by evaluating the reliability of non-repairable aircraft components. In addition, a reliability model was proposed to assess the subcomponents reliability. They also presented a new approach to determine the required amount of aircraft spare parts available in the inventory over a given period of time. Hou et al. [9] described the reliability relationship between components in complex systems. They developed a preventive maintenance action model for series and parallel systems to determine the optimal cost and timing. A case study is presented to prove the practicality of the developed model.

Vishnu and Regikumar [10] proposed a generalized method for reliability centered maintenance (RCM) where the maintenance is scheduled based on reliability of equipment in process plants. The maintenance strategy was developed based on two factors: criticality score and reliability parameters of the equipment. These factors derived from historical data of equipment failure and previous maintenance schedules. Consequently, a new maintenance strategy for each component was constructed using the analytical hierarchy process (AHP) method with the help of maintenance experts and practitioner opinions. Tee and Ekpiwhre [11] presented a reliability-centered maintenance analysis, carried out on key assets of a newly built Nigerian road junction infrastructure. The adaptation of the classical RCM succeeded in enforcing the different assets allocated to preventive maintenance (PM) strategies and showed that its implementation in the road industry can minimize unnecessary maintenance and regular reactive maintenance by optimizing its preventive maintenance interval in an effective way. In addition, Yssaad et al. [12] proposed a reliability-based centered maintenance for the distribution of a power system. The authors discussed the two major goals of RCM; ensuring protection through preventative maintenance measures; and preserving functionality in the most cost-effective way when protection is not important.

Li et al. [13] expressed the importance of reducing machine failures to have an uninterrupted operation and emphasized the poor efficiency of the conventional preventive maintenance method, which sometimes causes unnecessary costs because it does not consider the current state of the equipment. In this study, the researchers developed a dynamic preventive maintenance model based on the reliability of the current of the equipment rather than the conventional fixed maintenance. In addition, they proved in a case study that the proposed model can reduce the unavailability of the equipment and increase the efficiency of its operation throughout its useful life.

Rafael and Shamsi [14] considered the decision-making process in maintenance strategy for water distribution systems. The probability of having a functional network system upon demand was calculated based on the current level of components reliability based on which maintenance need could be prioritized in low reliability areas. They also took into account the severity of impact on the entire system from the failure of every component.

In addition, a report was published by Stiftelsen for industriell og teknisk forskning (SINTEF) along with the cooperation of multiple global organizations in the oil and gas industry [15]. It mainly provides an estimation with regards to the reliability data for selected equipment normally used in control systems within the oil and gas industry. The reliability data can support organizations to analyze reliability levels in line with the international standards IEC 61508 and IEC 61511. Generally, the study took the reliability approach to investigate the root causes of failures in equipment and to analyze the effect level in terms of failure on demand and critical safety failures.

The biggest documented and published project by oil and gas exploration and production organization evaluated, assessed, recorded, and analyzed failures of a huge number of piping components and mechanical machines used in these industries over a defined period of time. The project considered both the operational and calendar time in the assessment of the failure rates that took place in the normal operating lifecycle rather than the infant mortality or burn-in phase. The main purpose of this project is to provide the stated industry with a comprehensive reliable data to ensure cost-effective and safe process improvement. To minimize the effect of variation in the collected failure data, we considered a 90% uncertainty interval for the lower and upper limits. Additionally, the failure modes of each component are categorized based on their severity; allowing the readers to understand the likelihood of a critical failure in each of these components [16].

The previous two studies, SINTEF and Offshore Reliability Data (OREDA), are the best documented sources of oil and gas equipment reliability data due to their practical nature, which came from the collaboration of a number of large operation organizations in that field. The data are also reliable because of the wide range of involvement within the operation organizations and outside of these organizations. Some of the components in the oil and gas transportation systems are highly reliable, which makes it extremely hard to estimate a failure rate for them. These two studies fill this gap by applying long periods of tests to these particular components. Furthermore, a study has been conducted by Sun et al. [17] to improve the long-term preventive maintenance decisions using a reliability prediction approach. A case study of water pipelines was developed to collect real failure data of two main segments: exposed pipelines and buried pipelines. The case study was developed on the basis that each one of the two pipeline segments had its own failure distribution, which helps one come up with an overall system reliability. In addition, the study investigates the improvement of the preventive maintenance on the system reliability to develop a cost-effective system maintenance strategy. Subsequently, a mathematical reliability formula was derived from the exponential distribution to optimize the need for preventive maintenance.

The above study, unlike this research, focuses on water pipelines rather than oil and gas pipelines. However, its analytical approach of the preventive maintenance can be used in this research to recommend a planning framework for the preventive maintenance of the unloading system. Rimkevicius et al. [18] developed a comprehensive method for assessing reliability in energy pipeline systems scientifically, which can be applied to oil and gas systems. Basically, the failure in different mechanisms and the pipes structural integrity are considered as the base to assess system reliability. The study is centered around three types of analyses: mathematical, thermal-hydraulic, and pipeline structure analyses. The authors used homogeneous and non-homogeneous Poisson distributions with a constant failure rate assumption to give a better insight into the likelihood of future failures in the pipeline network system. Interestingly, they are confident that the developed method can be applied under any environmental condition. It was found that pipe corrosion constitutes around 30–40% of the total pipeline failures. Moreover, the frequency of failures in pipelines increases when they exceed 30 years in service. The assumption of a constant failure rate throughout the pipeline is the main assumption of this research. The approaches taken in this study assumed a constant failure rate and consideration of different mechanisms, like what we have done in the unloading application, that has enlightened the context of this research.

Shalay et al. [19] discussed the reliability distributions, such as exponential, normal, Rayleigh, logarithmic, and Weibull. The authors highlighted that despite the fact that the usage of exponential distribution under the assumption of constant failure rate of equipment is beneficial during the normal operation life, it is important to apply other distributions to further study the failure behavior of wearing and change of properties of the equipment by applying normal and Weibull distributions. Moreover, this gives a better understanding of the failure prediction under varying working conditions and changing loads on the equipment. An algorithm has been developed that analyzes the technological parameters, which helps in the selection of the appropriate distribution function. The main purpose of this study is to provide operating entities in the pipeline transportation industry with mathematical model to better predict the likelihood of failure of their equipment under dynamic operating conditions. Kong et al. [20] proposed a reliability-based maintenance approach for the deterioration of civil infrastructure systems addressing different uncertainties. These uncertainties include resistances and time loads changes, and interventions of the maintenance series that are used to save the systems. The authors used computer software for this analysis.

Wu et al. [21] introduced a new model of cost effective maintenance strategies to identify an acceptable period of the monitoring conditions and level of deterioration after incomplete preventive repairs. The models on maintenance optimization for the production system were developed by Barata and Tong respectively using the various methods [22,23].

Rao et al. [24] proposed an opportunity maintenance model for the production line to carry out preventive maintenance on several machines simultaneously, when one machine in the production line fails. Nonetheless, the maintenance technique is difficult to use for a complicated line consisting of a large number of machines. Zhang et al. [25] developed an opportunistic maintenance strategy for wind turbines addressing an inconsistent maintenance plan dependent on reliability. The authors characterized this maintenance and illustrated the consequences of corrective maintenance action using a hybrid model based on the hazard rate. The findings show the effect that different maintenance costs have on the economic benefits of an opportunistic maintenance strategy. Xie et al. [26] established an efficient strategy in view of accessibility for offshore wind turbines to decrease maintenance costs. A failure rate based on minimal data has been estimated using the Weibull distribution with a three-parameter process approach. The opportunistic maintenance plan was implemented to minimize maintenance costs by optimizing the preventive maintenance age and an acceptable maintenance age.

Arunraj et al. [27] used the analytical hierarchy method and the other planning approaches for implementing maintenance plans based on analysis of the failure rate of the system and maintenance costs, respectively. Zhao et al. [28] also suggested a delay-time model to prevent system failure during the preventive maintenance cycle when the machines were damaged. Li et al. [29] proposed a strategy based on preventive maintenance for the manufacturing plant. The authors measured the criticality of the machines in the line and then categorized the machines. Hadidi and Alkhaldi [30] presented several practical solutions for enhancing reliability in the systems of arm loading. The purpose of this study is to prevent oil shipment delay.

The previous study gives an insight about the normal behavior of similar equipment under the effect of aging. In other words, it shows the effect of wear out of such equipment. Additionally, it gives the impact of the changing operating conditions that might be experienced throughout the lifecycle of the unloading system. Reis et al. [31] simulated the tank truck loading operations in a fuel distribution terminal to generate a theoretical simulation of the tank truck loading operation, which helps in improving the efficiency and effectiveness of the tank truck loading systems. Probability theories have been used to simulate tanks queuing for different fuel types aiming to improve the queuing policy. A simulation of varying fuel flow rates and number of fuel loading arms has been tested to come up with the new queuing policy. The results show that the simulation of the new optimum queuing policy has a more efficient use of resources than the previous policy, which was based on first-come-first-served.

Jamshidi and Esfahani [32] dealt with a maintenance-based reliability approach for a machine-scheduling problem. The authors proposed a nonlinear mixed-integer model to simultaneously increase efficiency, and reduce maintenance, delay, and interrupting costs. Navarro et al. [33] introduced an optimization model based on a reliability-based maintenance approach for the preventive design of corrosion, addressing the environmental effects and costs of the life cycle. The findings of their study showed a reduction in environmental impacts and life cycle costs using the proposed model.

Based on the above literature review, it is clear that no study has assessed and evaluated the reliability of truck unloading systems. Even though international standards cover the minimum requirements for these systems, the design improvement and the maintenance scheduling to maintain uninterrupted operation needs to have a clear assessment of the effect of each component to the overall system reliability. Consequently, the aim of this study is to fill the gap of the previous works. In addition, this paper presents an operation strategy of the truck unloading system, the failure modes of the components within the system, and a bottom-up approach to the reliability of the system. It assesses the truck unloading system reliability during the operating lifecycle to the end of its useful life, investigates the effect of the preventive maintenance on the truck unloading system reliability, and identifies the major contributors to the system reliability for design improvement purposes.

3. Materials and Methods

In this section, the operation philosophy of the unloading system is explained and the followed approaches are summarized. In general, the truck unloading system is a piping system that absorbs liquid fuel from trucks and measures the quantity as accurately as possible before transferring it to the storage tanks. The characteristics of the devices are listed in Table 1.

Table 1. Device characteristics.

	Low pressure drop—less than 3.2 psi at extended maximum flow	
Features	Relatively low initial cost and minimum maintenance	
	High accuracy	
	Repeatability	Less than or equal to ±0.02%.
Accuracy	Linearity	±0.15% over the normal flow range
		±0.25% over the extended flow range
	Stability	Better than ±0.05% per 10 million gallons (38 million liters).
Range	Normal Range	$17\ \text{m}^3/\text{h}$–$170\ \text{m}^3/\text{h}$
	Extended Range	$10\ \text{m}^3/\text{h}$–$204\ \text{m}^3/\text{h}$

In the following section, a process sequential overview is discussed to understand the role of main components in the system. Table 2 shows the list of components of the truck unloading system and their associated symbols. Figure 1 shows the truck unloading system showing the symbols.

Table 2. Unloading system components list.

Component	Symbol
Unloading Arms	UA
Ball Valves	BV
Strainers	ST
Differential Pressure Transmitter	DPT
Positive Displacement Pump	PDP
Check Valve	CV
Air Eliminator	AE
Level Switch	LS
Gate Valves	GV
Pressure Relief Valve	PRV
Positive Displacement Flow Meter	FM
Pressure Transmitter	PT
Plug Valve	PV
Flow Control Valve	FCV

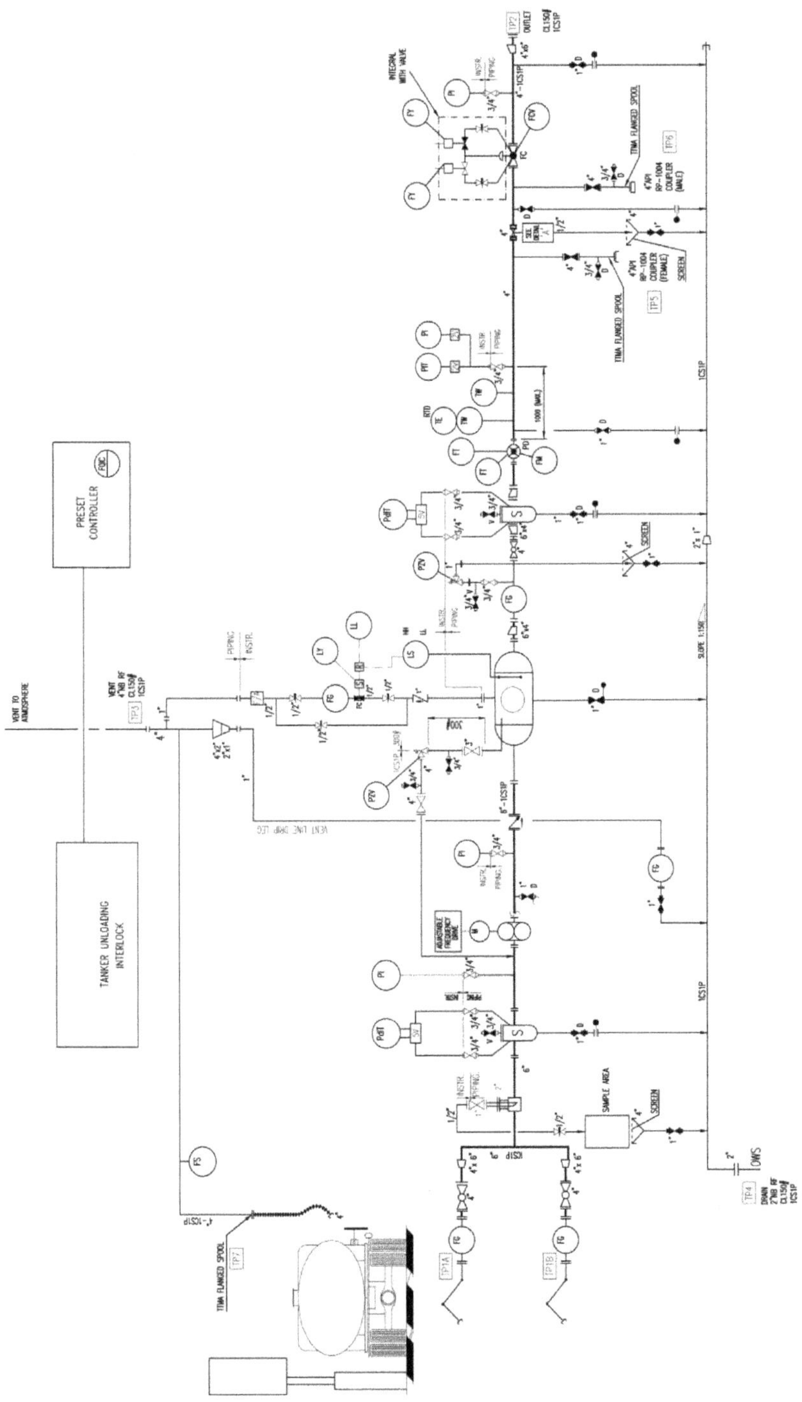

Figure 1. Unloading system piping and instrumentation diagram.

The unloading arms are connected to the bottom of a truck and the vapor arm is connected to the top of a truck to prevent a vacuum, which might cause the truck to collapse. The fluid starts flowing slowly from the truck to the unloading arms due to the gravitational force and difference in the elevation, which can be noticed through the visual flow glass in each of the two arms. The ball valves in each arm will be normally open so that fluid can pass through the two inlet streams. After that, the fluid will merge into the main streamline via a tee joint. The fluid then flows through a strainer that includes a differential pressure transmitter to monitor the difference in pressure between the inlet and outlet. This is mainly designed to filter the liquid in order to protect the rotating parts of the unloading pump against solid particles. Once the unloading pump starts working, the fluid will flow in the desired predefined rate in the pre-set controller, which is considered as the brain of the system. Downstream of the pump there is a check valve to prevent fluid from flowing in the opposite direction. Up to this point of the process, the fluid contains air, which needs to be removed to have an accurate measurement of the liquid quantity. Therefore, the fluid will pass through an air eliminator vessel where the air and liquid are separated. The air will flow back to the top of the truck through the vapor arm, while the liquid will continue its flow in the mainstream. There are two pressure relief valves; one valve is located in the vent line of the air eliminator and the other valve is available downstream of the air eliminator. The function of these valves is to protect the system from excess pressure with a predesigned pressure set point. Subsequently, the fluid will pass through another strainer equipped with a differential pressure transmitter for filtration purposes before entering the positive displacement meter to protect its rotating parts from solid particles, giving a more authentic measurement reading. The positive displacement meter then will measure the volumetric flow rate by displacement of the flowing fluid. The volume flow rate depends on the pressure of the fluid so that it passes into a pressure transmitter to ensure that the fluid being measured is within the desired pressure range. Otherwise, a signal will be sent to the system controller to notify any observed unwanted changes. Just before the outlet, there is a flow control valve through which the entire system flow is being controlled. The outlet point is directly connected to the storage tank.

The following are the assumptions and parameters of the system analysis:

- Piping spools failure rates are not considered as part of the system due to their short length and ease of replacement.
- Electrical boxes and terminations are not considered as part of the system due to the ease of replacement.
- System study is under the normal operating phase of the system lifecycle; burn-in and wear-out phases are not considered.
- System useful life is 20 years, which is common for such systems in the same field and under similar operating conditions.
- There is a constant failure rate (random failures) for each piece of equipment; proven to be very close to the actual failure rate during normal operating lifecycle.
- Difference in sample size in the determination of the failure rate is negligible.

3.1. System Reliability

The system reliability is the probability that the entire system will function successfully at any time. The system performance is dependent on the performance of its sub-systems (or its components). In other words, each component contributes in one way or another to the overall system performance. Consequently, the failure of any part of the system will either partially affect the overall system performance or it will cause a failure to the system. In this study, the focus is on the components of the system that have a major contribution to the system performance, and it is either impossible or not safe to operate without them. During the design stage of any system, it is necessary to decide which components are in series and which should be in parallel. A series system means that the whole system will fail if any components fail. A parallel system fails if all of its components fail. In the truck unloading system, there are two inlet streams, each one consists of an unloading arm and a ball valve connected in series, each one of these two streams are connected in parallel. In other words, if one stream fails, the other one can still be functional. The remaining components are connected in series. For instance, if the positive displacement pump fails to absorb the fluid, then the entire system will fail. A reliability block diagram of an unloading system is shown in Figure 2.

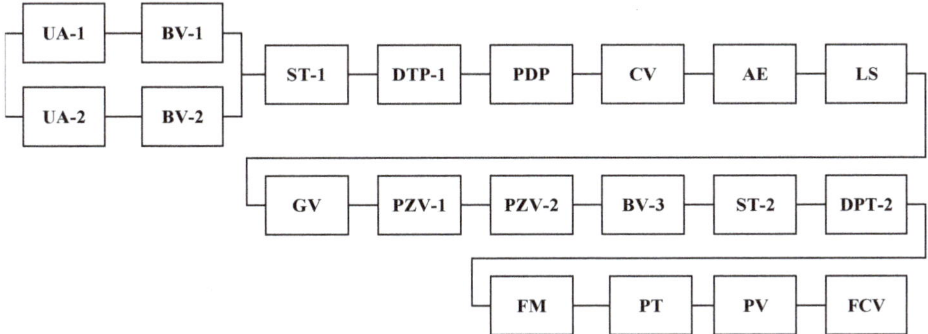

Figure 2. Unloading system reliability block diagram.

3.2. Data Collection

Reliability assessment studies are mainly dependent on failure rates data. The more reliable and valid the data, the better the reliability can be assessed. Specifically, in this case, it is necessary to have a large sample size to investigate random failures of a particular component, unlike other components where failures are not random. To satisfy this point, the components failure rates are extracted from the offshore and onshore reliability data (OREDA) [16] project, which is a large scale project under similar conditions in the same industry. For the remaining components, more valid samples were taken from one of the largest oil and gas operation companies in the Middle East. Some of the components have high reliability and failure rarely happens. For these components, using historical data from a truck unloading system was not feasible, instead, the data were taken from OREDA. These data include the number of tested units, number of failures and duration for a number of items such as ball valve, gate valve, pressure relief valve, flow control valve, differential pressure transmitter, pressure transmitter, level switch, and air eliminator. Table 3 shows the number of units tested, number of failures of tested units, test duration, and the mean time to failure (MTTF) for each component. Unloading arms, positive displacement pump, check valve, flow meter (positive displacement meter), plug valve, and strainer, where investigated under identical truck unloading system, were included, while the remaining were taken from the OREDA project. On the other hand, lambda (λ) and mean time to failure (MTTF) were calculated using Equations (1) and (2), respectively. Mean time between failures yielded clear figures on the random failure behavior of a particular component and ultimately it was essential for reliability calculations.

$$\lambda = \frac{Number\ of\ Failures}{Number\ of\ Units\ Tested\ \times\ Duration\ of\ Test} \tag{1}$$

$$MTTF = \frac{1}{\lambda} \tag{2}$$

Table 3. Unloading system component test details and results.

Component	Units Tested	Number of Failures	Duration (hours)	λ (1/hours)	MTTF (hours)
UA	14	1	25,920	2.75573×10^{-6}	362,880
BV	316	328	9,324,700	1.11315×10^{-7}	8,983,552
PDP	7	3	25,920	1.65344×10^{-5}	60,480
CV	14	2	25,920	5.51146×10^{-6}	181,440
GV	177	240	3,852,300	3.5198×10^{-7}	2,841,071
PZV	278	272	7,169,800	1.36464×10^{-7}	7,327,957
FM	7	4	25,920	2.20459×10^{-5}	45,360
PV	7	1	25,920	5.51146×10^{-6}	181,440
FCV	316	328	9,324,700	1.11315×10^{-7}	8,983,552
DPT	55	4	1,467,500	4.95586×10^{-8}	20,178,125
LS	76	37	1,862,400	2.61406×10^{-7}	3,825,470
PT	55	4	1,467,500	4.95586×10^{-8}	20,178,125
AE	50	174	3,725,900	9.34003×10^{-7}	1,070,661
ST	14	8	25,920	2.20459×10^{-5}	45,360

4. Data Analysis and Results

The aim of this study was to assess the truck unloading system reliability, which will allow the system users to know which part of the system is essential for preventive maintenance and the availability of spare parts to ensure continuous operation. Furthermore, the study will provide valuable insights to the system manufacturers, such as in which part of the system they should concentrate during the design improvement process, which will help produce a more reliable system that can be available and durable for a long period of time. In this section, mean time between failures, failure modes, and reliability of the system will be analyzed and discussed.

4.1. Mean Time between Failures

The mean time between failures plays a major role in the reliability investigations. It has a significant contribution to the maintenance scheduling and cost prediction. Figure 3 shows the mean time between failures of each component in the unloading system. Clearly, the differential pressure transmitter and the pressure transmitter have the longest time intervals without failures. This is of no surprise to the industry practitioners due to the nature of the instrumentation operation and material of construction. On the other hand, the positive displacement pump and the positive displacement flow meter have the shortest time intervals between consecutive failures due to the design complexity of the rotating mechanisms within the two components. Most importantly, any failure of the flow meter needs to be detected as early as possible to avoid uncertainty variation in fluid measurements. In addition, the strainer has a short period between consecutive failures because usually, such filtration systems need a continuous cleaning of its internal parts and the failures of these components can be observed without a regular frequent cleaning. Therefore, the frequent cleaning of the strainer will lead to an increase in the mean time between the failures. It is noted that all the valves have a medium time between failures, while piping spools have long time intervals between failures. According to some studies discussed earlier in the literature review section, the air eliminator has a relatively low mean time between failures due to the complex design in the separation process that takes place in this portion of the system. In addition, the unloading arms also have a relatively low mean time between failures compared to the piping spools due to the dynamic nature of the swivel joints that connects the arm parts together and the excessive movement they experience externally.

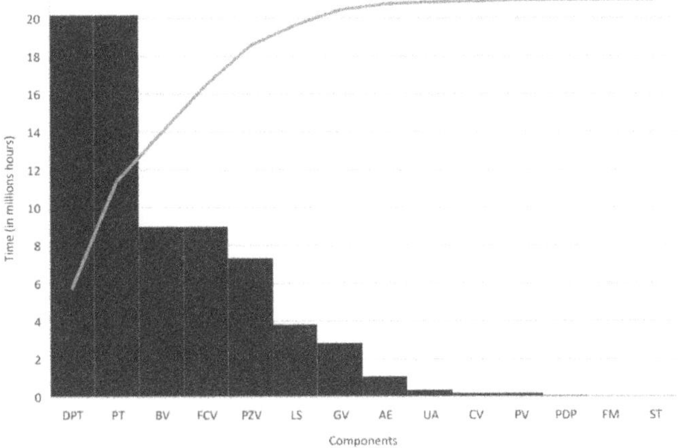

Figure 3. Mean time between failures of unloading system components.

4.2. Failure Modes

Table 4 shows the failure modes of each individual component that was observed in accordance with the same study of the failure rate of respective components. For instance, the three failures that usually happen in the positive displacement pump are vibration, parameter deviation, and failure to start on demand. Looking profoundly into the failure modes of the critical components, the positive displacement pump and positive displacement meter, it is noticed that these failures particularly, unlike other component failures, could be caused by deficiency of more than one part of the component construction, therefore the root cause of the failures could be very difficult to detect. On the other hand, unloading arm failures can easily be detected. Practically, valves are extremely important components in the system to either isolate or control the fluid flow. It has been found that almost all of the valve failures are a result of their construction integrity or excessive operation conditions; both of these failures cause an internal or an external leakage. Failure detection is out of the scope of this study; however, knowing the failure moods and their root causes are essential to reliability improvement.

Table 4. Unloading system component failure modes.

Component	Failure Mode
Unloading Arms	Rupture & Leak
	Leakage in closed position
	Internal leakage
Ball Valve	External leakage
	Low output
	Structural deficiency
	Vibration
Positive Displacement Pump	Parameter deviation
	Fail to start on demand
	External leakage
Check Valve	Opposite direction flow

Table 4. *Cont.*

Component	Failure Mode
Gate Valve	Internal leakage
	External leakage
Pressure Relief Valve	External leakage
	Fail to close on demand
	Fail to open on demand
	Spurious operation
	Valve leakage in closed position
	Delayed operation
Positive Displacement Flow Meter	Parameter deviation
	Low output
	Vibration
Plug Valve	External & Internal leakage
Flow Control Valve	Leakage in closed position
	Internal leakage
	External leakage
	Low output
	Structural deficiency
Differential Pressure Transmitter	Leakage
Level Switch	High output
	Low output
	Erratic output
	Fail to function on demand
	Spurious operation
Pressure Transmitter	Leakage
Air Eliminator—Separator	Parameter deviation
	Plugged/Choked
	Structural deficiency
Strainer	External leakage
	Plugging

4.3. Unloading System Reliability

Since the components and the whole system are assumed to have constant failure rates (random failures), the reliability figures must be investigated in multiple time intervals. For each component, the reliability of the system is calculated at the end of the first year, the fifth year, the tenth year, and the twentieth year using Equation (3). Table 5 shows the reliability of each component in the system.

$$R_i(t) = e^{-\lambda t} \tag{3}$$

$$R_p(t) = 1 - F(t) = 1 - \prod_{i=1}^{n_p} [1 - R_i(t)] \tag{4}$$

where $R_i(t)$ is the component i reliability, $R_p(t)$ is the reliability of components connected in parallel, $F(t)$ is the cumulative probability distribution, and n_p is the number of components connected in parallel.

For the other components connected in series, individual reliability was calculated using Equation (3). Considering each sub-system with its redundant as a single component that has a single reliability value as illustrated in the streams 1 and 2 combined row as given in Table 3 will validate the concept of a complete series system. Streams 1 and 2 combined reliability was calculated using Equation (4), considering stream 1 and stream 2 in parallel connection. As a result, the system's components shown in Figure 2 become one line of series components from start to end. Consequently, the total system reliability can be calculated using Equation (5). It is important to note that neither the unloading arms nor the ball valves were directly considered in the system reliability calculations, instead, their subsystem (stream 1 and 2 combined) was directly considered in the system reliability calculations and this will account for both unloading arms and the ball valves connected to them.

$$R_s(t) = r_1(t)r_2(t) \dots r_n(t) = \prod_{i=1}^{n} r_i(t) \tag{5}$$

where $r_i(t)$ is the reliability of component i and n is the number of components connected in series.

Figure 4 shows the total system reliability at different time periods (e.g., t = 0, 1, 5, 10, and 20 years) during the normal operating phase. During this phase, it is certain that the system is currently fully functional; therefore, the reliability equals 1 at time 0. Based on the system reliability plot, it is noticed that the reliability decreases dramatically with the highest slope between 0 and 2 years. Then, the slope (2–5 years) tends to be much smaller. After 5 years, as time increases, the system probability is very low and almost constant, which means that the probability that the system is functional after five years is close to the probability that the system is functional after ten and twenty years.

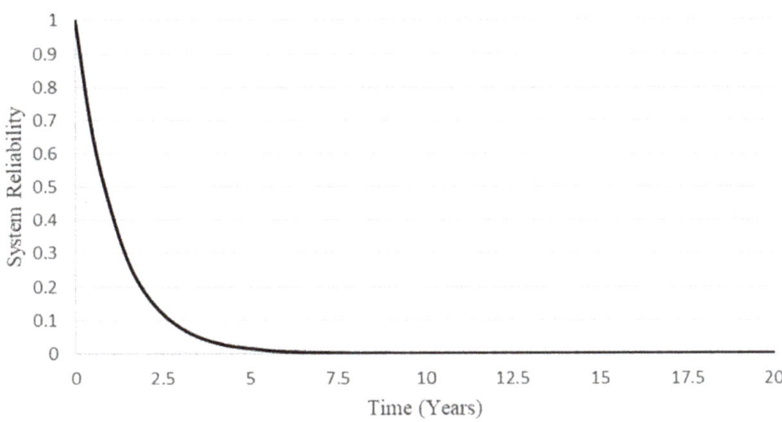

Figure 4. Unloading system reliability.

The reliability plot helps to determine the best schedule for preventive maintenance of the system, which will improve the system reliability. In this study, the unloading system reliability needs to be maintained above 90%, which means that the system components should be maintained every 45 days as shown in Figure 4. In the next section, the impact of critical component preventive maintenance on the whole system reliability will be investigated.

Table 5. Unloading system component reliability for different durations.

Component	Relation	Reliability (1 Year)	Reliability (5 Years)	Reliability (10 Years)	Reliability (20 Years)
Unloading Arm—1	Parallel	0.976471687	0.887765525	0.788127628	0.621145158
Unloading Arm—2	Parallel	0.976471687	0.887765525	0.788127628	0.621145158
Ball Valve—1	Parallel	0.999038705	0.995202756	0.990428525	0.980948663
Ball Valve—2	Parallel	0.999038705	0.995202756	0.990428525	0.980948663
Stream—1	Parallel	0.975533009	0.883506697	0.780584084	0.609311512
Stream—2	Parallel	0.975533009	0.883506697	0.780584084	0.609311512
Streams 1 & 2 Combined	Series	0.999401366	0.98642931	0.951856656	0.847362505
Ball Valve—3	Series	0.999038705	0.995202756	0.990428525	0.980948663
Positive Displacement Pump	Series	0.8668779	0.48954166	0.239651036	0.057432619
Check Valve—1	Series	0.953496955	0.788127628	0.621145158	0.385821307
Gate Valve—1	Series	0.996963513	0.984909488	0.9700467	0.940990601
Gate Valve—2	Series	0.996963513	0.984909488	0.9700467	0.940990601
Pressure Relief Valve—1	Series	0.998821649	0.994122112	0.988278773	0.976694934
Pressure Relief Valve—2	Series	0.998821649	0.994122112	0.988278773	0.976694934
Positive Displacement Flow Meter	Series	0.826565438	0.385821307	0.148858081	0.022158728
Plug Valve	Series	0.953496955	0.788127628	0.621145158	0.385821307
Flow Control Valve	Series	0.999038705	0.995202756	0.990428525	0.980948663
Differential Pressure Transmitter—1	Series	0.999571905	0.997861358	0.995727289	0.991472835
Differential Pressure Transmitter—2	Series	0.999571905	0.997861358	0.995727289	0.991472835
Level Switch	Series	0.997744003	0.988770794	0.977667684	0.95583410
Pressure Transmitter	Series	0.999571905	0.997861358	0.995727289	0.991472835
Air Eliminator—Separator	Series	0.991962691	0.96045427	0.922472404	0.850955336
Strainer—1	Series	0.826565438	0.385821307	0.148858081	0.022158728
Strainer—2	Series	0.826565438	0.385821307	0.148858081	0.022158728

4.4. Reliability Enhancement

From the system reliability plot, the preventive maintenance schedule for the unloading system is specified in order to maintain the reliability of the system above a certain level. In the previous sections, the critical components of the systems were identified. Therefore, preventive maintenance of these critical items will have a huge impact on the total system reliability. In this study, preventive maintenance of the strainer, positive displacement pump, and positive displacement meter will be considered due to their design nature. Practically, rotating part failures are relatively easier to detect before they occur, compared to static part failures. For example, if there is a slight alignment deviation in the pump coupling it will remain functional for a small period of time before the breakdown. While, for static components like valves, the failures happen within seconds without prior signals. In addition, the strainer is considered as a critical component because of the cleaning process that should take place during the maintenance, which will improve the system reliability.

Figure 5 shows the system reliability when preventive maintenance for the critical components is conducted every 45 days. It is noted that preventive maintenance of the three critical components significantly enhances the reliability of the unloading system. In other words, the probability that any of these components are functional for the next one and a half months is constant throughout the lifecycle of these components.

Figure 5 shows the improvements reflected on the system reliability due to conducting preventive maintenance. It is clear that the preventive maintenance has a large impact on the total system reliability improvement.

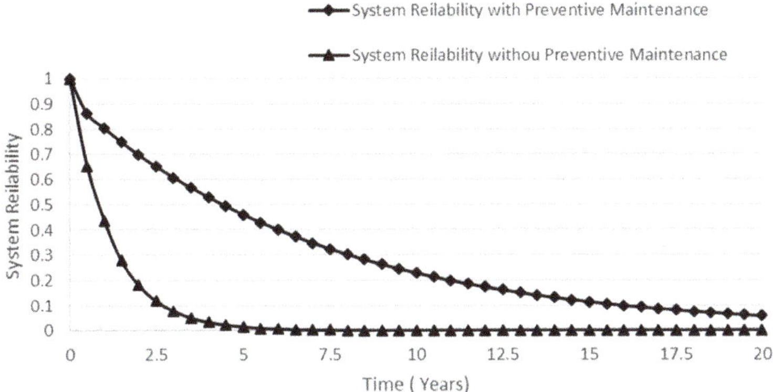

Figure 5. Unloading system reliability before and after preventive maintenance.

The analysis shows that the reliability of the unloading system reached approximately zero after 5 years of continuous operation, while after conducting the preventive maintenance, system reliability significantly improved. In addition, the results indicated that implementing preventive maintenance improves the system reliability by 80% in the first year, and this ratio increases dramatically as the period increases.

5. Conclusions and Future Work

The truck unloading system is a complicated integration of components with different levels of reliability. This study assessed and evaluated the reliability of the truck unloading system, explored and identified the components that have the lowest reliability within the system. The mean time between failures, failure modes of each component, and reliability of the system are analyzed. In addition, this study established a preventive maintenance strategy for improving the reliability of the truck unloading systems. Since the current level of the reliability of the truck unloading system is not satisfactory, it can be greatly enhanced through focusing on a few critical components.

According to reliability analysis, the best schedule for preventive maintenance to maintain system reliability above 90% is 45 days. Moreover, the positive displacement pump and the positive displacement meter are the most critical items. Therefore, it is recommended to avail spare parts for these two components and conduct preventive maintenance on regular basis; preferably every one and a half months as found from the reliability analysis. In addition, it is highly recommended that the operators should report any unusual observations at the earliest to avoid any further damage or poor function. Furthermore, the results indicated that implementing preventive maintenance improves the system reliability by almost 80% in the first year, and this ratio increases radically as the period increases.

Furthermore, the replacement of the positive displacement meter with a more advanced Coriolis meter should be considered. Comparatively, it has a lower failure rate due to the lower number of internal components and the simpler mechanism of its operation. Second, filtration components such as a strainer should be periodically inspected and cleaned to avoid plugging. Third, although valves have medium failure rates, it is important to keep spare parts for the frequently failing parts such as seals. Finally, further studies are highly recommended for design improvement of the truck unloading system to enhance the system reliability and having a longer uninterrupted safe operation.

Author Contributions: Conceptualization, A.M. and M.A.-S.; methodology, A.M. and M.A.-S.; software, A.G.; validation, A.G. and K.A.-O.; formal analysis, A.M., M.A.-S. and A.G.; investigation, A.M.; resources, A.G.; data curation, M.A.-S.; writing—original draft preparation, M.A.-S.; writing—review and editing, A.G. and A.M.; supervision, K.A.-O.; project administration, K.A.-O. All authors have read and agreed to the published version of the manuscript.

Funding: This research received no external funding.

Acknowledgments: The authors would like to thank King Fahd University of Petroleum and Minerals for the support that contributed to carrying out this research. They also thank the anonymous reviewers for their insightful comments.

Conflicts of Interest: The authors declare no conflict of interest.

References

1. Lee, D.; Pan, R. A nonparametric Bayesian network approach to assessing system reliability at early design stages. *Reliab. Eng. Syst. Saf.* **2018**, *171*, 57–66. [CrossRef]
2. Yoon, J.-H.; Yang, I.-H.; Jeong, J.-E.; Park, S.-G.; Oh, J.-E. Reliability improvement of a sound quality index for a vehicle HVAC system using a regression and neural network model. *Appl. Acoust.* **2012**, *73*, 1099–1103. [CrossRef]
3. Dalkilic, S. Improving aircraft safety and reliability by aircraft maintenance technician training. *Eng. Fail. Anal.* **2017**, *82*, 687–694. [CrossRef]
4. Corvaro, F.; Giacchetta, G.; Marchetti, B.; Recanati, M. Reliability, Availability, Maintainability (RAM) study, on reciprocating compressors API 618. *Petroleum* **2017**, *3*, 266–272. [CrossRef]
5. Catelani, M.; Ciani, L.; Venzi, M. Component Reliability Importance assessment on complex systems using Credible Improvement Potential. *Microelectron. Reliab.* **2016**, *64*, 113–119. [CrossRef]
6. Wu, J.; Yan, S.; Li, J.; Gu, Y. Mechanism reliability of bistable compliant mechanisms considering degradation and uncertainties: Modeling and evaluation method. *Appl. Math. Model.* **2016**, *40*, 10377–10388. [CrossRef]
7. Mi, J.; Li, Y.-F.; Yang, Y.-J.; Peng, W.; Huang, H.-Z. Reliability assessment of complex electromechanical systems under epistemic uncertainty. *Reliab. Eng. Syst. Saf.* **2016**, *152*, 1–15. [CrossRef]
8. Kontrec, N.; Milovanović, G.V.; Panic, S.R.; Milošević, H. A Reliability-Based Approach to Nonrepairable Spare Part Forecasting in Aircraft Maintenance System. *Math. Probl. Eng.* **2015**, *2015*, 1–7. [CrossRef]
9. Hou, W.; Jiang, Z.; Jin, Y. Reliability-Based Opportunistic Preventive Maintenance Model of Multi-units Serial-Parallel System. *J. Shanghai Jiaotong Univ.* **2009**, *4*, 658–662.
10. Vishnu, C.; Regikumar, V. Reliability Based Maintenance Strategy Selection in Process Plants: A Case Study. *Procedia Technol.* **2016**, *25*, 1080–1087. [CrossRef]
11. Tee, K.F.; Ekpiwhre, E. Reliability-based preventive maintenance strategies of road junction systems. *Int. J. Qual. Reliab. Manag.* **2019**, *36*, 752–781. [CrossRef]

12. Yssaad, B.; Khiat, M.; Chaker, A. Reliability centered maintenance optimization for power distribution systems. *Int. J. Electr. Power Energy Syst.* **2014**, *55*, 108–115. [CrossRef]
13. Li, L.; You, M.; Ni, J. Reliability-Based Dynamic Maintenance Threshold for Failure Prevention of Continuously Monitored Degrading Systems. *J. Manuf. Sci. Eng.* **2009**, *131*, 031010. [CrossRef]
14. Quimpo, R.G.; Shamsi, U.M. Reliability-Based Distribution System Maintenance. *J. Water Resour. Plan. Manag.* **1991**, *117*, 321–339. [CrossRef]
15. Stein, H. *PDS Data Handbook*, 2010th ed.; SINTEF: Trondheim, Norway, 2010. Available online: https://www.sintef.no/globalassets/project/pds/presentations/pds_data_handbook_2010_edition-selected_pages.pdf (accessed on 1 September 2020).
16. OREDA Participants. *Offshore Reliability Data Handbook*; DNV: Trondheim, Norway, 2002.
17. Sun, Y.; Ma, L.; Morris, J. A practical approach for reliability prediction of pipeline systems. *Eur. J. Oper. Res.* **2009**, *198*, 210–214. [CrossRef]
18. Rimkevičius, S.; Kaliatka, A.; Valinčius, M.; Dundulis, G.; Janulionis, R.; Grybenas, A.; Zutautaite, I. Development of approach for reliability assessment of pipeline network systems. *Appl. Energy* **2012**, *94*, 22–33. [CrossRef]
19. Shalay, V.; Zemenkova, M.; Zemenkov, Y.; Toropov, S. Modeling Parameters of Reliability of Technological Processes of Hydrocarbon Pipeline Transportation. In *MATEC Web of Conferences*; EDP Science: Les Ulis, France, 2016; Volume 73, p. 01029.
20. Kong, J.S.; Frangopol, D.M. Life-Cycle Reliability-Based Maintenance Cost Optimization of Deteriorating Structures with Emphasis on Bridges. *J. Struct. Eng.* **2003**, *129*, 818–828. [CrossRef]
21. Wu, F.; Niknam, S.A.; Kobza, J.E. A cost effective degradation-based maintenance strategy under imperfect repair. *Reliab. Eng. Syst. Saf.* **2015**, *144*, 234–243. [CrossRef]
22. Barata, J.M.; Soares, C.G.; Marseguerra, M.; Zio, E. Simulation modelling of repairable multi-component deteriorating systems for 'on condition' maintenance optimisation. *Reliab. Eng. Syst. Saf.* **2002**, *76*, 255–264. [CrossRef]
23. Tong, J.; Mao, D.; Xue, D.A. genetic algorithm solution for a nuclear power plant risk-cost maintenance model. *Nucl. Eng. Des.* **2004**, *229*, 81–89.
24. Rao, A.N.; Bhadury, B. Opportunistic maintenance of multi-equipment system: A case study. *Qual. Reliab. Eng. Int.* **2000**, *16*, 487–500. [CrossRef]
25. Zhang, C.; Gao, W.; Guo, S.; Li, Y.; Yang, T. Opportunistic maintenance for wind turbines considering imperfect, reliability-based maintenance. *Renew. Energy* **2017**, *103*, 606–612. [CrossRef]
26. Xie, L.; Rui, X.; Li, S.; Hu, X. Maintenance Optimization of Offshore Wind Turbines Based on an Opportunistic Maintenance Strategy. *Energies* **2019**, *12*, 2650. [CrossRef]
27. Arunraj, N.; Maiti, J. Risk-based maintenance policy selection using AHP and goal programming. *Saf. Sci.* **2010**, *48*, 238–247. [CrossRef]
28. Zhao, F.; Wang, W.; Peng, R. Delay-time-based preventive maintenance modelling for a production plant: A case study in a steel mill. *J. Oper. Res. Soc.* **2015**, *66*, 2015–2024. [CrossRef]
29. Li, G.; Li, Y.; Zhang, X.; Hou, C.; He, J.; Xu, B.; Chen, J. Development of a Preventive Maintenance Strategy for an Automatic Production Line Based on Group Maintenance Method. *Appl. Sci.* **2018**, *8*, 1781. [CrossRef]
30. Hadidi, L.A.; Alkhaldi, A.F. Reliability improvement for electrical pneumatic arm loading system. In Proceedings of the 2016 IEEE International Conference on Industrial Engineering and Engineering Management (IEEM), Bali, Indonesia, 4–7 December 2016; IEEE: Piscataway, NJ, USA, 2016.
31. Reis, A.N.; Neto, A.R.P.; Rolim, G.A. Simulation of Tank Truck Loading Operations in a Fuel Distribution Terminal. *Int. J. Simul. Model.* **2017**, *16*, 435–447. [CrossRef]
32. Jamshidi, R.; Esfahani, M.M.S. Reliability-based maintenance and job scheduling for identical parallel machines. *Int. J. Prod. Res.* **2014**, *53*, 1216–1227. [CrossRef]
33. Navarro, I.J.; Martí, J.V.; Yepes, V. Reliability-based maintenance optimization of corrosion preventive designs under a life cycle perspective. *Environ. Impact Assess. Rev.* **2019**, *74*, 23–34. [CrossRef]

Article

On the Use of Importance Measures in the Reliability of Inventory Systems, Considering the Cost

Liwei Chen [1,*], Meng Kou [2] and Songwei Wang [1,*]

[1]   School of Electrical Engineering, Zhengzhou University, Zhengzhou 450001, Henan, China
[2]   Beijing Institute of Astronautical System Engineering, Beijing 100076, China; dhynhnh@126.com
*   Correspondence: cliwei@zzu.edu.cn (L.C.); wangsongwei@zzu.edu.cn (S.W.)

Received: 25 August 2020; Accepted: 26 October 2020; Published: 9 November 2020

Abstract: In order to maximize inventory benefits or minimize costs, reliability and cost of inventory control models need to be identified and analyzed. These importance measures are one important approach to recognize and evaluate system weaknesses. However, importance measures have fewer applications in inventory systems' reliability. Considering the cost, this paper mainly discusses the reliability change of performance parameters with the importance measures in inventory systems. The calculation methods of differential importance and Birnbaum importance are studied in the inventory control model with shortages. By comparing the importance values of various parameters in the model, the optimization analysis of the inventory model can be used to identify the key parameters, so as to effectively reduce the total inventory cost. The importance order and the identification of key parameters are helpful to increase the operational efficiency of the inventory control and provide effective methods for improving the inventory management. Lastly, a case study with a shortage and limited inventory capacity is used to demonstrate the proposed model.

Keywords: reliability; importance measure; cost; inventory systems

1. Introduction

Importance measures refer to the influence on system reliability when a single or multiple components of a system fail or change state, which is a function of component reliability parameters and system structure. As one of the important branches and basic theories of reliability, importance measures penetrate all stages of products, including design, production, inspection, sale, maintenance, and so on. Identifying the factors influencing system reliability is most important [1,2]. In the phase of design, importance measures are used to identify weaknesses and support the improvement and optimization of system. During the system operation, importance measures can allocate enterprise resources reasonably to constituent part of a system to ensure that it is operating properly. By identifying and evaluating system weaknesses, importance measures have been widely applied in system reliability, decision making and risk analysis [3–7].

Since Birnbaum [8] firstly proposed the concept of importance analysis, importance measures have wide applications in the domains of fault analysis, model simulation, and network planning. Borgonovo and Apostolakis [9] introduced a new importance measure, the differential importance measure, for probabilistic safety assessment. Considering the transition rates of component states, Dui et al. [10–13] discussed the influence of importance measures on system performance and its applications in aviation and other fields. Kim and Song [14] proposed a generalized reliability importance measure that can deal with multiple critical failure regions, large curvatures of limit-state surfaces and the correlation between the input random variables. Li et al. [15] proposed a power flow element importance measure, which can improve cascading failure prevention, system backup setting, and overall resilience. Dui et al. [16] analyzed the applications of importance measures in the reliability of inventory systems and extended the importance measure to the three-echelon inventory systems. Dui et al. [17] studied the cascading failure in an inventory network from the perspective of the payoffs of nodes in a multi-strategy evolutionary game. Nguyen et al. [18] proposed opportunistic maintenance decision rules based on the criticality level of components and the availability of spare parts. Adak and Mahapatra [19] developed a cost-effective ordering inventory model where the increase in reliability of the item lead to a rise in demand and decreases the rate of deterioration. Maji et al. [20] found the optimal number of transportation cycles and components, which can maximize the total business profit and system reliability with volume, weight, and cost of the system as constraints. Huang [21] studied the system reliability of a stochastic delivery-flow distribution network with an inventory. Manna et al. [22] studied an imperfect production inventory model with production system reliability under two-layer supply chain management. Abdel-Aleem et al. [23] proposed an optimal solution of the reliability model by a generalized reduced gradient algorithm.

Currently, the competitive environment is drastically changing, which will lead to the adaptation of system resilience design and maintenance in dynamic environments [24,25]. There are too many uncertainty factors involved in inventory models. With the interference of various uncertainties, the problem of stock shortage occurs frequently. In situations where enterprises operate normally and minimize inventory, how to realize the optimization of inventory models out of stock, has become a hot topic for scholars from all circles [26]. Cárdenas-Barrón and Sana [27] proposed an economic order quantity inventory model of multi-items in a two-layer supply chain where demand was sensitive to promotional effort, and they compared collaborative and non-collaborative systems in terms of their average profits. Shekarian et al. [28] developed a reverse inventory model where the recoverable manufacturing process was affected by the learning theory. Chao et al. [29] characterized the optimal policies that simultaneously determine the optimal ordering and pricing decisions in each period over a finite planning horizon. Meanwhile, the impacts of supply source diversification and supplier reliability on the firm and on its customers are studied. Yu et al. [30] considered the optimal production, pricing, and substitution policies of a continuous-review production–inventory system with two products: a high-end product and a low-end product. Lee et al. [31] examined vendor-managed inventory systems with stockout-cost sharing between a supplier and a customer with shortages allowed under limited storage capacity, where a stockout penalty was charged to the supplier when stockouts occurred with the customer. Jia and Cui [32] analyzed the reliability of supply chain systems by using copulas. Flynn et al. [33] developed a theoretical conceptualization of supply chain uncertainty and reliability. He et al. [34] built a logistics service supply chain model under the stochastic demand to consider the feature of non-storage and reliability. Chen et al. [35] evaluated the supply chain reliability and resilience for the complexity of supply chain structures.

The inventory models provide an effective means for companies to carry out inventory management and reduce inventory costs. Scholars have continuously improved and optimized the models from various perspectives and fields, however, have neglected the application of importance measures to effectively identify and optimize systems in this field. Different parameters in the inventory system have different effects on the inventory system. Issues which have become the focus of research are: How to use the system reliability and importance measure to study the influence of the parameters in the inventory system on the inventory system? How to calculate the importance of different parameters and determine the parameters that have the greatest impact on the system? Therefore, this paper analyzes the reliability of the inventory system. According to the reliability of the inventory system, a cost-based inventory system importance model is proposed to study the importance of different parameters, which provides some support and reference for enriching importance measures and optimizing inventory models.

The rest of the paper is as follows. Section 2 analyses the reliability of inventory systems. The cost of inventory systems reliability is briefly described in Section 3. Section 3 also derives the computational methods of importance measures about various parameters of cost function. A numerical example is presented to illustrate importance values and changes in cost parameters in Section 4. Section 5 gives the conclusions and future work of this paper.

2. Reliability Analysis in Inventory Systems

Many logistics units constitute the logistics system organically according to a certain link mode. The premise of analyzing the reliability of the whole logistics system is to determine the calculation method of the reliability of a single logistics unit.

The reliability of a logistics unit refers to the probability that the service provided by logistics units remains within the specified error limit under certain conditions and time. In the service capacity curve of the logistics unit in Figure 1, the M curve represents the standard service curve that the logistics unit is expected to achieve, the N curve represents the error limit of the logistics service specified for the logistics unit, and the P curve represents the real logistics service curve provided by the logistics unit.

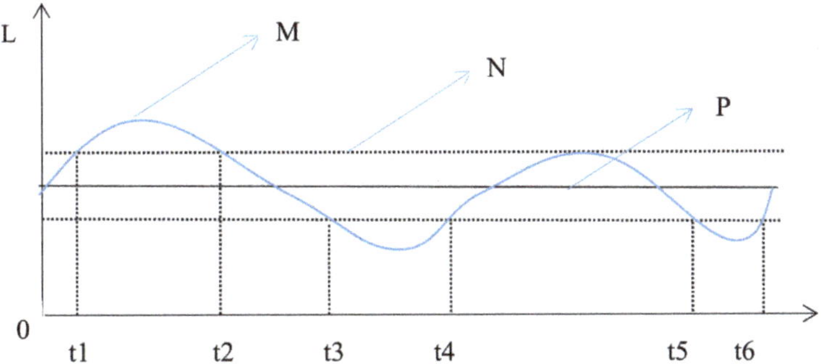

Figure 1. Service capacity curve of a logistics unit.

In Figure 1, L is the level of the service capacity of the logistics unit. In the intervals $[t1 - t2], [t3 - t4]$, $[t5 - t6]$, the logistics service provided by the logistics unit exceeds the specified error limit. Therefore, it is considered that the work of the logistics unit in this situation is unreliable. In the intervals $[t2 - t3]$, $[t4 - t5]$, the logistics service provided by the logistics unit does not exceed the specified error limit. Therefore, it is considered that the operation of the logistics unit is reliable here. Assuming that the reliability of the logistics unit is M_U, then

$$M_U = 1 - \frac{\sum\limits_{k=1}^{n} (t_{k2} - t_{k1})}{T}$$

where t_{k1}, t_{k2} are the starting and ending time, respectively, of the kth observation that the logistics service provided by the logistics unit exceeds the allowable deviation range, $(k = 1, 2, 3, \ldots, n)$. T is the total observation time, and n is the number of times that the logistics service provided by the logistics unit exceeds the allowable deviation range.

In a typical tandem logistics system, such as the one shown in Figure 2, the reliability of the five logistics units of transportation, storage, circulation processing, loading and unloading handling, and distribution are M_Y, M_C, M_L, M_Z and M_P, respectively.

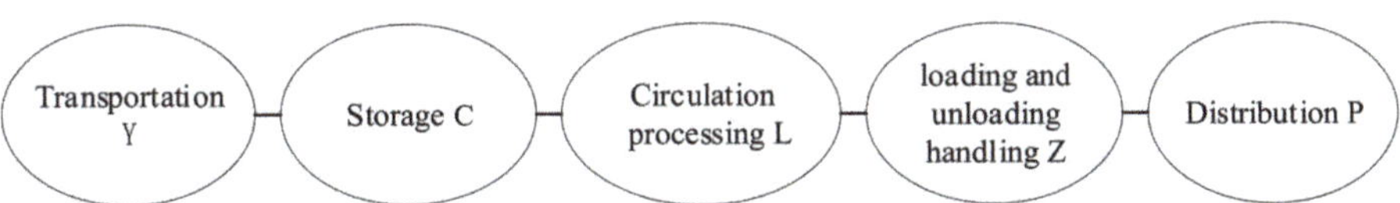

Figure 2. Structural diagram of a typical tandem logistics system.

This logistics system is composed of five logistics units in series. According to the mathematical reliability model of series system, the reliability of typical series logistics system is

$$M_T = M_Y M_C M_L M_Z M_P$$

The logistics system cannot be fully paralleled; therefore the reliability calculation can only be carried out for one parallel subsystem. The reliability of circulation processing, L1, and circulation processing, L2, are M_{L1} and M_{L2}, respectively, as shown in Figure 3.

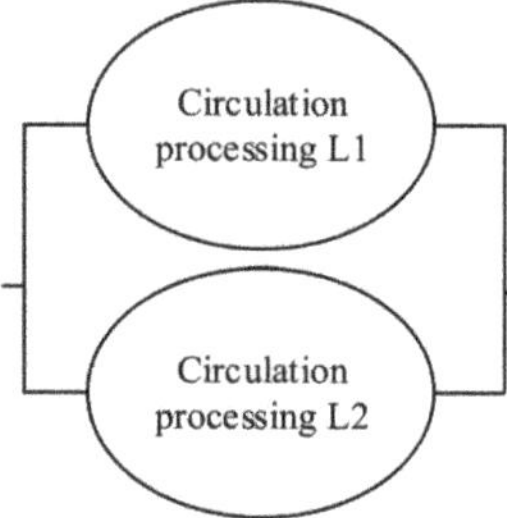

Figure 3. Structural diagram of a typical parallel logistics system.

Taking the parallel subsystem of circulation processing as an example, according to the mathematical reliability model of the parallel system, the reliability of the parallel subsystem is

$$M_{LT} = 1 - (1 - M_{L1})(1 - M_{L2}) = M_{L1} + M_{L2} - M_{L1}M_{L2}$$

In a typical series parallel logistics system, as shown in Figure 4, the reliability of transportation Y1, transportation Y2, storage C1, storage C2, circulation processing L1, circulation processing L2, loading and unloading handling Z1, loading and unloading handling Z2, distribution P1 and distribution P2 are M_{Y1}, M_{Y2}, M_{C1}, M_{c2}, M_{L1}, M_{L2}, M_{Z1}, M_{Z2}, M_{P1}, and M_{P2}, respectively.

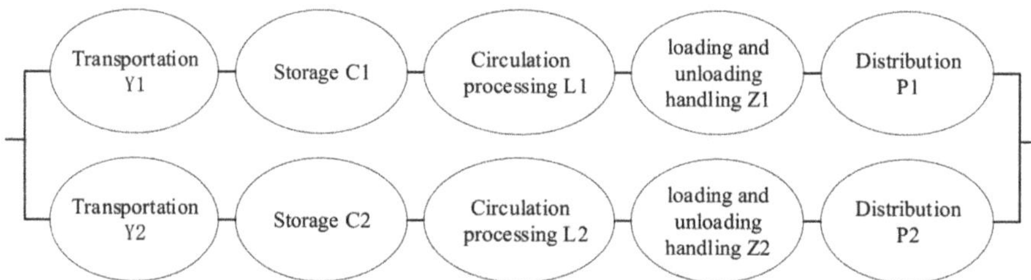

Figure 4. Structural diagram of a typical series parallel logistics system.

The first group series system consists of five logistics units: transportation Y1, storage C1, circulation processing L1, loading and unloading handling Z1, and distribution P1. According to the mathematical reliability model of series systems, the reliability of the first series system is

$$M_{T1} = M_{Y1} M_{C1} M_{L1} M_{Z1} M_{P1}$$

The second group of series systems consists of five logistics units: transportation Y2, storage C2, circulation processing L2, loading and unloading handling Z2 and distribution P2. Therefore, according to the mathematical reliability model of series system, the reliability of the second series system is

$$M_{T2} = M_{Y2} M_{C2} M_{L2} M_{Z2} M_{P2}$$

The reliability of two series systems in parallel is

$$M_T = 1 - (1 - M_{T1})(1 - M_{T2}) = M_{T1} + M_{T2} - M_{T1}M_{T2}$$
$$= M_{Y1}M_{C1}M_{L1}M_{Z1}M_{P1} + M_{Y2}M_{C2}M_{L2}M_{Z2}M_{P2} - M_{Y1}M_{C1}M_{L1}M_{Z1}M_{P1}M_{Y2}M_{C2}M_{L2}M_{Z2}M_{P2}$$

In a typical series parallel logistics system, as shown in Figure 5, the reliability of transportation $Y1$, transportation $Y2$, storage $C1$, storage $C2$, circulation processing $L1$, circulation processing $L2$, loading and unloading handling $Z1$, loading and unloading handling $Z2$, distribution $P1$ and distribution $P2$ are M_{Y1}, M_{Y2}, M_{C1}, M_{C2}, M_{L1}, M_{L2}, M_{Z1}, M_{Z2}, M_{P1}, and M_{P2}, respectively.

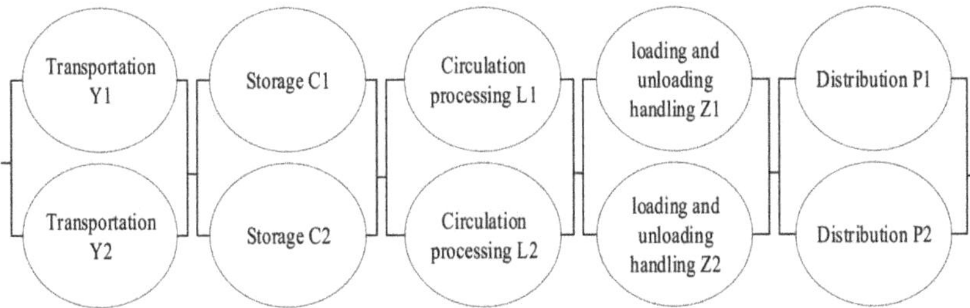

Figure 5. Structural diagram of a typical parallel series logistics system.

The first group of parallel subsystems is composed of transportation $Y1$ and transportation $Y2$ in parallel. According to the mathematical reliability model of the parallel system, the reliability of the first group of parallel subsystems is

$$M_Y = 1 - (1 - M_{Y1})(1 - M_{Y2}) = M_{Y1} + M_{Y2} - M_{Y1}M_{Y2}$$

Similarly, the reliability of the second group of parallel subsystems is

$$M_C = 1 - (1 - M_{C1})(1 - M_{C2}) = M_{C1} + M_{C2} - M_{C1}M_{C2}$$

The reliability of the third group of parallel subsystems is

$$M_L = 1 - (1 - M_{L1})(1 - M_{L2}) = M_{L1} + M_{L2} - M_{L1}M_{L2}$$

The reliability of the fourth group of parallel subsystems is

$$M_Z = 1 - (1 - M_{Z1})(1 - M_{Z2}) = M_{Z1} + M_{Z2} - M_{Z1}M_{Z2}$$

The reliability of the fifth group of parallel subsystems is

$$M_P = 1 - (1 - M_{P1})(1 - M_{P2}) = M_{P1} + M_{P2} - M_{P1}M_{P2}$$

The reliability of a typical parallel series logistics system is the reliability of the five parallel subsystems in series.

$$\begin{aligned} M_T &= M_Y M_C M_L M_Z M_P \\ &= (M_{Y1} + M_{Y2} - M_{Y1}M_{Y2})(M_{Y1} + M_{Y2} - M_{Y1}M_{Y2})(M_{L1} + M_{L2} - M_{L1}M_{L2}) \\ &\quad (M_{Z1} + M_{Z2} - M_{Z1}M_{Z2})(M_{P1} + M_{P2} - M_{P1}M_{P2}) \end{aligned}$$

3. Cost-Based Importance Measures of Inventory Systems Reliability

3.1. Birnbaum Importance Measure

From a mathematical perspective, Birnbaum importance [8] analyzes the influence of changes in variables on the whole function as $I(x_i) = \frac{\partial f(x_1, x_2, \dots, x_n)}{\partial x_i}$, where $f(x_1, x_2, \dots, x_n)$ means a function consisting of n variables $x_1, x_2, \dots, x_n$.

For various systems, x_i and $f(x_1, x_2, \dots, x_n)$ have different meanings. For example, x_i represents the reliability of component i, and $f(x_1, x_2, \dots, x_n)$ shows the reliability of whole system.

3.2. Differential Importance Measure

The differential importance measure [9] is $DIM(x_i) = \frac{\frac{\partial f(x_1, x_2, \dots, x_n)}{\partial x_i} dx_i}{\sum_j \frac{\partial f(x_1, x_2, \dots, x_n)}{\partial x_j} dx_j}$, in which, $f(x_1, x_2, \dots, x_n)$ represents the risk metrics associated with various parameters, and x_i means the parameters.

In inventory management, $f(x_1, x_2, \dots, x_n)$ means the total cost of the inventory control system, and x_i shows each parameter in inventory control, such as the order quantity, demand quantity, storage cost, shortage cost and so on.

3.3. Discussions on Importance Measures Based on the Inventory Systems Cost

In the inventory control model, the optimal lot size, Q^*, is obtained by minimizing the total cost. The function expression of the total cost is as follows.

$$\varphi(Q, \alpha) = \frac{(u + a/2)Q + \gamma}{1 - e^{-\rho Q/R}}$$

where Q is the selection variable, $\alpha = (u, a, R, \gamma, \rho)$ is a parameter variable, u is the unit price of the goods in stock, a is the unit holding cost, R is the demand speed, γ is the order cost, and ρ is the capital cost.

Among them, an increase in u, a, R and γ will cause an increase in the total cost. An increase in ρ will cause a decrease in total cost.

The total cost function takes the first-order derivative of Q and sets the result to 0, then

$$\varphi_Q(Q^*, \alpha) = \left(\frac{1}{2}a + u\right)\left(e^{(Q^*/R)\rho} - 1\right)R - \rho\left(\gamma + Q^*\left(\frac{1}{2}a + u\right)\right) = 0.$$

For the importance of several parameter variables with respect to function $Q^*(\alpha)$, the implicit importance can be applied to the optimization problem of the model. The final results are shown in Table 1. Assuming that $u = 10$ (yuan/piece), $a = 1$ (yuan/piece), $R = 8000$ (piece), $\gamma = 30$ (yuan/time), and $\rho = 8\%$, the comparative static (CS) can be used to analyze the results as follows.

Table 1. Results analysis of comparative static.

Parameter	CS	Expression	Mark	Value
u	Q_u	$-\dfrac{R\left(e^{(Q/R)\rho}-1\right)-Q\rho}{\rho\left(\frac{1}{2}a+u\right)\left(e^{(Q/R)\rho}-1\right)}$	$-$	36
a	Q_a	$-\dfrac{\frac{1}{2}R\left(e^{(Q/R)\rho}-1\right)-\frac{1}{2}Q\rho}{\rho\left(\frac{1}{2}a+u\right)\left(e^{(Q/R)\rho}-1\right)}$	$-$	18
R	Q_R	$-\dfrac{\frac{1}{R}\left(\frac{1}{2}a+u\right)\left(R\left(e^{(Q/R)\rho}-1\right)-Q\rho e^{(Q/R)\rho}\right)}{\rho\left(\frac{1}{2}a+u\right)\left(e^{(Q/R)\rho}-1\right)}$	$+$	0.047
γ	Q_γ	$-\dfrac{-\rho}{\rho\left(\frac{1}{2}a+u\right)\left(e^{(Q/R)\rho}-1\right)}$	$+$	13
ρ	Q_ρ	$-\dfrac{Q\left(\frac{1}{2}a+u\right)\left(e^{(Q/R)\rho}-1\right)-\gamma}{\rho\left(\frac{1}{2}a+u\right)\left(e^{(Q/R)\rho}-1\right)}$	$-$	4725

From Table 1, the results can be obtained as follows.

(1) The increase in u, a and ρ leads to a decrease in Q^* in the EOQ model.

(2) The increase in R and γ will cause an increase in Q^* in the EOQ model.

(3) It can be seen from the table that ρ is the most influential parameter in the EOQ results, and its influence degree is far greater than other parameters.

On the other hand, the partial derivatives of each parameter have different units of measurement, so they cannot be compared with each other. Similarly, because there are different measurement standards between parameters, assumption 1 of differential importance is not tenable in the application of differential importance. Even if the parameters have the same unit of measurement, but the direction of change is different, and the results of static analysis technology cannot be used as the standard of measurement.

According to the relationship between comparative static analysis technology and differential importance, the expression of differential importance measures in implicit models is proposed, and expressed as follows.

$$\Gamma(x^*, \alpha^*) = \left[\gamma_{j,s} : \gamma_{j,s} = \frac{|\Phi^{js}|}{|\Phi_j|} \right], \quad j = 1, \ldots, m, \quad s = 1, \ldots, n.$$

$\Gamma(x^*, \alpha^*)$ is a matrix, and the elements in the matrix represent the differential importance of the parameter α_s with respect to x_j. $\Phi^{js} = \left[J_x^1 J_x^2 \ldots J_x^{j-1} J_\alpha^s d\alpha_s J_x^{j+1} \ldots J_x^m \right]$, $\Phi_j = \left[J_x^1 J_x^2 \ldots J_x^{j-1} dJ_\alpha J_x^{j+1} \ldots J_x^m \right]$. According to the relationship between comparative static analysis technology and the expression of differential importance in the implicit model, the expression of Birnbaum importance is

$$\Gamma_B(x^*, \alpha^*) = \left[\begin{array}{c} \gamma_{j,s} : \gamma_{j,s} = \left\| \left[J_x^1 J_x^2 \ldots J_x^{j-1} J_\alpha^s \frac{\alpha}{x_s^j} J_x^{j+1} \ldots J_x^m \right] \right\| \\[4pt] = \dfrac{\left\| \left[J_x^1 J_x^2 \ldots J_x^{j-1} J_\alpha^s \frac{\alpha}{x_s^j} J_x^{j+1} \ldots J_x^m \right] \right\| \alpha}{x_s^j} \end{array} \right]$$

In an implicit multivariate function, the change of the whole function is caused by the change of one of the independent variables. The importance degree is used to rank the parameters, and the problems caused by different measurement units of each parameter are considered.

The change of the related variables needs to be calculated. According to the result of $\Gamma(x, \alpha)$, the optimal order quantity can be taken as the selection variable, and five parameters can be selected at the same time. In this case, $\Gamma(Q^*, \alpha) = \left[\gamma_{j,s} \right]$, $j = 1, s = 1, 2, \ldots, 5$. The results of the two importance analyses are presented in Tables 2 and 3. Columns 5 and 6 in Tables 2 and 3 represent the importance of the parameter and the resultant ranking, respectively.

Table 2. Analysis results of differential importance measure.

Parameter	Importance	Expression	Mark	Value	Order
u	$\Gamma_{2u}(Q^*, \alpha^*)$	$\dfrac{\left[R\left(e^{(Q/R)\rho}-1\right)-Q\rho\right]u}{\sum_{j=1}^{5} Q_j^* \alpha_j}$	$+$	8.02×10^3	4
a	$\Gamma_{2a}(Q^*, \alpha^*)$	$\dfrac{\left[\frac{1}{2}R\left(e^{(Q/R)\rho}-1\right)-\frac{1}{2}Q\rho\right]a}{\sum_{j=1}^{5} Q_j^* \alpha_j}$	$+$	4.01×10^2	5
R	$\Gamma_{2R}(Q^*, \alpha^*)$	$\dfrac{\left(\frac{1}{2}a+u\right)\left(R\left(e^{(Q/R)\rho}-1\right)-Q\rho e^{(Q/R)\rho}\right)}{\sum_{j=1}^{5} Q_j^* \alpha_j}$	$-$	8.4427×10^3	2
γ	$\Gamma_{2\gamma}(Q^*, \alpha^*)$	$\dfrac{-\rho\gamma}{\sum_{j=1}^{5} Q_j^* \alpha_j}$	$-$	8.421×10^3	3
ρ	$\Gamma_{2\rho}(Q^*, \alpha^*)$	$\dfrac{\left[Q\left(\frac{1}{2}a+u\right)\left(e^{(Q/R)\rho}-1\right)-\gamma\right]\rho}{\sum_{j=1}^{5} Q_j^* \alpha_j}$	$+$	8.4433×10^3	1

Table 3. Analysis results of Birnbaum importance measure.

Parameter	Importance	Expression	Mark	Value	Order
u	$\Gamma_{Bu}(Q^*,\alpha^*)$	$\dfrac{\left[R\left(e^{(Q/R)\rho}-1\right)-Q\rho\right]u}{Q^*}$	$+$	3.028×10^{-3}	4
a	$\Gamma_{Ba}(Q^*,\alpha^*)$	$\dfrac{\left[\frac{1}{2}R\left(e^{(Q/R)\rho}-1\right)-\frac{1}{2}Q\rho\right]a}{Q^*}$	$+$	1.514×10^{-4}	5
R	$\Gamma_{BR}(Q^*,\alpha^*)$	$\dfrac{\left(\frac{1}{2}a+u\right)\left(R\left(e^{(Q/R)\rho}-1\right)-Q\rho e^{(Q/R)\rho}\right)}{Q^*}$	$-$	3.187×10^{-3}	2
γ	$\Gamma_{B\gamma}(Q^*,\alpha^*)$	$\dfrac{-\rho\gamma}{Q^*}$	$-$	3.179×10^{-3}	3
ρ	$\Gamma_{B\rho}(Q^*,\alpha^*)$	$\dfrac{\left[Q\left(\frac{1}{2}a+u\right)\left(e^{(Q/R)\rho}-1\right)-\gamma\right]\rho}{Q^*}$	$+$	3.1872×10^{-3}	1

It can be seen from Table 2 that the proportional increase in parameters R, γ or ρ has almost the same effect on the results. At the same time, the importance of u is only slightly lower than R, γ and ρ, while a is almost an unimportant parameter relative to the other parameters. Although the results of Birnbaum importance obtained in Table 3 are different from those in Table 2, the relationship between the values is roughly the same in both tables, that is, the ranking results of differential importance and Birnbaum importance are the same. That is to say, compared with the influence of other parameters on the EOQ considering financing, the change of EOQ considering financing caused by a is almost negligible.

4. Numerical Example

In this section, a shortage and limited inventory capacity is used to demonstrate the proposed model. The inventory models are assumed, and displayed in Table 4.

Table 4. Case assumptions of the inventory models.

Case Assumptions	Case Assumption Contents
Assumption 1	A small amount of stockout will not cause much damage to customers and companies. The loss of unit goods per unit time is C_4.
Assumption 2	Companies can store products in their own warehouses or leased warehouses. C_2 is the storage fee of unit goods per unit time stored in their own warehouses, C_3 is the storage fee in leased warehouses, and $C_2 < C_3$.
Assumption 3	The capacity of their own warehouses is Q_0.
Assumption 4	When storing, companies firstly store products in their own warehouses until they are full, and then in leased warehouses.
Assumption 5	When selling, companies will firstly sell the products in leased warehouses until they are empty, and then from their own warehouses.

Assuming that $[0,T]$ is a time cycle, and when $t=0$, the instant purchase is Q_1, the inventory capacity is Q_0, so the capacity of leased warehouses is $Q_1 - Q_0$. Q_2 is the allowable shortage quantity, R is the demand rate, and it is decreasing constantly during $[0,T]$.

Therefore, the change of inventory volume during $[0,T]$ is shown in Figure 6.

Assuming the total cost of an inventory is C, then

$$C = \frac{C_3(Q_1-Q_0)^2}{2Q} + \frac{C_2Q_0(2Q_1-Q_0)}{2Q} + \frac{C_4(Q-Q_1)^2}{2Q} + \frac{C_1R}{Q}$$

where R means the demand speed, Q is the order quantity, and Q_1 and Q_0 represent the total inventory volume, so the inventory capacity of leased warehouses is $Q_1 - Q_0$. Q_2 is the shortage quantity, so $Q = Q_1 + Q_2$. C_1 is the order cost, C_2 is the inventory cost of unit goods per unit time when using their own warehouses, C_3 means the inventory cost of unit goods per unit time when using leased warehouses, and C_4 represents the shortage cost of unit goods.

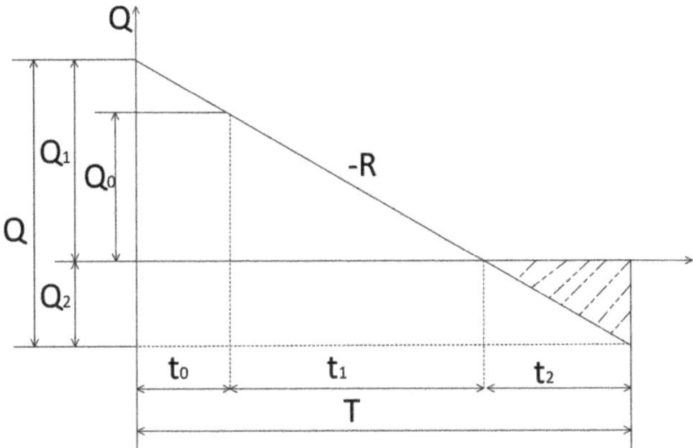

Figure 6. The storage volume changes over time.

In order to obtain the optimal solution of total cost, that is the minimal cost C^* and the optimal ordering quantity Q^*, taking the partial derivatives of Q, Q_1, Q_2 separately, then Q^*, Q_1^*, and Q_2^* can be obtained as follows:

$$\begin{cases} Q^* = \left[\dfrac{2C_1R(C_3+C_4)}{C_3C_4} + \dfrac{(C_3-C_2)(C_2+C_4)Q_0^2}{C_3C_4} \right]^{\frac{1}{2}} \\[3mm] Q_1^* = \left[\dfrac{2C_1C_4R}{C_3(C_3+C_4)} + \dfrac{C_4(C_3-C_2)(C_2+C_4)Q_0^2}{C_3(C_3+C_4)^2} \right]^{\frac{1}{2}} + \dfrac{(C_3-C_2)Q_0}{C_3+C_4} \\[3mm] Q_2^* = \left[\dfrac{2C_1C_3R}{C_3(C_3+C_4)} + \dfrac{C_4(C_3-C_2)(C_2+C_4)Q_0^2}{C_4(C_3+C_4)^2} \right]^{\frac{1}{2}} + \dfrac{(C_3-C_2)Q_0}{C_3+C_4} \end{cases}$$

Birnbaum importance can be used to analyze the importance of parameters in this model. The expressions are as follows:

$$I(Q_0) = \frac{\left[\dfrac{(C_3-C_2)(C_2+C_4)Q_0}{C_3C_4} \left(\dfrac{2C_1R(C_3+C_4)+(C_3-C_2)(C_2+C_4)Q_0^2}{C_3C_4} \right)^{-\frac{1}{2}} \right] Q_0}{Q^*}$$

$$I(R) = \frac{\left[\dfrac{C_1(C_2+C_4)}{C_3C_4} \left(\dfrac{2C_1R(C_3+C_4)+(C_3-C_2)(C_2+C_4)Q_0^2}{C_3C_4} \right)^{-\frac{1}{2}} \right] R}{Q^*}$$

$$I(C_2) = \frac{\left[\dfrac{(C_3-2C_2-C_4)Q_0^2}{2C_3C_4} \left(\dfrac{2C_1R(C_3+C_4)+(C_3-C_2)(C_2+C_4)Q_0^2}{C_3C_4} \right)^{-\frac{1}{2}} \right] C_2}{Q^*}$$

$$I(C_3) = \frac{\left[\dfrac{C_2C_4(C_2+C_4)Q_0^2-2C_1RC_4^2}{2C_3^2C_4^2} \left(\dfrac{2C_1R(C_3+C_4)+(C_3-C_2)(C_2+C_4)Q_0^2}{C_3C_4} \right)^{-\frac{1}{2}} \right] C_3}{Q^*}$$

$$I(C_4) = \frac{-\left[\dfrac{2C_1RC_3^2+C_2C_3(C_3-C_2)Q_0^2}{2C_3^2C_4^2} \left(\dfrac{2C_1R(C_3+C_4)+(C_3-C_2)(C_2+C_4)Q_0^2}{C_3C_4} \right)^{-\frac{1}{2}} \right] C_4}{Q^*}$$

The parameters (Q_0, R, C_2, C_3, C_4) can be selected for analysis in the inventory models with stockout. There are different metrics among the parameters, therefore differential importance measures can be applied to identify the importance of each parameter in this model. The expressions are as follows:

$$DIM(Q_0) = \frac{\left[\frac{(C_3-C_2)(C_2+C_4)Q_0}{C_3C_4}\left(\frac{2C_1R(C_3+C_4)+(C_3-C_2)(C_2+C_4)Q_0^2}{C_3C_4}\right)^{-\frac{1}{2}}\right]Q_0}{\sum_{j=1}^{6}Q_j^*\alpha_j}$$

$$DIM(R) = \frac{\left[\frac{C_1(C_2+C_4)}{C_3C_4}\left(\frac{2C_1R(C_3+C_4)+(C_3-C_2)(C_2+C_4)Q_0^2}{C_3C_4}\right)^{-\frac{1}{2}}\right]R}{\sum_{j=1}^{6}Q_j^*\alpha_j}$$

$$DIM(C_2) = \frac{\left[\frac{(C_3-2C_2-C_4)Q_0^2}{2C_3C_4}\left(\frac{2C_1R(C_3+C_4)+(C_3-C_2)(C_2+C_4)Q_0^2}{C_3C_4}\right)^{-\frac{1}{2}}\right]C_2}{\sum_{j=1}^{6}Q_j^*\alpha_j}$$

$$DIM(C_3) = \frac{\left[\frac{C_2C_4(C_2+C_4)Q_0^2-2C_1RC_4^2}{2C_3^2C_4^2}\left(\frac{2C_1R(C_3+C_4)+(C_3-C_2)(C_2+C_4)Q_0^2}{C_3C_4}\right)^{-\frac{1}{2}}\right]C_3}{\sum_{j=1}^{6}Q_j^*\alpha_j}$$

$$DIM(C_4) = \frac{-\left[\frac{2C_1RC_3^2+C_2C_3(C_3-C_2)Q_0^2}{2C_3^2C_4^2}\left(\frac{2C_1R(C_3+C_4)+(C_3-C_2)(C_2+C_4)Q_0^2}{C_3C_4}\right)^{-\frac{1}{2}}\right]C_4}{\sum_{j=1}^{6}Q_j^*\alpha_j}$$

Assuming that $C_1 = 30$, $C_2 = 1$, $C_3 = 2$, $C_4 = 3$, $Q_0 = 500$, and $R = 8000$, when combining the inventory models with shortage and the formula of each parameter above in the model, the results of Birnbaum importance and the differential importance measures of various parameters are as shown in Table 5.

Table 5. The results of importance measures.

Parameter	Birnbaum Importance	Order	Differential Importance	Order
Q_0	0.2940	2	0.4545	2
R	0.3528	1	0.5455	1
C_2	−0.1103	4	−0.1705	4
C_3	−0.0647	5	−0.1000	5
C_4	−0.1779	3	−0.2750	3

From Table 5, firstly, although the Birnbaum importance and differential importance measure of parameters have a certain difference in values, the sorting results are identical. Secondly, the values of importance are positive or negative, but the parameters are sorted by their absolute value. When the value is positive, it indicates that the optimal order quantity increases when the parameter is increasing, otherwise, it decreases with the increase in the parameter. Thirdly, according to the magnitude of values, the order of R, Q_0, C_4, C_2, C_3 decreases in sequence. In other words, when these parameters change proportionally, the optimal order quantity has the largest change caused by R, and C_3 is the least important among these variables. Therefore, the changes of R and Q_0 are most important. The demand speed and inventory capacity are the key parameters for reducing the total cost and optimizing the inventory model.

In order to improve analysis of the dynamic effects caused by various parameters on Birnbaum importance and differential importance measure, each parameter can be set for the changes of Birnbaum importance and the differential importance measure, as parameters change within the intervals. For each parameter, the variation of Birnbaum importance and the differential importance measure are shown in Figure 7.

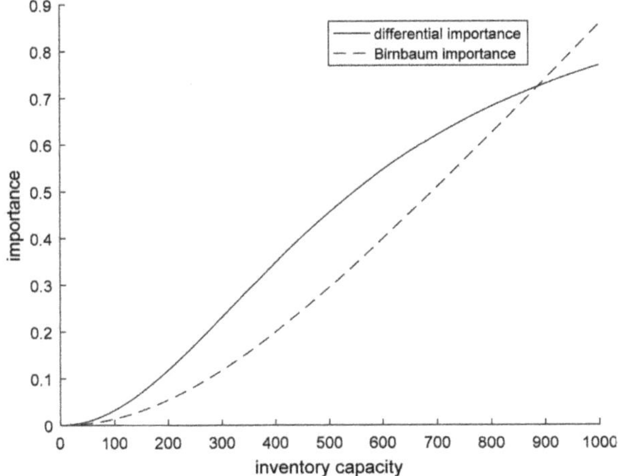

(**a**) Importance changes with inventory capacity.

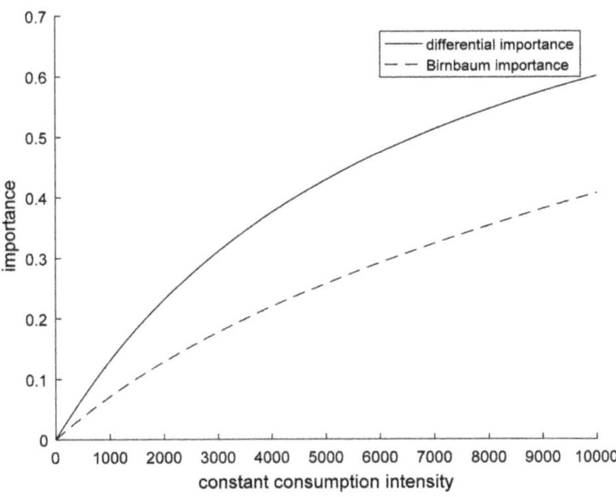

(**b**) Importance changes with constant consumption intensity.

Figure 7. *Cont.*

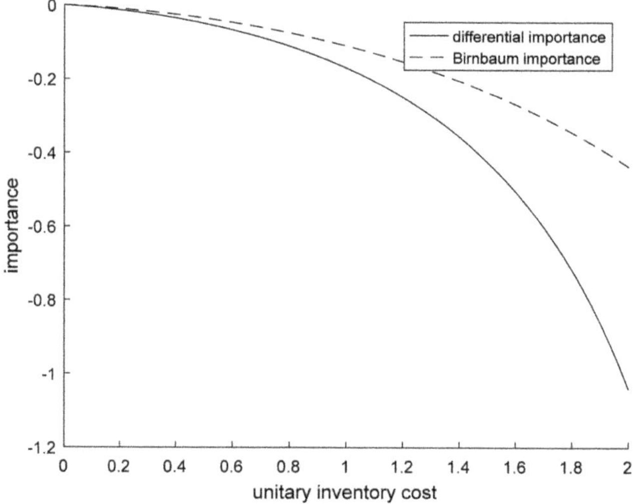

(**c**) Importance changes with unitary inventory cost.

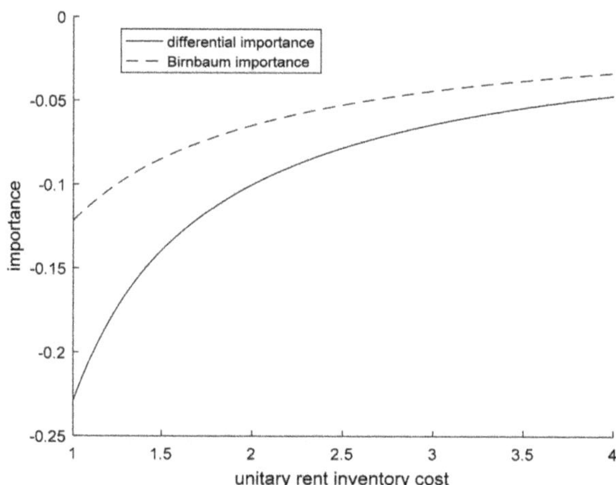

(**d**) Importance changes with unitary rent inventory cost.

Figure 7. *Cont.*

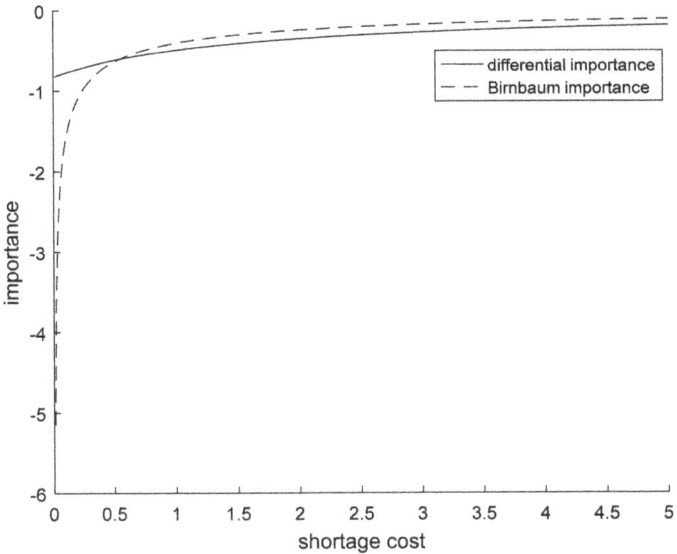

(**e**) Importance changes with shortage cost.

Figure 7. Importance changes with various parameters.

Figure 7 shows that Birnbaum importance and differential importance measures have similarities in the variation law. The plus-minus and direction of curves are consistent, but the slopes are not the same. In Figure 7e, when the shortage cost is less than 0.5, Birnbaum importance changes almost linearly, otherwise its change is quite slow and tends to be stable. However, the curve of the differential importance measure keeps a steady change rate within the interval. In Figure 7a,e, the two curves even appear to be intersected. At the intersection, the values of Birnbaum importance and the differential importance measure are equivalent. In addition, the differential importance measure is above the Birnbaum importance in Figure 7b, while the exact reverse is the case in Figure 7c,d; the Birnbaum importance of the parameter lies above the differential importance measure. The two importance measures show different changes with the varying parameters, which suggests that there are certain differences between the two importance measures in practice. Meanwhile, when applying the additivity of the differential importance measure, let $S = \{R, C_2, C_3, C_4\}$, then

$$DIM_S(Q^*, \alpha^*) = DIM_R(Q^*, \alpha^*) + DIM_{C_2}(Q^*, \alpha^*) + DIM_{C_3}(Q^*, \alpha^*) + DIM_{C_4}(Q^*, \alpha^*) = 0$$

Based on the analysis above, in the optimization analysis of inventory models with stockout, Birnbaum importance and the differential importance measure are different in calculation method and values, but the orders of importance are ultimately identical. This hints that Birnbaum importance and differential importance measures can effectively identify the importance of each parameter in a model, however, due to their own advantages and disadvantages, it is better to combine them together to provide a better solution for decision-makers in practice.

5. Conclusions and Future Work

Based on the theory of importance measures and inventory models, this paper mainly discusses the diagnosis and recognition of performance parameters in inventory control models with stockout. After a brief introduction about the inventory models that allows stockout, the concepts and calculation method of importance measures were applied into the models to analyze the application of Birnbaum importance and the differential importance measure. By comparing and analyzing the calculations and results of the two importance measures, the importance order was obtained, and the key parameters were identified to optimize the inventory control and management from the view of system reliability. The main contributions of this article are as follows.

(1) Based on the research of inventory systems, it was found that there was almost no literature on the reliability of an inventory system. Combining the concept of reliability with the inventory system, an inventory system reliability model was proposed in this paper. It could enrich the research in the field of inventory system reliability.

(2) Based on the inventory system reliability model, cost-based importance measures of inventory systems' reliability were proposed. The purpose was to study the impact of the changes of different parameters in the inventory system on the inventory system.

(3) Based on the analysis of numerical examples, it was concluded that Birnbaum importance and differential importance measures can effectively determine the importance of each parameter in the inventory system. According to the calculation result, the order of parameter importance is R, Q_0, C_4, C_2, C_3.

In future work, we will consider the impact of the relationship between different parameters on the inventory system.

Author Contributions: Conceptualization, L.C. and S.W.; methodology, L.C.; software, S.W.; writing—original draft preparation, L.C. and M.K.; writing—review and editing, L.C. and S.W.; supervision, L.C. and S.W.; funding acquisition, L.C. All authors have read and agreed to the published version of the manuscript.

Funding: This research was funded by the financial supports for this research from the National Natural Science Foundation of China (No. 61807031), Key Science and Technology Program of Henan Province, China (Nos. 132102210560, 162102210004).

Conflicts of Interest: The authors declare no conflict of interest. The funders had no role in the design of the study; in the collection, analysis, or interpretation of data; in the writing of the manuscript, or in the decision to publish the results.

References

1. Kozłowski, E.; Kowalska, B.; Kowalski, D.; Mazurkiewicz, D. Survival Function in the Analysis of the Factors Influencing the Reliability of Water Wells Operation. *Water Res. Manag.* **2019**, *33*, 4909–4921. [CrossRef]

2. Kozłowski, E.; Mazurkiewicz, D.; Kowalska, B.; Kowalski, D. Application of a multidimensional scaling method to identify the factors influencing on reliability of deep wells. In *Advances in Intelligent Systems and Computing, Proceedings of the Intelligent Systems in Production Engineering and Maintenance, ISPEM 2018, Wrocław, Poland, 17–18 September 2018*; Burduk, A., Chlebus, E., Nowakowski, T., Tubis, A., Eds.; Springer: Cham, Switzerland, 2018; Volume 835, pp. 56–65. [CrossRef]

3. Zhang, C.; Zhang, Y. Common cause and load-sharing failures-based reliability analysis for parallel systems. *Eksploatacja i Niezawodnosc Maint. Reliab.* **2019**, *22*, 26–34. [CrossRef]

4. Cai, B.; Liu, Y.-H.; Fan, Q.; Zhang, Y.; Liu, Z.; Yu, S.; Ji, R. Multi-source information fusion based fault diagnosis of ground-source heat pump using Bayesian network. *Appl. Energy* **2014**, *114*, 1–9. [CrossRef]

5. Cui, J.; Ren, Y.; Xu, B.; Yang, D.; Zeng, S. Reliability analysis of a multi-eso based control strategy for level adjustment control system of quadruped robot under disturbances and failures. *Eksploatacja i Niezawodnosc Maint. Reliab.* **2019**, *22*, 42–51. [CrossRef]

6. Sun, B.; Li, Y.; Wang, Z.; Li, Z.; Xia, Q.; Ren, Y.; Feng, Q.; Yang, D.; Qian, C. Physics-of-failure and computer-aided simulation fusion approach with a software system for electronics reliability analysis. *Eksploatacja i Niezawodnosc Maint. Reliab.* **2020**, *22*, 340–351. [CrossRef]

7. Liu, D.; Wang, S.; Tomovic, M. Degradation modeling method for rotary lip seal based on failure mechanism analysis and stochastic process. *Eksploatacja i Niezawodnosc Maint. Reliab.* **2020**, *22*, 381–390. [CrossRef]

8. Birnbaum, Z.W. *On the Importance of Different Components in a Multi-Component System*; Academic Press: New York, NY, USA, 1969.

9. Borgonovo, E.; Apostolakis, G. A new importance measure for risk-informed decision making. *Reliab. Eng. Syst. Saf.* **2001**, *72*, 193–212. [CrossRef]

10. Dui, H.; Li, S.; Xing, L.; Liu, H. System performance-based joint importance analysis guided maintenance for repairable systems. *Reliab. Eng. Syst. Saf.* **2019**, *186*, 162–175. [CrossRef]

11. Dui, H.; Si, S.; Zuo, M.J.; Sun, S. Semi-Markov Process-Based Integrated Importance Measure for Multi-State Systems. *IEEE Trans. Reliab.* **2015**, *64*, 1–12. [CrossRef]

12. Dui, H.; Si, S.; Wu, S.; Yam, R.C. An importance measure for multistate systems with external factors. *Reliab. Eng. Syst. Saf.* **2017**, *167*, 49–57. [CrossRef]

13. Dui, H.; Si, S.; Yam, R.C. Importance measures for optimal structure in linear consecutive-k-out-of-n systems. *Reliab. Eng. Syst. Saf.* **2018**, *169*, 339–350. [CrossRef]

14. Kim, T.; Song, J. Generalized Reliability Importance Measure (GRIM) using Gaussian mixture. *Reliab. Eng. Syst. Saf.* **2018**, *173*, 105–115. [CrossRef]

15. Li, J.; Dueñas-Osorio, L.; Chen, C.; Shi, C.-L. AC power flow importance measures considering multi-element failures. *Reliab. Eng. Syst. Saf.* **2017**, *160*, 89–97. [CrossRef]

16. Dui, H.; Li, C.; Chen, L. Reliability-oriented extended importance measures in uncertain inventory systems for mechanical products. *Adv. Mech. Eng.* **2019**, *11*, 1687814019828438. [CrossRef]

17. Dui, H.; Meng, X.; Xiao, H.; Guo, J. Analysis of the cascading failure for scale-free networks based on a multi-strategy evolutionary game. *Reliab. Eng. Syst. Saf.* **2020**, *199*, 106919. [CrossRef]

18. Nguyen, K.-A.; Do, P.; Grall, A. Joint predictive maintenance and inventory strategy for multi-component systems using Birnbaum's structural importance. *Reliab. Eng. Syst. Saf.* **2017**, *168*, 249–261. [CrossRef]

19. Adak, S.; Rani, P. Effect of reliability on multi-item inventory system with shortages and partial backlog incorporating time dependent demand and deterioration. *Ann. Oper. Res.* **2020**, 1–21. [CrossRef]

20. Maji, A.; Bhunia, A.K.; Mondal, S.K. Exploring a production-inventory model with optimal reliability of the production in a parallel-series system. *J. Ind. Prod. Eng.* **2020**, *37*, 71–86. [CrossRef]

21. Huang, C.-F. Evaluation of system reliability for a stochastic delivery-flow distribution network with inventory. *Ann. Oper. Res.* **2017**, *277*, 33–45. [CrossRef]

22. Manna, A.K.; Dey, J.K.; Mondal, S.K. Two layers supply chain in an imperfect production inventory model with two storage facilities under reliability consideration. *J. Ind. Prod. Eng.* **2017**, *35*, 57–73. [CrossRef]

23. Abdel-Aleem, A.; El-Sharief, M.A.; Hassan, M.A.; El-Sebaie, M.G. Optimization of reliability based model for production inventory system. *Int. J. Manag. Sci. Eng. Manag.* **2017**, *13*, 54–64. [CrossRef]

24. Daniewski, K.; Kosicka, E.; Mazurkiewicz, D. Analysis of the correctness of determination of the effectiveness of maintenance service actions. *Manag. Prod. Eng. Rev.* **2018**, *9*, 20–25. [CrossRef]

25. Feng, Q.; Zhao, X.; Fan, D.; Cai, B.; Liu, Y.; Ren, Y. Resilience design method based on meta-structure: A case study of offshore wind farm. *Reliab. Eng. Syst. Saf.* **2019**, *186*, 232–244. [CrossRef]

26. Zhao, X.; Chen, M.; Nakagawa, T. Periodic replacement policies with shortage and excess costs. *Ann. Oper. Res.* **2020**, 1–19. [CrossRef]

27. Cárdenas-Barrón, L.E.; Sana, S.S. Multi-item EOQ inventory model in a two-layer supply chain while demand varies with promotional effort. *Appl. Math. Model.* **2015**, *39*, 6725–6737. [CrossRef]

28. Shekarian, E.; Olugu, E.U.; Abdul-Rashid, S.H.; Bottani, E. A Fuzzy Reverse Logistics Inventory System Integrating Economic Order/Production Quantity Models. *Int. J. Fuzzy Syst.* **2016**, *18*, 1141–1161. [CrossRef]

29. Chao, X.; Gong, X.; Zheng, S. Optimal pricing and inventory policies with reliable and random-yield suppliers: Characterization and comparison. *Ann. Oper. Res.* **2014**, *241*, 35–51. [CrossRef]

30. Yu, Y.; Shou, B.; Ni, Y.; Chen, L. Optimal production, pricing, and substitution policies in continuous review production-inventory systems. *Eur. J. Oper. Res.* **2017**, *260*, 631–649. [CrossRef]

31. Lee, J.-Y.; Cho, R.K.; Paik, S.-K. Supply chain coordination in vendor-managed inventory systems with stockout-cost sharing under limited storage capacity. *Eur. J. Oper. Res.* **2016**, *248*, 95–106. [CrossRef]

32. Jia, X.; Cui, L. Reliability Research of k-out-of-n: G Supply Chain System Based on Copula. *Commun. Stat. Theory Methods* **2012**, *41*, 4023–4033. [CrossRef]

33. Flynn, B.B.; Koufteros, X.; Lu, G. On Theory in Supply Chain Uncertainty and its Implications for Supply Chain Integration. *J. Supply Chain Manag.* **2016**, *52*, 3–27. [CrossRef]

34. He, M.; Xie, J.; Wu, X.; Hu, Q.; Dai, Y. Capability Coordination in Automobile Logistics Service Supply Chain Based on Reliability. *Procedia Eng.* **2016**, *137*, 325–333. [CrossRef]
35. Chen, X.; Xi, Z.; Jing, P. A Unified Framework for Evaluating Supply Chain Reliability and Resilience. *IEEE Trans. Reliab.* **2017**, *66*, 1144–1156. [CrossRef]

Publisher's Note: MDPI stays neutral with regard to jurisdictional claims in published maps and institutional affiliations.

Article

Application of Predictive Maintenance Concepts Using Artificial Intelligence Tools

Diogo Cardoso and Luís Ferreira *

Faculty of Engineering, University of Porto, 4200-465 Porto, Portugal; up201505446@fe.up.pt
* Correspondence: lferreir@fe.up.pt

Abstract: The growing competitiveness of the market, coupled with the increase in automation driven with the advent of Industry 4.0, highlights the importance of maintenance within organizations. At the same time, the amount of data capable of being extracted from industrial systems has increased exponentially due to the proliferation of sensors, transmission devices and data storage via Internet of Things. These data, when processed and analyzed, can provide valuable information and knowledge about the equipment, allowing a move towards predictive maintenance. Maintenance is fundamental to a company's competitiveness, since actions taken at this level have a direct impact on aspects such as cost and quality of products. Hence, equipment failures need to be identified and resolved. Artificial Intelligence tools, in particular Machine Learning, exhibit enormous potential in the analysis of large amounts of data, now readily available, thus aiming to improve the availability of systems, reducing maintenance costs, and increasing operational performance and support in decision making. In this dissertation, Artificial Intelligence tools, more specifically Machine Learning, are applied to a set of data made available online and the specifics of this implementation are analyzed as well as the definition of methodologies, in order to provide information and tools to the maintenance area.

Keywords: predictive maintenance; Industry 4.0; Internet of Things; artificial intelligence; machine learning

Citation: Cardoso, D.; Ferreira, L. Application of Predictive Maintenance Concepts Using Artificial Intelligence Tools. *Appl. Sci.* **2021**, *11*, 18. https://dx.doi.org/10.3390/app11010018

Received: 19 November 2020
Accepted: 18 December 2020
Published: 22 December 2020

1. Introduction

Maintenance is a relevant factor for the competitiveness of an organization, since the actions carried out at this level have a direct impact on aspects such as the cost, deadlines and quality of the products produced or services provided [1,2]. Maintenance is a support to the operational area of a company and cannot be dissociated from it, given the implication it has in terms of the efficiency of productive assets. These two areas, operation and maintenance, must operate in parallel in order to guarantee the availability and the rapid response of human and material resources to operational problems, thus ensuring the achievement of objectives with the maximization of available resources. Thus, it becomes important not only to achieve the proposed objectives, but also to achieve them with the minimum consumption or use of resources [2].

It is in this context of constant transformation that Industry 4.0 arises [1]. Industry 4.0 implements the tools provided by advances in information and communication technologies in order to increase the levels of automation and digitalization in industrial and production processes [1,3]. The objective is to manage the entire value chain process, improving production efficiency and creating superior products and services. One of the key points of this technological evolution is the data, which is now more easily read, processed, stored, analyzed and shared between machines and human beings [4]. Additionally, the Internet of Things (IoT) is defined as an ecosystem in which the objects and equipment inserted in it are equipped with sensors and other digital devices, thus being able to gather and exchange information with each other, in a networked system [1].

In recent years, a drop in cost and an increase in the reliability of sensors, data transmission and storage devices have promoted the emergence of condition monitoring systems for industrial equipment. Simultaneously, the IoT allows real-time transmission of this information about the conditions of the systems captured by different monitoring devices. This development offers an excellent opportunity to use condition monitoring data intelligently within predictive maintenance, combining the ability to collect data with an effective and integrated analysis of it [5].

In this sense, the potential of Artificial Intelligence tools, more specifically Machine Learning, allows us to aim for an improvement in the availability of systems, reducing maintenance costs, increasing operational performance and safety, and the ability to support decision making in relation to the ideal time and the ideal action for carrying out the maintenance intervention [4–7].

Machine Learning can be defined as "the field of study that gives the computer the ability to learn without being explicitly programmed". It can be said that "Machine Learning algorithms use computational methods to learn information directly from data without using predefined equations as a model" [8].

The main objective of this work is to apply Artificial Intelligence tools, more specifically Machine Learning, to a set of data, coming from different sources, available online [9]. Furthermore, we seek to analyze the specificities of this implementation and the definition of methodologies, in order to provide information and tools to the maintenance area.

2. Machine Learning Process Workflow and Techniques

One of the main difficulties of applying a Machine Learning process to maintenance data is the choice of the right workflow, as in literature there are many different approaches to this problem, depending on the origin of the data and the objectives of the analysis [10–13]. As the different applications are difficult to compare, in this work it was decided to explore a simple but complete framework and to use a set of data that can be used by other researchers, as the data set is publicly available.

In the present work, the workflow presented in [14], represented in Figure 1, will be followed. From our point of view, a Machine Learning project must always start with the establishment of rigorous and clear definition of the objectives, since such a system fulfills a very specific task and the definition of vague objectives can mean that the model developed is not able to predict exactly what it is intended to.

Quite possibly, the most important part of a Machine Learning project is the ability to understand the data used and how it relates to the task we want to solve. It will not be effective to randomly choose an algorithm, use the data set we have available and expect good results [15]. It is necessary to understand what is happening in the data set before starting to build a model. When building a Machine Learning solution, we must answer or at least keep in mind the following questions: What questions are we trying to answer? Does the available data set allow you to answer these questions? What is the best way to paraphrase my question as a Machine Learning problem? Is the available data set sufficient to represent the problem we are trying to solve? What features (or attributes) have been extracted and will they be able to lead to the correct predictions? How to measure the success of the application of Machine Learning? How will the Machine Learning solution interact with the rest of the process?

Machine Learning algorithms and methods are only part of a larger process for solving a specific problem, and it is important to keep that in mind. Sometimes, a lot of time is spent building complex Machine Learning solutions, only to discover in the end that they do not solve the problem that we were waiting for [16]. By deepening the technical aspects of Machine Learning, it is easy to lose sight of the final goals. It is important to keep in mind all the assumptions created, either explicitly or implicitly, when building Machine Learning models.

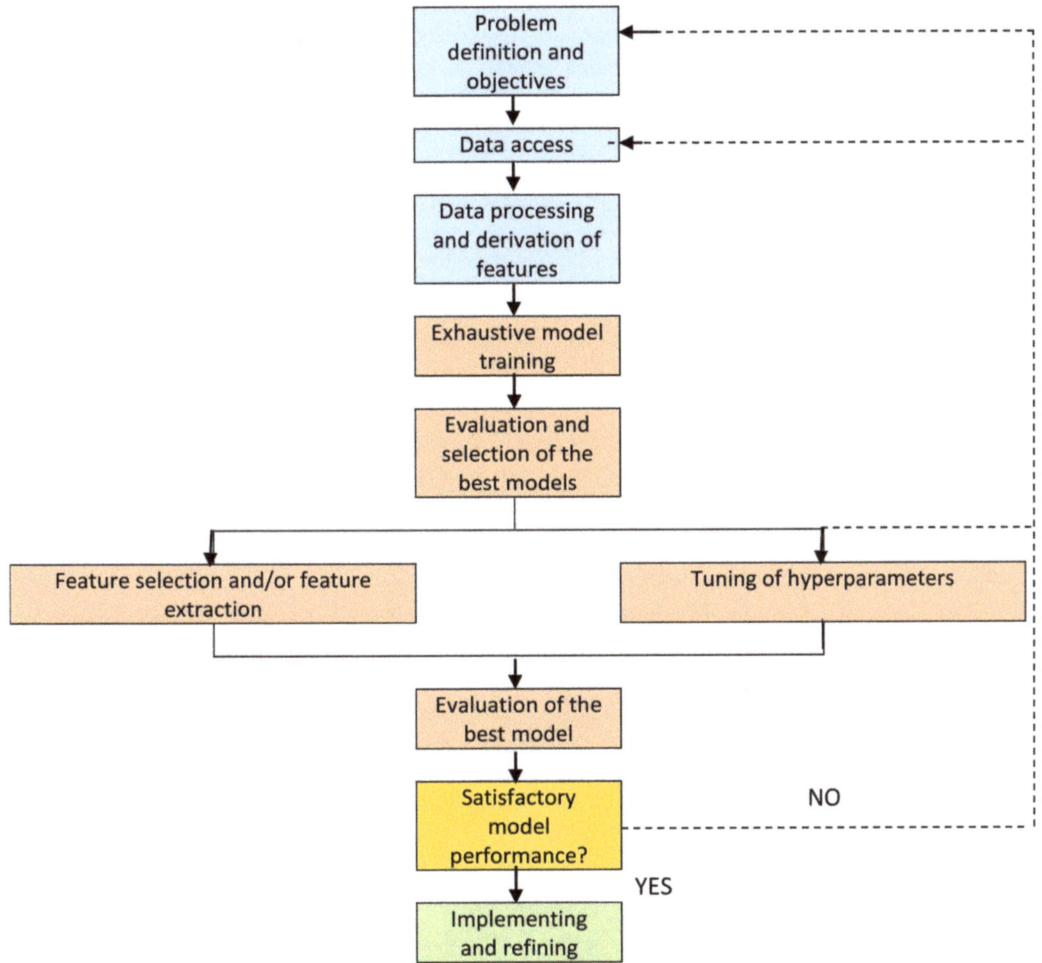

Figure 1. Machine Learning process workflow [14].

Typical machine learning algorithms, such as the hidden Markov models [17], hidden semi-Markov models [18], self-organizing neural network [19], SVM [20], multimodal deep support vector classification [21], deep random forest [22], genetic algorithms [23], blind source separation [24], fuzzy logic [25], k-nearest neighbor algorithms [26] and Bayesian algorithms [27], etc., have been applied in fault diagnosis of dynamic equipment. To the best of our knowledge, there are two main categories of approach for fault diagnosis of gearboxes: Data-driven and physical model-based methods. Although these methods have been successfully applied in many applications, it is very difficult to know what is the best algorithm to apply to a particular data set.

A systematic review of the scientific literature was carried out in [28], from which it is possible to draw several conclusions. Predictive maintenance strategies are being applied to the most diverse equipment, in multiple areas. The equipment where these methods are applied include, but are not limited to, turbines, engines, compressors, pumps. About 89% of the published papers use a set of real data, with 11% using synthetic data. Regarding the use of Machine Learning algorithms in scientific publications, the most used is Random Forest (RF)—33%, followed by methods based on Neural Networks (NN), such as Artificial

NN, Convolution NN, Long Short-Term Memory Metwork (LSTM) and Deep Learning—27%, Support Vector Machine (SVM)—25% and k-means—13%. There was also a greater tendency to use vibration signals.

3. Application Example

3.1. Data Applied

Throughout this paragraph, the process of implementing Machine Learning algorithms to a set of maintenance data will be detailed and explained.

First, the data set is presented and the choice is justified. Then, the objectives of this Machine Learning application are rigorously and clearly established. Subsequently, the data set is processed through feature engineering, creating new features in order to seek better performance from the models. The data set used is key to solving Machine Learning problems. A sensible choice of what data to use and how to handle it is crucial to improving the performance of the algorithms. According to Domingos [29], feature engineering is the key to Machine Learning projects and that, often, the measured signals are not suitable for the learning process, and it is necessary to build features from those that are.

Then, the data set is divided into training, validation and testing subsets and the first application of Machine Learning models was carried out, where a variety of algorithms were trained and evaluated. The training process of the algorithms is carried out in the training subset and the validation subset provides an impartial assessment of the fit of the models to the training data, while simultaneously fine-tuning the model and its hyper-parameters in order to seek better performance. Finally, the test set is used to obtain an estimate of the model's performance, simulating its behavior for future data.

The implementation described in this chapter will be carried out using the Python programming language, using the packages Matplotlib, Numpy, Pandas and Scikit-Learn [30,31].

3.2. Data Sources

Despite the growth of this area, due to business competitiveness, sharing sensitive information of this nature is rare, which means that the number of publicly available datasets (relevant to this application) is very scarce.

Within an industrial environment, there is a very complete data set, made available by Microsoft, published in [9], of relevance to the present project.

This dataset contains data from five different sources:

- real-time telemetry;
- error log;
- maintenance history;
- fault history; and
- information about machines

The data were acquired over a year (2015) for one hundred machines, except for the maintenance history, which also contains records for the year 2014.

For a total of one hundred machines, from four different models, the data set contains 876,100 hourly telemetry records, that is, 8761 records per machine. The failure records contain 3919 entries and the maintenance history 3286. The failure history has 761 records, that is, on average, about 8 failure records per machine, throughout 2015. Each machine has 4 components of interest for analysis and also 4 sensors, which measure tension, pressure, vibration and rotation. A controller monitors the system and is able to alert you to the occurrence of 5 types of errors.

Thus, real-time telemetry data consists of measurements from different sensors (4 per machine), with the associated date and time. The measurements of voltage ("volt"), rotation ("rotate"), pressure ("pressure") and vibration ("vibration") are acquired in real time and the average of these measurements over an hour is recorded in Table 1.

Table 1. Typical example of real-time telemetry recording.

	Datetime	machineID	Volt	Rotate	Pressure	Vibration
0	2015-01-01 06:00:00	1	176.217853	418.504078	113.077935	45.087686
1	2015-01-01 07:00:00	1	162.879223	402.747490	95.460525	402.747490
2	2015-01-01 08:00:00	1	170.989902	527.349825	75.237905	34.178847
3	2015-01-01 09:00:00	1	162.462833	346.149335	109.248561	41.122144
4	2015-01-01 10:00:00	1	157.610021	435.376873	111.886648	25.990511

To better understand the behavior of each sensor, a simple statistical analysis is performed in Table 2, where the mean, the standard deviation, and the minimum and maximum values are calculated for the parameters voltage ("volt"), rotation ("rotate"), pressure ("pressure") and vibration ("vibration") during 2015. As an example, Figure 2 shows the graphical evolutions of the Tension signals (Figure 2a), Rotation (Figure 2b), Pressure (Figure 2c) and Vibration (Figure 2d), over the first fifteen days of January 2015, for machine 1 (machineID = 1).

Table 2. Statistical analysis of telemetry data in real time.

	Volt	Rotate	Pressure	Vibration
count	876100	876100	876100	876100
mean	170.777736	446.605119	100.858668	40.385007
std	15.509114	52.673886	11.048679	5.370361
min	97.333604	138.432075	51.237106	14.877054
25%	160.304927	412.305714	93.498181	36.777299
50%	170.607338	447.558150	100.425559	40.237247
75%	181.004493	482.176600	107.555231	43.784938
max	255.124717	695.0209841	185.951998	76.791072

The second source of information is the error log. These are errors that did not immediately lead to a failure, as the machine remained operational. There are 5 types of errors: error1, error2, error3, error4, error5. The date and time are rounded to the nearest time. Each record consists of a date/time, machine and type of error—Table 3. The total number of error records over the year 2015 is 3919. In Figure 3 it is possible to observe the number of errors per type over the year 2015.

The maintenance records contain data of component replacements resulting from a scheduled or unscheduled maintenance intervention, periodic inspections, or performance degradation. In case of maintenance intervention due to the failure of a component, a fault record is also generated, see next paragraph. For each machine, this data set contains information about 4 types of components: comp1, comp2, comp3, comp4. The date and time are rounded to the nearest time. Each record consists of a date/time, machine and the type of component replaced—Table 4. The total number of maintenance records throughout 2015 is 3286. As previously mentioned, the maintenance records also contain entries for 2014. Figure 4 shows the number of components replaced, by type. It is possible to observe that in this case the number of substitutions is similar for the 4 types of components.

The fault records contain the component replacement records, resulting from the maintenance intervention, due to the occurrence of a fault. The data is for the 4 types of components: comp1, comp2, comp3, comp4. The date and time are rounded to the nearest time. Each record consists of a date/time, machine and the type of replaced component—Table 5. The total number of failure records during 2015 is 761. In Figure 5, it is possible to observe the number of replacements, by type of component, during 2015, due to the occurrence of a failure.

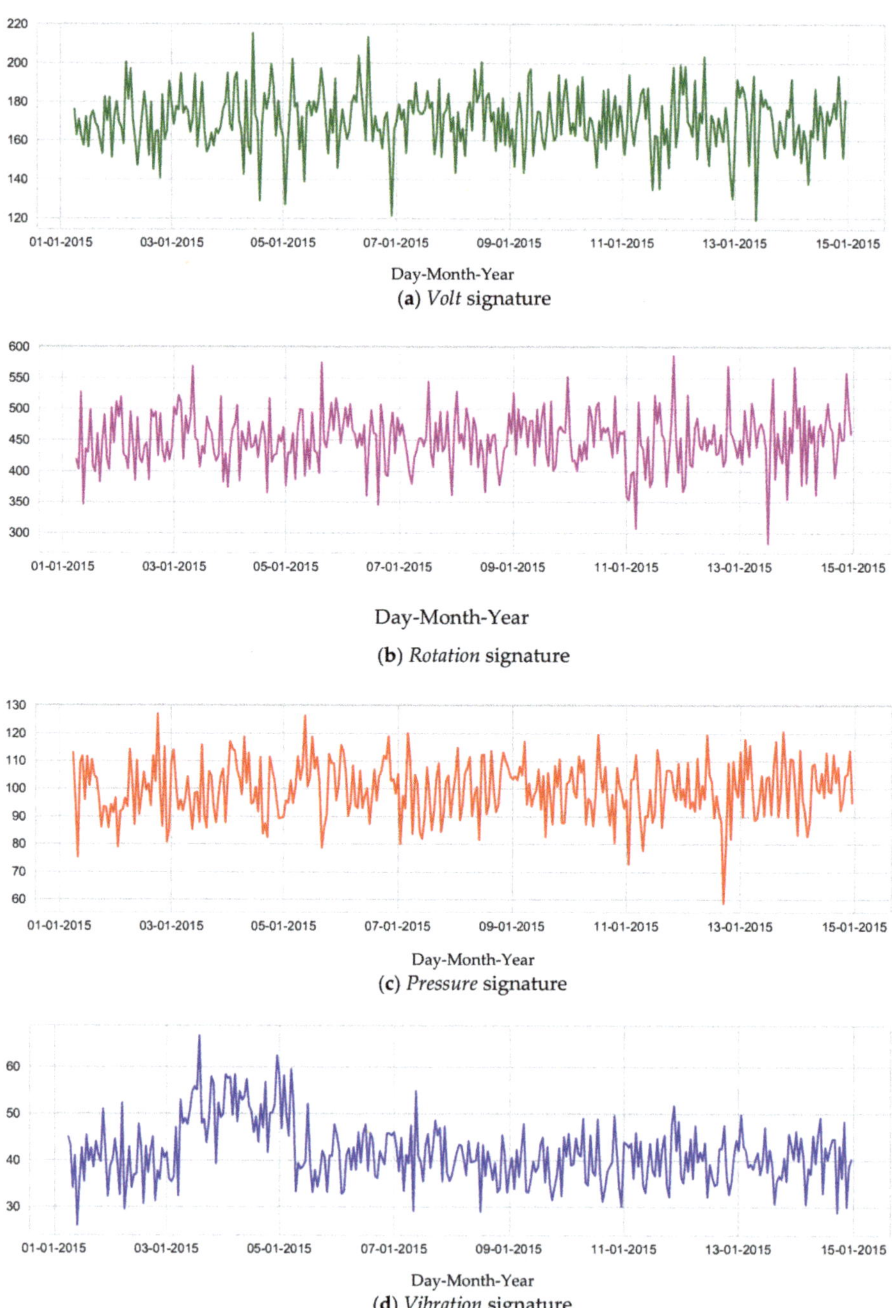

Figure 2. Evolution of Telemetry data over the first fifteen days of January 2015, for the machine 1.

Table 3. Typical error logging example.

	Datetime	machineID	errorID
0	2015-01-03 07:00:00	1	error1
1	2015-01-03 20:00:00	1	error3
2	2015-01-04 06:00:00	1	error5
3	2015-01-10 15:00:00	1	error4
4	2015-01-22 10:00:00	1	error4

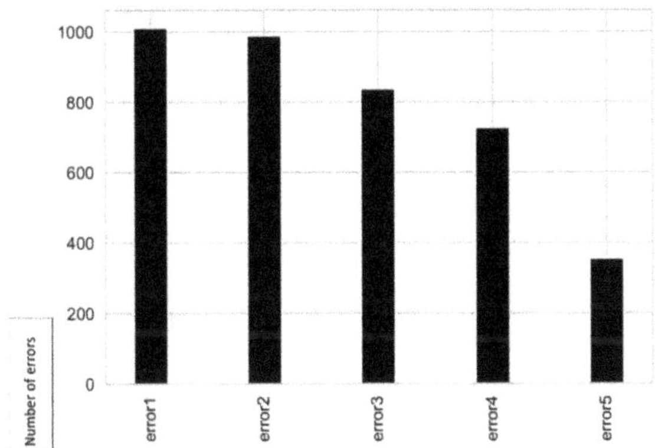

Figure 3. Representation of the number of errors by type.

Table 4. Typical example of maintenance records.

	Datetime	machineID	comp
0	2014-06-01 06:00:00	1	comp2
1	2014-07-16 06:00:00	1	comp4
2	2014-07-31 06:00:00	1	comp3
3	2014-12-13 06:00:00	1	comp1
4	2015-01-05 06:00:00	1	comp 4

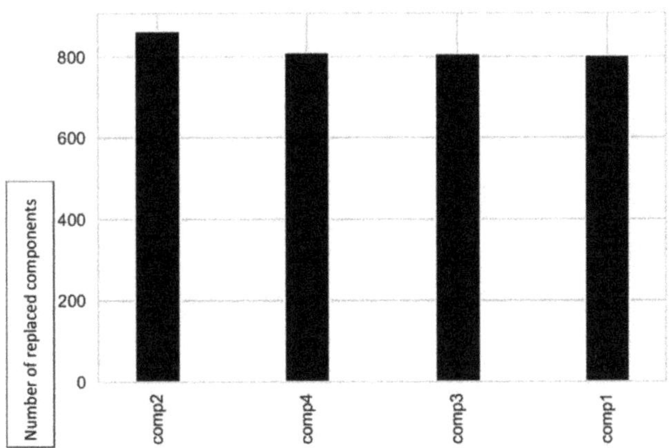

Figure 4. Representation of the number of components replaced, by type.

Table 5. Typical example of failure records.

	Datetime	machineID	Failure
0	2015-01-05 06:00:00	1	comp 4
1	2015-03-06 06:00:00	1	comp 1
2	2015-04-20 06:00:00	1	comp2
3	2015-06-19 06:00:00	1	comp4
4	2015-09-02 06:00:00	1	comp 4

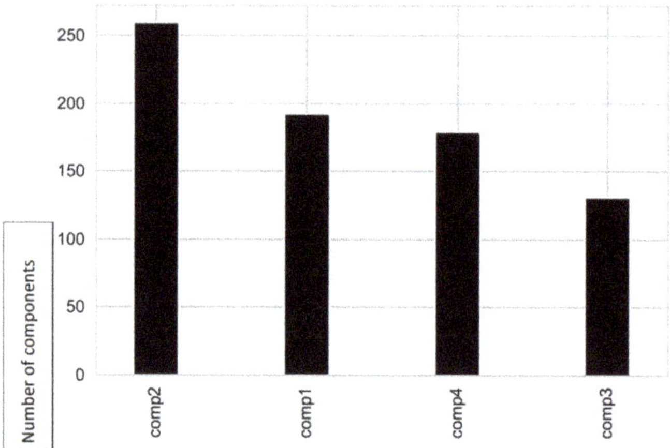

Figure 5. Representation of the number of components replaced, by type, due to the occurrence of a failure.

Finally, this data set contains information about the model and number of years of service for each of the 100 machines—Table 6. Figure 6 shows a histogram showing the distribution of the number of machines and service time, by model.

Table 6. Typical example of information for each machine.

	machineID	Model	Age
0	1	model3	18
1	2	model4	7
2	3	model3	8
3	4	model3	7
4	5	model3	2
5	6	model3	7
6	7	model3	20
7	8	model3	16

3.3. Definition of Objectives

As already mentioned, a Machine Learning project must start with the rigorous and clear establishment of objectives. In this case, the main objective of the models used will be to predict the probability of a failure occurring within the defined time window. More specifically, the probability of a machine failure occurring in the next 24 h (duration of the time window chosen for this application) related to one of the components (components 1, 2, 3 or 4).

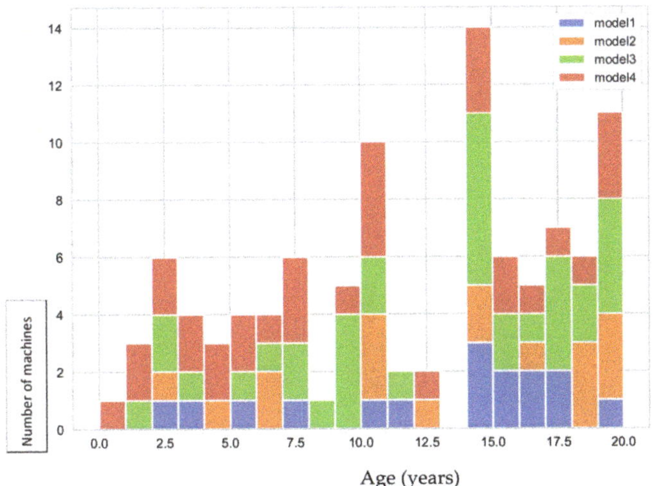

Figure 6. Histogram representing the number of machines and service time, by model.

Then, given that a particular and clear objective has already been set, more specific questions can be asked about Machine Learning itself: (1) Should supervised, unsupervised, or reinforcement learning models be chosen or, possibly, combinations of learning modes? (2) Whether supervised learning, classification, or regression? (3) Are models intended to train immediately as new data is obtained (*batch learning* or *online learning*)?

After analyzing the problem, and bearing in mind the proposed objective, we opted for supervised learning and, in particular, classification. Furthermore, taking into account the existence of 4 different components under analysis, the problem will be multi-class classification ("Multiclass Classification"). It was also considered that it will not be necessary, given the scope of the problem and the nature of the data, for models to train immediately as new data is obtained. Therefore, we are facing a problem of *batch learning*.

3.4. Feature Engineering

A feature is a predictive attribute for the model. The purpose of feature engineering is to seek to increase the predictive power of Machine Learning algorithms, creating new features from the available data. As a rule, feature engineering is carried out in the first place and then the selection of features occurs, eliminating irrelevant, redundant or highly correlated features. Starting from the different sources of information presented in the previous paragraphs, a single dataset will be created, which will be used for the application of predictive models.

The historical data that models have access to are individual moments in the past. In particular, for telemetry data, disturbances resulting from measurements, such as noise, are possible, thus making the predictive task more difficult. In this way, the data can be aggregated in time windows, thus allowing to "smooth" the values, minimizing the effects of noise on the features used by the models.

Bearing in mind how far in the future the model should be able to predict, according to the requirements of the project, it is important to define how far it should "look" to make these predictions. This interval of time passed to where the model should "look back" is called lag. Several features can be extracted from these time intervals—lag features. The data set used to generate lag features usually has a date/time associated with it.

For each record, a time window of dimension N is created and the lag features are calculated for the period N before the date/time of that record. Figure 7 shows an example of this application for a t_i measurement with N = 3. The value of N is typically in minutes or hours, depending on the nature of the data.

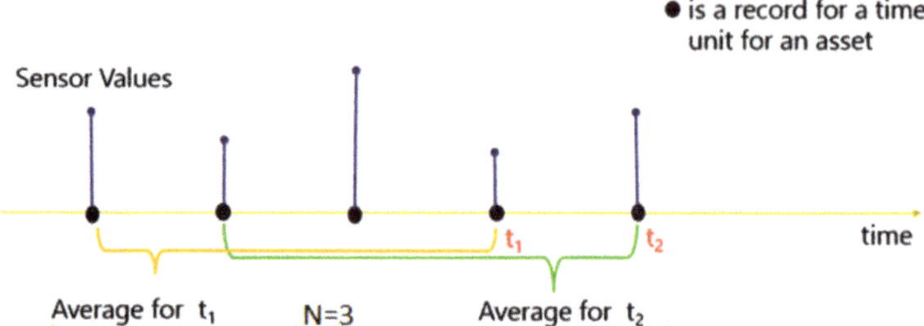

Figure 7. Lag Features [9].

Thus, two temporal windows were created. The first, of 3 h, in order to allow us to portray the behavior of telemetry data in the short term (Table 7) and the second, of 24 h, in order to represent the long-term evolution (Table 8). In each of these time intervals, two new parameters were calculated every 3 h for each of the features: the moving average and the standard deviation. Note that, in the case of N = 24 h (Table 8), naturally the two new parameters are not available for the initial moments (first 24 h).

Table 7. Example of *Lag Features* for telemetry data in real time, with N = 3.

	machineID	Datetime	Voltmean_3h	Rotatemean_3h	Pressuremean_3h	Vibrationmean_3h
0	1	2015-01-01 09:00:00	170.028993	449.533798	94.592122	40.893502
1	1	2015-01-01 12:00:00	164.192565	403.949857	105.687417	34.255891
2	1	2015-01-01 15:00:00	168.134445	435.781707	107.793709	41.239405
3	1	2015-01-01 18:00:00	165.514453	430.472823	101.703289	40.373739
4	1	2015-01-01 21:00:00	168.809347	437.111120	90.911060	41.738542

Table 8. Example of *Lag Features* for telemetry data in real time, with N = 24.

	machineID	Datetime	Voltmean_24h	Rotatemean_24h	Pressuremean_24h	Vibrationmean_24h
7	1	2015-01-02 06:00:00	169.733809	445.179865	96.797113	40.385160
8	1	2015-01-02 09:00:00	170.614862	446.364859	96.849785	39.736826
9	1	2015-01-02 12:00:00	169.893965	447.009407	97.715600	39.498374
10	1	2015-01-02 15:00:00	171.243444	444.233563	96.666060	40.229370
11	1	2015-01-02 18:00:00	170.792486	448.440437	95.766838	40.055214

As with telemetry data, the error log also has a date/time associated with it. However, these data are categorical and not numerical. In this case, the number of errors of each type is added, every 3 h, for the time window N = 24 (Table 9). Each line in the table represents the sum of the number of errors of each type in the 24 h prior to the indicated *datetime*.

Table 9. Example of *Lag Features* for error logging.

	machineID	Datetime	Error1count	Error2count	Error3count	Error4count	Error5count
15	1	2015-01-03 06:00:00	0.0	0.0	0.0	0.0	0.0
16	1	2015-01-03 09:00:00	0.0	0.0	0.0	0.0	0.0
17	1	2015-01-03 12:00:00	1.0	0.0	0.0	0.0	0.0
18	1	2015-01-03 15:00:00	1.0	0.0	0.0	0.0	0.0
19	1	2015-01-03 18:00:00	1.0	0.0	0.0	0.0	0.0
20	1	2015-01-03 21:00:00	1.0	0.0	0.0	0.0	0.0
21	1	2015-01-04 00:00:00	1.0	0.0	1.0	0.0	0.0
22	1	2015-01-04 03:00:00	1.0	0.0	1.0	0.0	0.0
23	1	2015-01-04 06:00:00	1.0	0.0	1.0	0.0	0.0
24	1	2015-01-04 09:00:00	1.0	0.0	1.0	0.0	1.0

The maintenance log, which contains information related to the replacement of components, allows the generation of new potentially important features, such as, for example, how long ago a component was last replaced—Table 10. It is expected that this feature relates well to the possible failures of the components, since, the longer the time of use of a component, the greater the expected degradation.

Table 10. Time since the last replacement, by type of component.

	Datetime	machineID	comp1	comp2	comp3	comp4
0	2015-01-01 06:00:00	1	19.000000	214.000000	154.000000	169.000000
1	2015-01-01 07:00:00	1	19.041667	214.041667	154.041667	169.041667
2	2015-01-01 08:00:00	1	19.083333	214.083333	154.083333	169.083333
3	2015-01-01 09:00:00	1	19.125000	214.125000	154.125000	169.125000
4	2015-01-01 10:00:00	1	19.166667	214.166667	154.166667	169.166667

It is relevant to note that the creation of features based on maintenance data is not as linear as in the previous cases. However, this type of case-specific feature engineering is very common in predictive maintenance, where domain knowledge and experience play a crucial role in understanding and creating relevant features.

Finally, information about the machines can be used without further modifications, that is, information related to the model and number of years in service of each machine—Table 6.

3.5. Feature Selection

An analysis of the linear correlation between the variables was performed (Figure 8). The correlation coefficient varies between −1 and 1. This coefficient makes it possible to see whether one variable justifies the linear variation of another. When it is close to 1, it means that there is a strong positive correlation, that is, if a given feature A increases, then feature B also increases and if A decreases, B it also decreases.

In this case, it appears that the correlation between the features is mostly low or nonexistent (correlation coefficient close to zero). Even so, in the case of the features pressuremean_3h and pressuremean_24h, the value of the correlation coefficient is approximately 0.5 and a more detailed analysis will be relevant.

Thus, Figure 9 shows the failures, by type of components, according to the evolution of the features pressuremean_3h and pressuremean_24h. It is possible to observe that, for components 1 and 3, there are clusters of points.

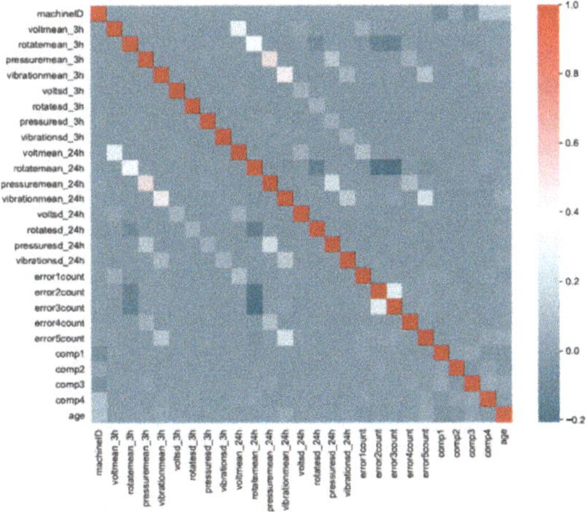

Figure 8. Features correlation.

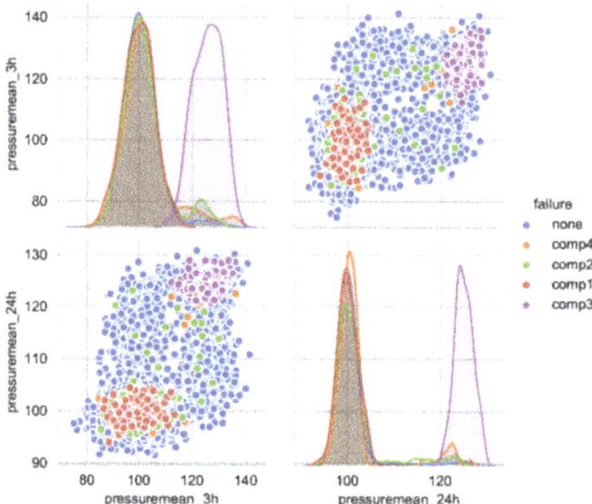

Figure 9. Detailed analysis of the correlation between features pressuremean_3h and pressuremean_24h.

However, the same is not true for components 2 and 4. It is likewise possible to observe that most failures for component 3 occur for higher pressuremean_3h and pressuremean_24h values, when compared with the other components. Thus, it was decided to keep both features, since there is a clear relationship between these and the occurrence of failures in at least some of the components

3.6. Classification of Data and Construction of Labels

As previously mentioned, the problem of predictive maintenance under analysis is a case of Supervised Learning. In order to train a model to predict failures, it is necessary for not only examples of failure but also a time series of observations that led to that failure. Furthermore, the model needs examples of "normal" operating periods in order to be able to see the difference between the two. The classification between these two states is binary (stable or without failure/unstable or with failure). With this information available (stable/unstable), the model is only useful if it is able to promptly alert you to the imminence of a failure.

In order to fulfill this early warning criterion, it is necessary to modify the definition of the failure event label, which occurs at a specific time, to a time interval where the failure event may occur. The time until the failure occurs, which delimits the boundary between the two categories, must be chosen according to operational criteria. Is the knowledge that a failure will occur within 12 h sufficient to prevent it? And what about 24 h? And two weeks? The model's ability to predict a failure will also depend on the duration of this time window.

This process is illustrated in Figure 10. In order to achieve the reset from unstable to pre-unstable, observations within the time window (represented by "X" in Figure 9) before the occurrence of a failure have been labeled as *pre-instable*, while records outside this time interval X have been labeled as stable.

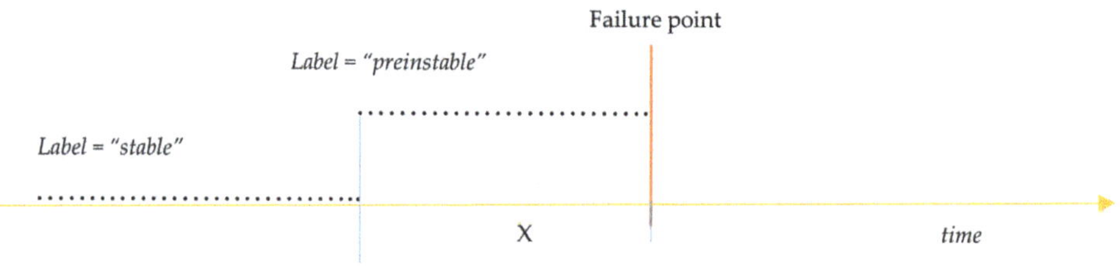

Figure 10. Data classification and label construction—adapted from [9].

The main objective of the Machine Learning models used will be to predict the probability of a failure occurring within this time window. In this case, more specifically, the probability of a machine failure occurring in the next 24 h (duration of the time window chosen for this application) related to one of the components (components 1,2,3 or 4). Thus, a new categorical feature "failure" was created, where all records in the 24 h prior to the occurrence of a failure in component 1 have the value failure = comp1 and so on for components 2,3 and 4. Records that do not check these conditions have the value of failure = none. This leads to the problem turning from a binary problem (stable/pre-unstable) to a multi-class classification problem (stable/n pre-unstable component). It should also be noted that, henceforth, due to this redefinition of failure event, when it is mentioned that a certain algorithm predicts a failure, in fact, what is being mentioned is that the algorithm predicts the occurrence of a failure within this time window.

Table 11 shows examples of failure in component 2. Note that the first 8 records occur in the 24 h prior to the occurrence of the first failure of component 2. The next 8 records in the 24 h prior to another failure of component 2.

3.7. Data Splitting

When working with associated day and time data, as is the case here, the division between the training, validation and test sets must be carried out carefully, in order to ensure that the evaluations obtained correspond to the actual performance that should be expected of the models, since there is an inherent temporal correlation between observations (high similarity between temporally close data). This validation technique is called Holdout.

In problems of predictive maintenance, in most situations, the best option is to perform a division based on time, that is, choose a point in time, train the model with all records prior to that point, using the later records to validate the model. This methodology also allows to simulate how the model will actually behave in practice.

Table 11. Example of failure representation in component 2.

	machineID	Datetime	Model	Age	Failure
857	1	2015-04-19 09:00:00	model3	18	comp2
858	1	2015-04-19 12:00:00	model3	18	comp2
859	1	2015-04-19 15:00:00	model3	18	comp2
860	1	2015-04-19 18:00:00	model3	18	comp2
861	1	2015-04-19 21:00:00	model3	18	comp2
862	1	2015-04-20 00:00:00	model3	18	comp2
863	1	2015-04-20 03:00:00	model3	18	comp2
864	1	2015-04-20 06:00:00	model3	18	comp2
2297	1	2015-10-16 09:00:00	model3	18	comp2
2298	1	2015-10-16 12:00:00	model3	18	comp2
2299	1	2015-10-16 15:00:00	model3	18	comp2
2300	1	2015-10-16 18:00:00	model3	18	comp2
2301	1	2015-10-16 21:00:00	model3	18	comp2
2302	1	2015-10-17 00:00:00	model3	18	comp2
2303	1	2015-10-17 03:00:00	model3	18	comp2
2304	1	2015-10-17 06:00:00	model3	18	comp2

Thus, in the present application, the registrations until 08/31/2015 1:00:00 were assigned to the test set, the registrations between 01/09/2015 1:00:00 and 31/10/2015 1:00:00 to the validation set and registration from 01-11-2015 1:00:00 to the test set. In order to guarantee that the data in different sets do not share time windows, the records at the borders, that is, the records of the 24 h preceding the date of division, have been removed. Thus, Table 12 shows the amount of data that was attributed to each of the sets and the percentage that corresponds to failures.

Table 12. Amount of data attributed to each of the sets and the percentage corresponding to failures.

Quantity	/%	Failure/%
Training	66.52	2.02
Validation	16.57	1.89
Test	16.91	1.92

3.8. Class Imbalance in Maintenance Problem Applications

Something to take into account in predictive maintenance is the fact that the occurrence of failures is rare during the life cycle of a given machine, when compared to normal operation. This leads to an imbalance between classes (Table 13), which usually leads to an illusory performance on the part of the algorithms, which tend to classify the most common example more often at the expense of the less common, since the total number of incorrect classifications are thus less. Therefore, the Recall and Precision values can be low, although the Accuracy value is high. A clear example of this phenomenon is, in the validation set (where most of the evaluation metrics will be calculated), where 98.11% (Table 12) of the data correspond to the Stable category (failure = none), that is, a model (without any use) that provides functioning stable values at all times would have an Accuracy of 98.11%. It is therefore essential to look at other evaluation metrics.

Table 13. Example of the imbalance between the different classes for the 'failure' feature in the total data set.

	Failure	%
none	285,684	98.06
comp2	1985	0.68
comp1	1464	0.50
comp4	1240	0.43
comp3	968	0.33

For a considerable number of critical equipment applications, the model's inability to predict a failure can be exorbitantly expensive. In predictive maintenance, as a general rule, the most important is the number of real failures that the model is capable of predicting, that is, the model's Recall3. This parameter becomes even more important as the consequences of false negatives, that is, true failures that the model was unable to predict, exceed the consequences of false positives, that is, a false prediction of a failure. This phenomenon is known as "incorrect classification cost" and can be estimated by companies according to the cost of repair, parts and labor. Generally, it is preferable that the model errs as a precaution, since it will be more economical to carry out a maintenance check than a partial or total interruption of the operation. However, the wrong prediction of a failure, that is, a false positive, can also lead to a loss of time and resources. In this case, the model must be adjusted to a high Precision. However, as mentioned earlier, the Recall and Precision metrics are not independent: Increasing one implies decreasing the other.

3.9. Application of Models in the Validation Set

In this first application, the validation set is used in order to understand how a wide variety of models behave, as well as to look for the tuning of the hyper-parameters of certain models. Such an approach is due to the fact that it is not possible, at the outset, to determine which algorithm is most suitable for a given problem. When training and evaluating a wide variety of models at an early stage, it is possible to see which ones have the greatest potential, however, for this step to be successful, it is necessary that the metrics for evaluating the models have been chosen in accordance with the established objectives.

The models tested were K-Nearest Neighbors, Decision Tree, Random Forest, Naïve Bayes and Artificial Neural Networks. The best results were obtained by Random Forest and Artificial Neural Networks models [32].

3.10. Test Set Behavior

The validation set has been used, so far, to fine-tune the models and respective hyper-parameters, in order to seek better performance. It is now important to check how the models behave in the test set. Although, in a real case, it is advisable to evaluate only the model that is intended to be implemented [8], this section presents the results obtained for the evaluation of the two best models (in the validation set), in this case, Random Forest and Artificial Neural Networks, with min-max scaling normalization.

Tables 14 and 15 show the values obtained for Precision, Recall and F1 Score, for the Random Forest and Artificial Neural Network model, respectively, in the validation and test sets.

Table 14. Performance for the Random Forest model in the validation and test sets, with 'n_estimators = 70.

	None	comp1	comp2	comp3	comp4
Precision	0.9997	0.9244	0.9916	1.0000	0.9812
Conj. Develop. Recall	0.9999	0.9578	0.9861	0.9514	0.9543
F1 Score	0.9998	0.9408	0.9889	0.9751	0.9676
Precision	0.9988	0.9718	0.9711	0.9855	0.9830
Conj. Test Recall	0.9998	0.8150	0.9882	0.9189	0.9611
F1 Score	0.9993	0.8865	0.9796	0.9510	0.9719

Table 15. Performance for the Artificial Neural Network model in the validation and test sets, with 100 hidden layers (hidden layers = 100) and min-max scaling normalization.

	None	comp1	comp2	comp3	comp4
Precision	0.9997	0.9451	0.9917	1.0000	0.9520
Conj. Develop.Recall	0.9998	0.9337	0.9972	0.9167	0.9954
F1 Score	0.9997	0.9394	0.9945	0.9565	0.9732
Precision	0.9990	0.9030	0.9941	0.9858	0.9725
Conj. Test Recall	0.9995	0.8425	0.9853	0.9392	0.9833
F1 Score	0.9993	0.8717	0.9897	0.9619	0.9779

As expected, there is a generalized decrease in performance of the evaluation metrics in the test set. Still, the results remain satisfactory. As previously mentioned, in predictive maintenance, as a general rule, the most important is the number of real failures that the model is capable of predicting, that is, the value of the model's Recall parameter [28]. This parameter becomes even more important as the consequences of false negatives, that is, true failures that the model was unable to predict, exceed the consequences of false positives, that is, a false prediction of a failure [33,34].

For both models, there is a drop in the value of Recall (and, consequently, of F1Score) to values below 90% for component 1 in the test set. In the present application, the four components were considered to be of equal importance.

In a real application, where it may be possible to know more information about each one of them (such as cost, importance in the process, location in the equipment, ease of replacement), the analysis may involve trying to optimize certain metrics that are considered to be of greater relevance.

4. Conclusions

In this paper, Machine Learning models were applied to a dataset available online. The data set used was published by Microsoft, in [9], in a Notebook for Predictive Maintenance and Machine Learning. The use of this data set was justified. In the implementation carried out in the present project, until the final phase of feature engineering, the steps presented in that Notebook were followed. However, from that moment on, as it is considered that the approach presented in [9] is too simplistic (no validation technique is used and only a single model is applied), it was decided to deepen the analysis with the implementation of the Holdout validation, which divides the data set into three subsets (Training, Validation and Test), as well as various Machine Learning models, thus showing how to fine-tune the models and respective hyper-parameters using the validation set.

The fact that it is a multi-class classification problem added complexity to the analysis and, perhaps, starting with a binary classification problem may be advisable for a better understanding of the basic concepts of Machine Learning, fundamental to the success of any application.

It is possible to address the imbalance between classes, very common in maintenance applications, since the occurrence of failures is rare during the life cycle of a given machine, when compared to its normal operation.

At the outset, and knowing that a sensible choice of which data to use and how to handle it is crucial for the performance of Machine Learning algorithms, good results would be expected based on the result obtained. However, more important than any result was the demonstration of a methodology, starting from data of different types and sources (very common in maintenance applications), that allowed us to show how it is possible to visualize and treat them in order to apply Artificial Intelligence tools in the analysis of maintenance data, in this case, Machine Learning.

Although the results obtained compare well with those presented so far in the literature, the biggest disadvantage in using the presented methodology lies in the definition of the features. If the selection of features is not the most correct, the results obtained can lead to wrong predictions. For future work, the application of feature learning concepts will be considered instead of feature engineering, which appears to be promising to improve the results obtained [35,36].

Author Contributions: D.C.: developed and performed the work during is M.Sc. thesis; L.F.: supervised the work and prepared and edited the manuscript. All authors have read and agreed to the published version of the manuscript.

Funding: This research received no external funding.

Conflicts of Interest: The authors declare no conflict of interest.

References

1. Gilchrist, A. *Industry 4.0: The Industrial Internet of Things*; Springer: Berlin/Heidelberg, Germany, 2016; ISBN 978-1-4842-2047-4.
2. Kobbacy, K.A.H.; Murthy, D.N.P. (Eds.) *Complex System Maintenance Handbook*; Springer Science & Business Media: Berlin/Heidelberg, Germany, 2008; ISBN 978-1849967006.
3. Kagermann, H.; Lukas, W.-D.; Wahlster, W. *Securing the Future of German Manufacturing Industry Recommendations for Implementing the Strategic Initiative INDUSTRIE 4.0*; Final report of the Industrie 4.0 Working Group; Acatech-National Academy of Science and Engineering: Frankfurt, Germany, 2013.
4. Ribau, J. Afinal, o que é isto da Indústria 4.0? Available online: https://visao.sapo (accessed on 7 June 2020). (In Portuguese)

5. Paolanti, M.; Romeo, L.; Felicetti, A.; Mancini, A.; Frontoni, E.; Loncarski, J. Machine learning approach for predictive maintenance in industry 4.0. In Proceedings of the 14th IEEE/ASME International Conference on Mechatronic and Embedded Systems and Applications (MESA), Oulu, Finland, 2–4 July 2018; pp. 1–6.

6. Li, Y.; Wang, K. Modified convolutional neural network with global average pooling for intelligent fault diagnosis of industrial gearbox. *Eksploat. Niezawodn. Maint. Reliab.* **2020**, *22*, 63–72. [CrossRef]

7. Rodrigues, J.; Costa, I.; Torres Farinha, J.; Mendes, M.; Margalho, L. Predicting motor oil condition using artificial neural networks and principal component analysis. *Eksploat. Niezawodn. Maint. Reliab.* **2020**, *22*, 440–448. [CrossRef]

8. Müller, A.C.; Guido, S. *Introduction to Machine Learning with Python: A Guide for Data Scientists*; O'Reilly Media Inc.: Sebastopol, CA, USA, 2016; ISBN 1449369901.

9. Microsoft. Predictive Maintenance Modelling Guide. 2018. Available online: https://notebooks.azure.com/Microsoft/projects/PredictiveMaintenance (accessed on 1 May 2020).

10. Çinar, Z.M.; Abdussalam Nuhu, A.; Zeeshan, Q.; Korhan, O.; Asmael, M.; Safaei, B. Machine Learning in Predictive Maintenance towards Sustainable Smart Manufacturing in Industry 4.0. *Sustainability* **2020**, *12*, 8211. [CrossRef]

11. Cheng, J.C.; Chen, W.; Chen, K.; Wang, Q. Data-driven predictive maintenance planning framework for MEP components based on BIM and IoT using machine learning algorithms. *Autom. Constr.* **2020**, *112*, 103087. [CrossRef]

12. Ran, Y.; Zhou, X.; Lin, P.; Wen, Y.; Deng, R. A Survey of Predictive Maintenance: Systems, Purposes and Approaches. *IEEE Commun. Surv. Tutor* **2019**, arXiv:191207383R.

13. Florian, E.; Sgarbossa, F.; Zennaro, I. Machine learning for predictive maintenance: A methodological framework. In Proceedings of the XXIV Summer School "Francesco Turco"—Industrial Systems Engineering, Bergamo, Italy, 9–11 September 2020.

14. Filipe Gomes Pereira, L. Previsão de Falhas em Empanques Mecânicos da Refinaria de Matosinhos Usando Modelos de Machine Learning. Master's Thesis, FEUP, University of Porto, Porto, Portugal, 2018.

15. Kanawaday, A.; Sane, A. Machine learning for predictive maintenance of industrial machines using IoT sensor data. In Proceedings of the 8th IEEE International Conference on Software Engineering and Service Science (ICSESS), Beijing, China, 24–26 November 2017; pp. 87–90, ISBN 1538604973.

16. Ali, Y.H. Artificial Intelligence Application in Machine Condition Monitoring and Fault Diagnosis. *Artif. Intell. Emerg. Trends Appl.* **2018**, *14*, 275–291. [CrossRef]

17. Mba, C.U.; Makis, V.; Marchesiello, S.; Fasana, A.; Garibaldi, L. Condition monitoring and state classification of gearboxes using stochastic resonance and hidden markov models. *Measurement* **2018**, *126*, 76–95. [CrossRef]

18. Li, X.; Makis, V.; Zuo, H.; Cai, J. Optimal bayesian control policy for gear shaft fault detection using hidden semi-markov model. *Comput. Ind. Eng.* **2018**, *119*, 21–35. [CrossRef]

19. Zuber, N.; Bajrićb, R. Gearbox faults feature selection and severity classification using machine learning. *Eksploat. Niezawodn. Maint. Reliab.* **2020**, *22*, 748–756. [CrossRef]

20. Zaeri, R.; Ghanbarzadeh, A.; Attaran, B.; Moradi, S. Artificial neural network based fault diagnostics of rolling element bearings using continuous wavelet transform. In Proceedings of the 2nd International Conference on Control, Instrumentation and Automation, Shiraz, Iran, 27–29 December 2011; pp. 753–758. [CrossRef]

21. Li, C.; Zurita, G.; Cerrada, M.; Cabrera, D. Multimodal deep support vector classification with homologous features and its application to gearbox fault diagnosis. *Neurocomputing* **2015**, *168*, 119–127. [CrossRef]

22. Saufi, S.R.; Ahmad, Z.A.B.; Leong, M.S.; Lim, M.H. An intelligent bearing fault diagnosis system: A review. *MATEC Web Conf.* **2019**, *255*, 06005. [CrossRef]

23. Chen, F.; Tang, B.; Chen, R. A novel fault diagnosis model for gearbox based on wavelet support vector machine with immune genetic algorithm. *Measurement* **2013**, *46*, 220–232. [CrossRef]

24. Li, Z.; Peng, Z. A new nonlinear blind source separation method with chaos indicators for decoupling diagnosis of hybrid failures: A marine propulsion gearbox case with a large speed variation. *Chaos Solitons Fractals* **2016**, *89*, 27–39. [CrossRef]

25. Cerrada, M.; Li, C.; Sánchez, R.V.; Pacheco, F.; Cabrera, D.; Oliveira, J.V.D. A fuzzy transition based approach for fault severity prediction in helical gearboxes. *Fuzzy Sets Syst.* **2018**, *337*, 52–73. [CrossRef]

26. Lei, Y.; Zuo, M.J. Gear crack level identification based on weighted k-nearest neighbor classification algorithm. *Mech. Syst. Sig. Process.* **2009**, *23*, 1535–1547. [CrossRef]

27. Jiang, R.; Yu, J.; Makis, V. Optimal bayesian estimation and control scheme for gear shaft fault detection. *Comput. Ind. Eng.* **2012**, *63*, 754–762. [CrossRef]

28. Thyago, P.; Carvalho, A.; Fabrízzio, M.N.; Soares, R.V.; Roberto Francisco, P.; Basto, S.A. A systematic literature review of machine learning methods applied to predictive maintenance. *Comput. Ind. Eng.* **2019**, *137*, 106024.

29. Domingos, P. A Few Useful Things to Know About Machine Learning. *Commun. ACM* **2012**, *55*, 78–87. [CrossRef]

30. Géron, A. *Hands-On Machine Learning with Scikit-Learn and Tensor-Flow: Concepts, Tools, and Techniques to Build Intelligent Systems*; O'Reilly Media Inc.: Sebastopol, CA, USA, 2019; ISBN 978-1491962299.

31. Pedregosa, F.; Varoquaux, G.; Gramfort, A.; Michel, V.; Thirion, B.; Grisel, O.; Blondel, M.; Prettenhofer, P.; Weiss, R.; Dubourg, V.; et al. Scikit-learn: Machine Learning in Python. *J. Mach. Learn. Res.* **2011**, *12*, 2825–2830.

32. Cardoso, D. Application of Predictive Maintenance Concepts with Application of Artificial Intelligence Tools. Master's Thesis, FEUP, University of Porto, Porto, Portugal, 2020.

33. Microsoft. Azure AI Guide for Predictive Maintenance Solutions. Available online: https://docs.microsoft.com/en-us/azure/machine-learning/team-data-science-process/predictive-maintenance-playbook (accessed on 20 April 2020).
34. Poosapati, V.; Katneni, V.; Manda, V.K.; Ramesh, T.L.V. Enabling Cognitive Predictive Maintenance Using Machine Learning: Approaches and Design Methodologies. In *Soft Computing and Signal Processing*; Springer: Berlin/Heidelberg, Germany, 2019; pp. 37–45.
35. LeCun, Y.; Bengio, Y.; Hinton, G. Deep learning. *Nature* **2015**, *521*, 436–444. [CrossRef]
36. González-Muñiz, A.; Díaz, I.; Cuadrado, A.A. DCNN for condition monitoring and fault detection in rotating machines and its contribution to the understanding of machine nature. *Heliyon* **2020**, *6*, e03395. [CrossRef] [PubMed]

Article

The Method of Production Scheduling with Uncertainties Using the Ants Colony Optimisation

Iwona Paprocka [1,*], Damian Krenczyk [1] and Anna Burduk [2]

[1] Department of Engineering Processes Automation and Integrated Manufacturing Systems, Faculty of Mechanical Engineering, Silesian University of Technology, Konarskiego 18A str., 44-100 Gliwice, Poland; damian.krenczyk@polsl.pl

[2] Faculty of Mechanical Engineering, Wroclaw University of Science and Technology, 50-370 Wrocław, Poland; anna.burduk@pwr.edu.pl

* Correspondence: iwona.paprocka@polsl.pl; Tel.: +48-32-237-16-57

Abstract: Production and maintenance tasks apply for access to the same resources. Maintenance-related machine downtime reduces productivity, but the costs incurred due to unplanned machine failures often outweigh the costs associated with predictive maintenance. Costs incurred due to unplanned machine failure include corrective maintenance, reworks, delays in deliveries, breaks in the work of employees and machines. Therefore, scheduling of production and maintenance tasks should be considered jointly. The problem of generating a predictive schedule with given constrains is considered. The objective of the paper is to develop a scheduling method that reflects the operation of the production system and nature of disturbances. The original value of the paper is the development of the method of a basic schedule generation with the application of the Ant Colony Optimisation. A predictive schedule is built by planning the technical inspection of the machine at time of the predicted failure-free time. The numerical simulations are performed for job/flow shop systems.

Keywords: maintenance; predictive scheduling; flow shop; job shop; ant colony optimisation

Citation: Paprocka, I.; Krenczyk, D.; Burduk, A. The Method of Production Scheduling with Uncertainties Using the Ants Colony Optimisation. *Appl. Sci.* **2021**, *11*, 171. https://dx.doi.org/10.3390/app11010171

Received: 19 November 2020
Accepted: 23 December 2020
Published: 27 December 2020

1. Introduction

The criteria of cost, quality and time availability are always contradictory. Entrepreneurs look for solutions that will not be reflected in the loss of quality or extension of deadlines for the implementation of tasks. Entrepreneurs are looking for organizational, technological and IT solutions that will allow for improvements in these areas.

Consider the problem of scheduling production tasks and planning technical inspections of machines. Production and maintenance tasks apply for access to the same resources, machines. Production and maintenance managers have divergent goals. Machines immobilization for maintenance decrease productivity. Boudjelida [1] investigated the robustness of joint production and maintenance scheduling in the problem of flow permutation and proved that the loss of efficiency due to the insertion of maintenance tasks into the production schedule increases. But costs incurred due to unplanned machine failure often outweigh the costs associated with predictive maintenance. Costs incurred due to unplanned machine failure include corrective maintenance, reworks, delays in deliveries, breaks in the work of employees and machines. Therefore, scheduling of production and maintenance tasks should be considered jointly.

The related literature distinguishes three approaches to production and maintenance planning in disturbance conditions: predictive, proactive and reactive. The goal of the predictive approach is to obtain a schedule that can absorb the disturbance without affecting planned external activities, while maintaining high system efficiency [2,3]. The proactive approach examines the influence of the disturbance on a schedule, using the criteria of stability. The schedule obtained for the best sequence of tasks related to maintenance and production is assumed for implementation [4]. The objective of the reactive approach is to adapt the schedule to the current situation [5].

There are two proactive approaches: proactive without or with prediction (Figure 1). In the first approach, only the impact of a disturbance on the schedule using robustness measures is examined. Researchers search for the best sequence of idle times between production tasks or batches taking the advantage of the simulation process [6,7]. Considering only the relationship between production and maintenance tasks as a conflict in management decisions may cause unmet demand or unexpected machine failures. A common objective is to maximize a system productivity and efficiency. Usually, the time interval for maintenance task and the number of maintenance tasks are fixed in advance. The mentioned deficiencies of proactive-reactive approaches are eliminated in predictive-reactive approaches.

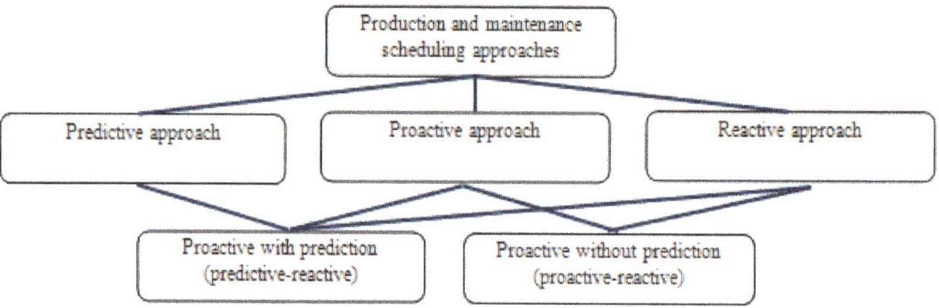

Figure 1. Classification of production and maintenance scheduling approaches.

The predictive-reactive approach is regarded as a combination of predictive and proactive scheduling techniques. Researchers predict maintenance time and then evaluate the effect of a disturbance on the predictive schedule using robustness measures [8,9]. Using the probability theory to describe machine conditions allows for more reliable maintenance planning. However, accepting the assumption that machine conditions are observable at the beginning of each period is not sufficient. Popular maintenance strategies are based on the periodic inspection of a machine and age dependent inspection and are also not sufficient. Attributes to describe the machine age and the influence of maintenance should be drawn from analysis of historical data on failure-free times and observation of dynamic machine conditions. The predictive-reactive method is considered in the presented paper.

Benbouzid-Sitayeb et al. [10] propose the joint production and preventive maintenance scheduling problem in permutation flowshops with the objective of minimizing the makespan. The insertion of the maintenance tasks is done according to several heuristics. Fei and Ma [11] propose a joint optimization on a hybrid flow shop system. A preventive maintenance strategy is based on reliability. The multi-objective is to minimize the makespan and total production cost. The authors proved that joint optimization is superior compared with independent decision-making. Nourelfath and Châtelet [12] present the integrating preventive maintenance and tactical production planning method for a parallel production system. The authors assume two possible causes for system failure: the independent failure of single components, and the simultaneous common cause failure of all components. The objective is to minimize the sum of preventive and corrective maintenance costs, setup costs, holding costs, backorder costs and production costs. Berrichi et al. [13] propose the Ant Colony Optimization algorithm to solve the joint production and maintenance scheduling problem. The trade-off solutions between objectives of production and maintenance is searched. Reliability models are used to take into account the maintenance aspect.

This paper faces the problem of generating a predictive schedule with given constrains in the conditions of disturbances for job shop/flow shop systems. The objective of the article is to develop an effective method of task scheduling, reflecting the operation of the production system and the nature of the disturbances. The method of estimation unknown system parameters such as Mean Time to Failure, Mean Time of Repaier is based on the theory of probability. The original value of the paper is the development of the method of a basic schedule generation with the application of the Ants Colony Optimisation (ACO). A predictive schedule is built by planning the technical inspection of the machine at time of the predicted failure-free time. Flexible operations are allocated to the machine during an increased risk of a failure. Three algorithms: genetic (GA) [14], immune (MOIA) [15] and clonal selection (CSA) [16] have been developed and compared for the presented problem of predictive schedules generation.

In this paper the concept of the ACO is presented and numerical examples are given for predictive scheduling. The ant colony optimization algorithm is applied to the problem of makespan minimization and schedule stability maksimisation. Comparative analyses of parameter variants of the ant colony optimization algorithm are performed.

The paper is organized as follows: The job shop scheduling problem for experimental study is presented in Section 2. The general concept of ACO is presented in Section 3. The application of ACO for the problem of production and maintenance task scheduling is described in Section 4. Section 5 contains numerical simulations and experimental test results related to the research. The paper concludes with a brief summary of the results (Section 6).

2. Production and Manitenance Scheduling Model

The scheduling problem in a job shop system where production tasks are allocated to resources with performance constraints due to maintenance is considered. Production systems are described by: (a) production tasks, (b) machines, (c) routes of production tasks, (d) operation times, (e) task completion dates. Production tasks are executed in exclusive-like mode and operations are not preempted. After a machine failure, the disrupted operations can be performed on parallel machines.

Data on the failure-free operation of the machine is collected. Knowledge about the machine reliability characteristics for the future planning horizon is acquired in five stages:

1. Adoption of the hypothesis that the time of failure-free operation is described by the reliability distribution depending on the phase of the machine's life cycle.
2. Application of methods for estimating distribution parameters.
3. Prediction of distribution parameters for the future planning period.
4. Calculation of the reliability characteristics (e.g., mean time to failure (MTTF)) for the future planning period.
5. Assessment of the impact of variable dates of failure on the values of stability and robustness criteria for given values of machine reliability characteristics.

The analyzed historical period is divided into i equal scheduling periods, $[(i-1)T, iT)$, $i = 1, ..., m+1$. For each of them, N_i events are observed, i.e., machine failures with failure-free times $X_{i,1}, ..., X_{i,N_i}$. For each historical period i, the distribution parameters are estimated in order to describe the phenomenon of failure rate. Let us assume the hypothesis that the failure-free times $X_{i,1}, ..., X_{i,N_i}$ in period i $[(i-1)T, iT), i = 1, ..., m+1$ are described by the exponential distribution with parameter $\mu_i > 0$ with a density function:

$$f_i(t) = \begin{cases} \mu_i \exp(-\mu_i t), t > 0, \\ 0, t \leq 0, \end{cases} \tag{1}$$

Parameters μ_i are estimated in the second stage. Values of μ_i generally differ in subsequent historical periods i. Using the maximum likelihood method, the parameter $\tilde{\mu}_1$ for the first period is estimated:

$$\tilde{\mu}_1 = \frac{n_1}{\sum_{k=1}^{n_1} x_{1,k}} \tag{2}$$

In the empirical moment method, value $\tilde{\mu}_1$ is determined comparing the equations:

$$m_1(x_1) = \frac{1}{n_1} \sum_{k=1}^{n_1} x_{1,k} \tag{3}$$

where:

$$m_1(x_1) = \frac{1}{\mu_1} \tag{4}$$

and the formula for the estimated parameter $\tilde{\mu}_1$ (2) is obtained.

After obtaining the estimated values of the distribution parameters for each historical period, parameter $\tilde{\mu}_{m+1}$ is predicted for the future planning period using classical regression technique. Defining the function describing the parameters consists in eliminating fluctuations and identifying trends of the analyzed data on failure-free times. The least squares method is used for smoothing time series in linear and quadratic functions. To confirm the hypothesis that the scatter plot of a given function is the most reliable, two coefficients are calculated: (a) coefficient of determination (R2) which measures the trend fit to the failure-free data and (b) the function of losses (SSE) which is the sum of squared deviations residues. The hypothesis with the function with the highest R2 value and the lowest SSE value is selected.

In the fourth stage, we determine the reliability characteristics, such as [16]:

- Mean Time Between Failures = Mean Time To Failure + Mean Time of Repair, (*MTBF = MTTF + MTTR*),

$$MTBF = E\{X_{m+1,1} + Y_{m+1,1}\} = \frac{1}{\mu_{m+1}} + \frac{1}{\alpha_{m+1}}, \tag{5}$$

  where: $\alpha_{m+1} > 0$ is predefined.
- Probability P that in the interval $[f, g] \in [mT, (m+1)T)$ there occurs at least one failure,
- Period of increased risk of failure $[a, b + MTTR]$, where: a on the assumption that the probability of the failure-free time of the bottleneck is higher than a equalling 30%, b on the assumption that the probability of the failure-free time of the bottleneck is less than b equalling 70%.

In the fifth stage, the predictive schedule is generated for the reliability characteristics using the ant colony optimisation algorithm. The procedure of generating predictive schedules is presented in Section 4. The stability of schedule k is measured using the quality robustness and solution robustness criteria. The reactive schedule k^* is generated in a situation where the predictive schedule k can not absorb the impact of the disturbance. The newly generated schedule should reproduce the previous one as much as possible according to the stability criterion:

$$SR(k*) = \sum_{j=1}^{J} \sum_{v_j=1}^{V_j} \left| st_{j,v_j}(k) - st_{j,v_j}(k*) \right|, \tag{6}$$

where: $st_{j,v_j}(k)$ is start time of operation v_j of task j in predictive schedule k; $st_{j,v_j}(k*)$ is start time of operation v_j of task j in reactive schedule k^*.

After the disturbance, the value of the criterion used to evaluate the predictive schedule should not be significantly influenced. The quality robustness of schedule k is assessed by calculating the difference between the makespan criterion C before and after the machine failure:

$$QR(k*) = |C(k) - C(k*)|. \tag{7}$$

3. Basics on Ant Colony Optimization

Modeling how ants behave and interact helps solve many optimization problems. The first ant algorithm (Ant Colony Optimization) was presented by Marco Dore in 1992 [8]. The strength of ants lies in their numbers and the cooperation. The cooperation between individuals ensures the survival of the entire community. Each ant can find the shortest path from the anthill to the food source without analyzing the visible terrain that surrounds it. Ants easily adapt to new conditions. When the road is blocked by an obstacle, they can avoid it, when the place where the food was located becomes inaccessible, they will start looking for a new source of food.

An ant that has reached the food and returns to the anthill leaves a pheromone trail behind. Depending on what signal the ant wants to send to others, the smell and the intensity of the pheromone varies. Any other ant, sensing the pheromone in its immediate neighbourhood and analyzing its intensity, is able to determine which direction to go in order to reach the food. The more ants pass along the path from food to the anthill, the stronger the smell of the pheromone will remain on that path, making it the most attractive path. The paths that are less traveled are forgotten over time, and even if they led to food, the pheromone will not be enough to guide the ants to their destination.

The structure of the ant algorithm consists of three parts: main transition rule, global update rule, local update rule.

3.1. Main Transition Rule

Each ant follows the pseudo—random—proportional rule taking the next step. The rule determines whether the ant is focused on exploration (random path selection) or exploitation (determinism) moving from point r to point s (Equation (8)).

If an ant is focused on exploration, it does not react to the pheromone trace in its environment. This makes it more likely that the ant will pass over to an area that may be more attractive. If an ant is focused on exploitation, it only goes where it senses the pheromone trail, which makes the paths from the anthill to the food more abundant in pheromone.

$$
S = \begin{cases} \arg\max_{u \in N_k(r)} \left\{ [\tau(r,u)] \cdot [\eta(r,u)]^\beta \right\}, & \text{for } q \le q_0 \\ p_k(r,s), & \text{for } q > q_0, \end{cases}
\tag{8}
$$

where: q_0—a parameter, $q_0 \in \langle 0, 1 \rangle$, q—a random number from $\langle 0, 1 \rangle$, τ—size of the pheromone trace on the edge u, between points r and s, $\eta = \frac{1}{\delta}$—reciprocal of the distance $\delta(r, u)$ representing heuristics, β—the parameter of the relative importance between the pheromone trace and the reciprocal of the distance, $N_k(r)$—the set of those points that ant k (located at point r) has not yet visited, $p_k(r, s)$—a random variable selected according to the formula:

$$
p_k(r,s) = \begin{cases} \dfrac{[\tau(r,s)] \cdot [\eta(r,s)]^\beta}{\sum\limits_{u \in N_k(r)} [\tau(r,s)] \cdot [\eta(r,s)]^\beta}, & \text{for } s \in N_k(r) \\ 0, & \text{for } s \notin N_k(r) \end{cases}
\tag{9}
$$

If parameter $q \le q_0$, the ant is driven by the desire to exploit already discovered areas. The most attractive point s for an ant is the one to which the distance from r is the shortest, and the pheromone value on the path from r to s is the highest.

If parameter $q > q_0$, the ant is driven by the desire to discover new areas—exploration. In this case point s is a random point from all available points connected to point r. Each ant exploring a new area learns about it, which, if useful, passes on to other ants by means of the pheromone left behind. Any available point can be chosen, not only the best one.

Appropriate selection of parameter q_0 results in the quality improvement of solutions generated by the algorithm. By lowering parameter q_0, ants may start to pay too much attention to explore new areas. Already discovered routes leading to the target are quickly forgotten by other ants.

On the other hand, when parameter q_0 is overestimated, it is likely that ants focus on the suboptimal solution. There is not enough ants to explore new areas in search of a new, perhaps better solution.

3.2. Local Update of the Pheromone Trace

Local updating of the pheromone trace takes place every iteration, for each ant [8]. Looking for solutions, ants move between points on the edges connecting these points. At the same time, ants update the value of the pheromone, even if they have not found the best solution. Updating the pheromone trace locally aims to reduce the value of the pheromone on each visited edge in each iteration. Updating the pheromone trace locally prevents ants from accumulating on one path only, and introduces some variation in the results obtained:

$$\tau(r,s) \leftarrow (1-\rho) \cdot \tau(r,s) + \rho \cdot \Delta\tau(r,s), \tag{10}$$

where: ρ—pheromone evaporation factor $\rho \in \langle 0,1 \rangle$, $\tau(r,s)$—the amount of pheromone on the way from point r do s, $\Delta\tau_K(r,s)$—reduction of the pheromone trace:

$$\Delta\tau_K(r,s) = \tau_0 = \frac{1}{n \cdot L_{nn}} \tag{11}$$

where: n—number of possible points to visit from the point r, L_{nn}—minimum distance between two adjacent points.

3.3. Global Pheromone Update

The global update of the pheromone consists in updating the pheromone value at the edges of the relatively optimal path from the anthill to the food. The relatively optimal path is the best solution to the problem from the beginning of the algorithm's operation or determined for each iteration [8]:

$$\tau(r,s) \leftarrow (1-\alpha) \cdot \tau(r,s) + \alpha \cdot \sum_{k=1}^{m} \Delta\tau_K(r,s), \tag{12}$$

where: α—pheromone evaporation rate, $(1-\alpha) \in \text{<}0,1\text{>}$ is the glow of the pheromone, $\tau(r,s)$—the amount of pheromone on the way from point r to s, m—the number of ants that have passed from point r to point s, $\Delta\tau_K(r,s)$—the increase of the pheromone trace is calculated from:

$$\Delta\tau_K(r,s) = \begin{cases} \frac{1}{L_K}, & \text{for } (r,s) \in L_K \\ 0, & \text{for } (r,s) \notin L_K \end{cases} \tag{13}$$

where: $(r,s) \in L_K$—edge belonging to the global best solution, K—index of the ant that discovered the best solution, L_K—the length of the globally best solution.

4. ACO for Scheduling Production and Maintenance Tasks

The presented predictive-reactive method uses the advantage of computer simulation by repeating three steps:

(1) generating a population of best ants,
(2) conversion of basic schedules (represented by ants) into predictive schedules using the Minimal Impact of Disrupted Operation on the Schedule (MIDOS) rule.
(3) assessment of the impact of a disruption on reactive schedule/s using criteria: solution robustness (SR) and quality robustness (QR) [15].

In the following the ACO implementation for generating basic schedules (first step) in job shop scheduling problems is presented.

Pheromone and heurisitic information initialization is inspired by Boudjelida [1]. The same ant coding procedure was presented in [1] as in this article. But the ACO algorithms differ in the procedure for improving the solution and the number of parameters controlling the intensity of the pheromone and the visibility of the pheromone. The main difference is also the approach to scheduling maintenance tasks. In this paper a predictive-reactive approach is considered, the author [1] proposes a proactive-reactive approach. Both articles also consider different types of scheduling problems.

4.1. Ants Coding

Ant k is positioned on a randomly selected task j from a randomly selected vector of tasks V_k. The selected task is placed on the ant taboo list T_k. The size of the taboo list is equal to the number of tasks J ($j = 1, 2, ..., J$) in a scheduling problem. The neighbourhood size for each selected task is two, $n = 2$. In other words, the ant can select two adjacent tasks of j from list V_k in the next step.

4.2. Solution Construction

Ant k selects a task to schedule by selecting parameter q and calculating a transition probability for exploration or exploitation (1 and 2). The task sekected from the neighbourhood is inserted in the Tabu list T_k. The selected task is scheduled. Ant k moving from task r to task s reduces the value of the pheromone information on the track (r, s) (3). In the scheduling problem, L_{nn} is the minimum deadline for completing a task after scheduling all neighborhood tasks and n is the number of tasks in the neighborhood. The process of a task selection is repeated until vector V_k is empty. The final solution achieved by ant k is presented by the production task sequence in Tabu list T_k.

4.3. The Best Solution Selection

The best solution selection is repeated after each ant has constructed a production task sequence. The best solution obtains the minimum value of makespan criterion C. Makespan represents the end time of the last operation in a schedule. The pheromone information is updated for each track that the best ant has followed (5). L_K is the value of criterion (C_{max}) in the presented scheduling problem.

4.4. The Predictive Schedule Generation

Predictive schedules take the advantage of prognostic analysis in the Minimal Impact of Disrupted Operation on the Schedule (MIDOS) rule. The MIDOS rule transforms schedules to be more robust and stable in the event of disruptions. In the MIDOS rule, the job which is predicted to be disturbed is rescheduled. The most flexible operation of the job is assigned to the bottleneck. The backward and forward scheduling are applied for remaining operations [12].

4.5. The Predictive-Reactive Schedule Generation

The predictive and reactive schedules are generated for the basic schedules achieved by the ACO. Predictive schedules are generated using the MIDOS rule. The MIDOS rule modifies the basic schedules so that they are more reliable and stable when there is a risk of disruption. Following the MIDOS rule, a task that is predicted to be disrupted is analyzed for the flexibility of its operations. The most flexible operations are assigned to the critical machine. For the remaining operations, the back and forth scheduling rule applies. There are two variants of the MIDOS rule. The MIDOS I rule uses a left-shifting heuristic of operations preceding a critical operation, and a right-shifting of operations following a critical operation. In the MIDOS II rule, forward and backward scheduling depends on the availability of parallel machines. Operations are scheduled appropriately on the earliest available parallel machines for the upstream and downstream operations of the critical operation, respectively.

After the disturbance, two rescheduling procedures are applied for disrupted operations: Right Shifting (RS) and Reschedule on Parallel Machines (RPM). SR assesses how much the current schedule differs from the previously adopted one. QR assesses how much the current value of the quality indicator differs from the value of the previously adopted schedule.

4.6. Update of the Pheromone Trace for Makespan Optimisation

The formula for updating the pheromone trace locally is modified in order to perform makespan optimisation. The reduction of the pheromone trace is calculated using.

$$\Delta\tau_K(r,s) = \tau_0 = \frac{1}{n \cdot C(nn)} \tag{14}$$

where: $C(nn)$ is the end date of the last task in the schedule (makespan):

$$C(nn) = \max\left[tz_{V_j} \right] \tag{15}$$

tz_{v_j} is the completion time of operation v_j of job j, $v_j = 1, \ldots, V_j$, $j = 1, \ldots, J$.
The increase of the pheromone trace is calculated from

$$\Delta\tau_{K**}(r,s) = \begin{cases} \frac{1}{C(nn)}, & \text{for } (r,s) \in L_{K**} \\ 0, & \text{for } (r,s) \notin L_{K**} \end{cases} \tag{16}$$

where: $(r,s) \in L_{K**}$—job sequence belonging to the global best schedule, $K**$—index of the ant that discovered the best schedule.

The steps of the ACO are presented in Figure 2. The next Section presents a job shop (JS) scheduling problem for experimental study to better understood the steps of the ACO.

```
While a number of iterations η > E
for each ant k
select vector of tasks Vₖ
For each passage of ant
    •      if (passage =1)
    •        select task jfrom a vector of tasks Vₖ,
    •      else if (passage !=1)
    •        select parameter q∈[0,1]
            -   If (q<=q₀), go to exploitation step
            -   If (q>q₀), go to exploration step
    •   Exploitation: for each task r belonging to the neighborhood of the scheduled task s
        determine the probability of its selection (8)
            -   select the destination point s * with maximum probability,
            -   ant k goes to s *, point s * becomes the new r point for the ant
            -   Local pheromone update (10)
            -   schedule task j on machines, put task j on the ant taboo list Tₖ
    •   Exploration: for each task r belonging to the neighborhood of the scheduled task s
        determine the probability of its selection (18)
            -   determine numerical intervals proportional to probabilities,
            -   generate a random number in the range [0,1]
            -   Selection of the destination point s* that has been assigned to the interval
            -   Local pheromone update (10)
            -   schedule task j on machines, put task j on the ant taboo list Tₖ
Select the best schedule represented by ant K
Global pheromone update (12)
Apply MIDOS (I or II) rule in order to transform the best basic schedule into predictive schedule
After the disturbance, generate reactive schedules using heuristics: RS or RPM
Select the best predictive-reactive schedule using criteria: SR and QR
```

Figure 2. The steps of the ACO.

5. Predictive-Reactive Scheduling Case Study

This section introduces various ACO parameter data sets to verify the performance of ACO and MIDOS I or MIDOS II for predictive scheduling in various job shop sizes.

Job shop scheduling problems are investigated to fine-tune the parameters where 9 jobs have to be performed on 8 machines (9 × 8) and 11 jobs have to be performed on 10 machines (11 × 10). The first machine is the most heavily loaded. The failure-free time of the bottleneck *MTTF* equals 66. The repair time of the bottleneck *MTTR* equals 6. The increased probability of the bottleneck failure occurs in time horizon [*a, b* + *MTTR*] where: *a* = 60 and *b* = 72. The objective is to find an approach which is able to generate stable and robust schedules in the event of the bottleneck failure. The objective is to achieve a robust and stable schedule for the problem, $C_{max}(k) \rightarrow$ min (15).

Computer simulation of the Ant Colony Optimisation is run for the parameter of the relative importance between the pheromone trace and the reciprocal of the distance $\beta = 1$; pheromone evaporation factor $\alpha = \rho = \{0.2, 0.4, 0.6, 0.8\}$; number of ants, $K = \{10, 15, 20, 25\}$; number of iterations, $E = \{10, 20, 30, 40\}$ and parameter q_0 which decides abot exploration or exploatation selection by an ant, $q_0 = \{0.3, 0.4, 0.5, 0.6\}$. The ACO is run 10 times for each set of input parameters $\{\rho, K, E, q_0\}$.

First, the influence of the number of iterations, $E = \{10, 20, 30, 40\}$ over the quality of basic schedules generation for single criterion problem is investigated and for unchanging pheromone evaporation factor $\alpha = \rho = \{0.6\}$ and number of ants $K = \{10\}$. The parameter q_0 is equal to 0.5 to get an equal chance of choosing to explore and exploit by ants. By observing the first and third quartiles of C_{max} and the best schedules achieved the following conclusion can be drawn that a larger number of iteration is, the higher chances of achieving a better solution are in scheduling problem (11 × 10) (Figure 3b). By observing the first and third quartiles of C_{max} and the best schedules achieved for the scheduling problem (9 × 8), the opposite phenomenon can be observed. The smaller the number of iterations is, the greater the chances of achieving a better solution are (Figure 3a).

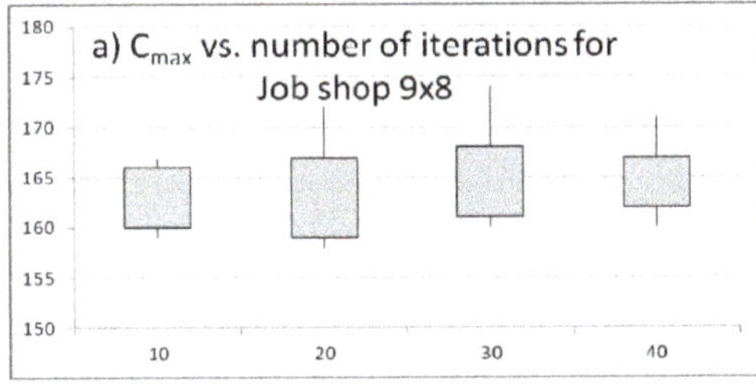

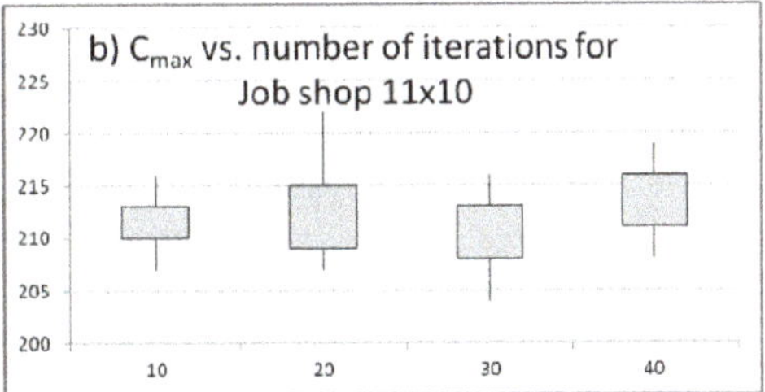

Figure 3. The C_{max} of the basic schedules achieved by the ACO for different iteration numbers.

Next, the influence of the number of ants, $K = \{10, 15, 20, 25\}$ over the quality of basic schedules generation for single criterion problem is investigated for unchanging pheromone evaporation factor $\alpha = \rho = \{0.6\}$, number of iterations $E = \{20\}$, parameter q_0 is equal to 0.5. Observing the results of the achieved value of the makespan criterion for the basic schedules (Figure 4), the following conclusion can be drawn that a larger the ant population is, the greater the chances of achieving a better solution are. This phenomenon is noticed for both sizes of scheduling problems (9 × 8) and (11 × 10).

Then, the simulations are continued for the number of ants, $K = \{15\}$, iteration size, $E = \{20\}$, parameter $q_0 = 0.5$ and changing values of pheromone evaporation factor $\alpha = \rho = \{0.2, 0.4, 0.6, 0.8\}$. Observing average values of C_{max} and the best schedules achieved (Figure 5) the following conclusion can be drawn that the higher values of pheromone evaporation factors $\alpha = \rho$ are, the higer chances of achieving a better solution are. Algthough the average quality of population does not increase with the parameter values, better solutions are achieved for scheduling problem (9 × 8). The best scheduel is achieved for C_{max} equals 152 for scheduling problem (9 × 8), for the pheromone evaporation factor $\alpha = \rho = 0.8$ (Figure 5a). The best scheduel is achieved for C_{max} equals 203 for scheduling problem (11 × 10), for the pheromone evaporation factor $\alpha = \rho = \{0.4, 0.6, 0.8\}$ (Figure 5b).

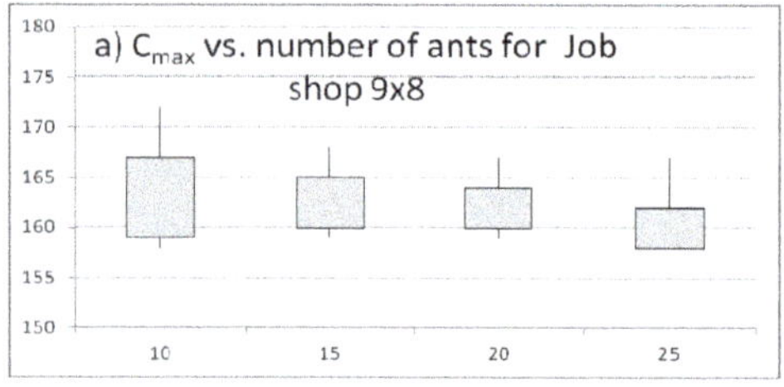
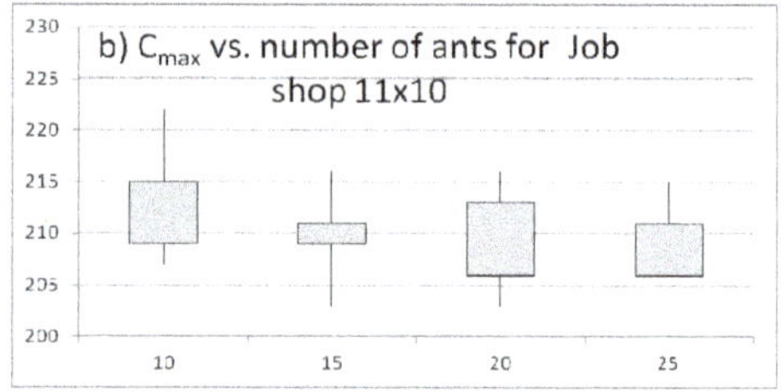

Figure 4. The C_{max} of the basic schedules achieved by the ACO for different size of ant population.

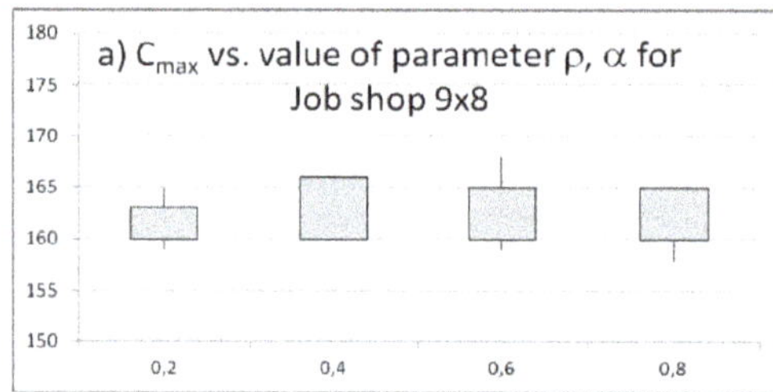
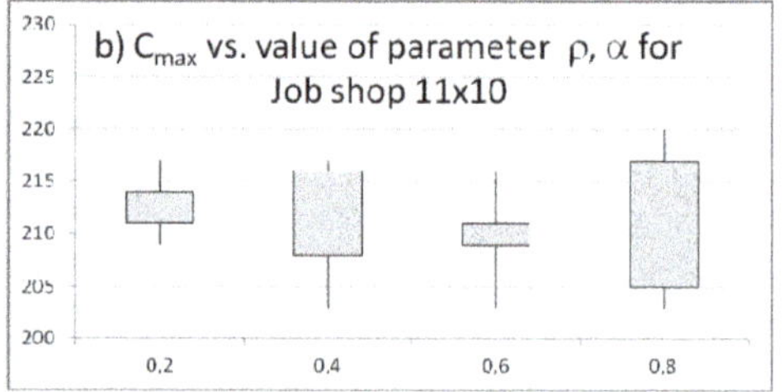

Figure 5. The C_{max} of the best basic schedules achieved by the ACO for different values of parameter $\alpha = \rho$.

Then, the simulations are continued for the number of ants, $K = \{15\}$, iteration size, $E = \{20\}$, parameter $\alpha = \rho = \{0.2\}$ and changing values of parameter q_0 which decides abot exploration or exploatation selection by an ant, $q_0 = \{0.3, 0.4, 0.5, 0.6\}$. Observing average values of C_{max} and the best schedules achieved (Figure 6) the following conclusion can be drawn that the lower values of parameter q_0 are, the higer chances of achieving a better solution are. The average quality of population does not increase with the parameter values for scheduling problems (9 × 8) and (11 × 10) (Figure 6a,b).

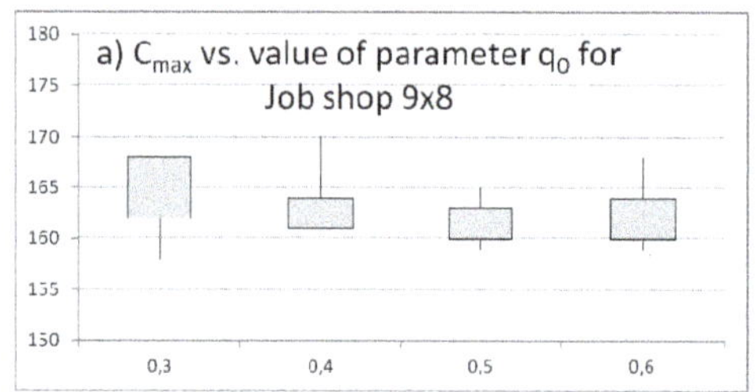
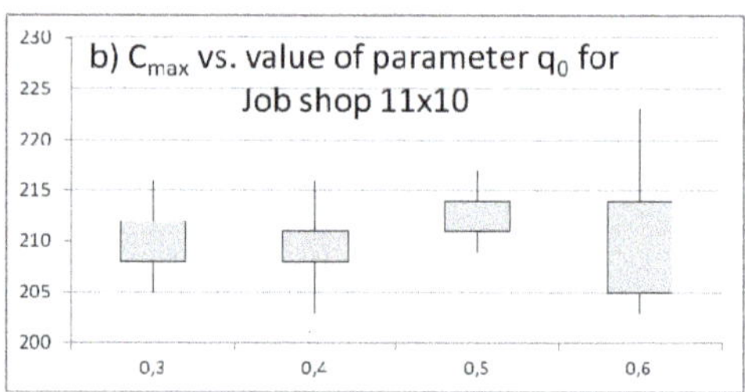

Figure 6. The C_{max} of the basic schedules achieved by the ACO for different values of parameter q_0.

Next, the performance of the ACO and MIDOS I or MIDOS II is verified for predictive scheduling for different datasets of job shops. The predictive and reactive schedules are generated for the basic schedule achieved by the ACO for each set of input parameters $\{\rho, K, E, q_0\}$. Predictive schedules are generated using rules: the MIDOS I or MIDOS II.

For example, in the first simulation, the predictive and reactive schedules were generated for the basic schedule obtained by the ACO and MIDOS I for the sequence of tasks: {7 8 6 9 5 2 4 3 1} for scheduling problem (9 × 8) and {10 8 7 6 11 9 3 4 1 5 2} for scheduling problem (11 × 10) (Table 1). The makespan function of the predictive schedule generated using the MIDOS I was $C_{max}(1) = 141$. The makespan function of the reactive schedule generated using the MIROS was also $C_{max}(1^*) = 141$. The solution robustness was SR(1) = 48 and the quality robustness was QR(1) = 0 for the first scheduling problem (9 × 8). Quality of the task sequences achieved for the remaining ants for scheduling problems (9 × 8) and (11 × 10) is described in Table 1. Also, computer simulations were run for generating predictive schedules using the MIDOS II. Quality of the predictive and reactive schedules for scheduling problems (9 × 8) and (11 × 10) is described in Table 2. The average solution robustness of predictive schedules generated using the ACO and MIDOS I was 32.69 for scheduling problem (9 × 8) and 42.07 for scheduling problem (11 × 10) (Table 1). The average solution robustness of predictive schedules generated using the ACO and MIDOS II was 31.92 for scheduling problem (9 × 8) and 27.46 for scheduling problem (11 × 10) (Table 2). All achieved schedules are robust taking into account the quality robustness criterion for both scheduling problems (9 × 8) (Table 1) and (11 × 10) (Table 2). By analyzing the minimum, maximum, first quantile, third quantile and the mean values of solution and quality robustness, the following conclusion can be drawn: the MIDOS II heuristic is better to apply to the basic schedules generated by ACO (Figure 7).

Table 1. The schedules generated using the MIDOS I and MIROS for the best basic schedules achieved by the ACO and input parameters $\{\rho, K, E, q_0\}$.

$\{E, K, \rho, q_0\}$	Scheduling Problem (9 × 8)					Scheduling Problem (11 × 10)				
	C_{max} of Basic Schedule	C_{max} of Predictive Schedule	C_{max} of Reactive Schedule	SR	QR	C_{max} of Basic Schedule	C_{max} of Predictive Schedule	C_{max} of Reactive Schedule	SR	QR
{10, 10, 0.6, 0.5}	159	141	141	48	0	207	177	177	0	0
{20, 10, 0.6, 0.5}	158	138	138	100	0	207	169	169	41	0
{30, 10, 0.6, 0.5}	160	144	144	75	0	204	168	168	97	0
{40, 10, 0.6, 0.5}	160	134	134	4	0	208	166	166	9	0
{20, 15, 0.6, 0.5}	159	140	140	2	0	207	175	175	8	0
{20, 20, 0.6, 0.5}	159	122	122	57	0	203	167	167	116	0
{20, 25, 0.6, 0.5}	158	136	136	2	0	206	174	174	14	0
{20, 15, 0.4, 0.5}	160	138	138	4	0	209	172	172	86	0
{20, 15, 0.6, 0.5}	159	130	130	61	0	203	165	165	156	0
{20, 15, 0.8, 0.5}	158	140	140	2	0	203	165	165	4	0
{20, 15, 0.2, 0.3}	158	147	147	23	0	205	174	174	12	0
{20, 15, 0.2, 0.4}	161	141	141	45	0	203	172	172	0	0
{20, 15, 0.2, 0.6}	159	140	140	2	0	203	174	174	4	0
average				32.69					42.077	

Table 2. The schedules generated using the MIDOS II and MIROS for the best basic schedules achieved by the ACO and input parameters $\{\rho, K, E, q_0\}$.

$\{E, K, \rho, q_0\}$	Scheduling Problem (9 × 8)					Scheduling Problem (11 × 10)				
	C_{max} of Basic Schedule	C_{max} of Predictive Schedule	C_{max} of Reactive Schedule	SR	QR	C_{max} of Basic Schedule	C_{max} of Predictive Schedule	C_{max} of Reactive Schedule	SR	QR
{10, 10, 0.6, 0.5}	159	141	141	47	0	207	177	177	0	0
{20, 10, 0.6, 0.5}	158	138	138	100	0	207	169	169	6	0
{30, 10, 0.6, 0.5}	160	144	144	62	0	204	174	174	0	0
{40, 10, 0.6, 0.5}	160	134	134	4	0	208	166	166	9	0
{20, 15, 0.6, 0.5}	159	137	137	14	0	207	175	175	2	0
{20, 20, 0.6, 0.5}	159	122	122	56	0	203	167	167	182	0
{20, 25, 0.6, 0.5}	158	136	136	6	0	206	173	173	3	0
{20, 15, 0.4, 0.5}	159	138	138	4	0	209	172	172	13	0
{20, 15, 0.6, 0.5}	160	130	130	19	0	203	165	165	131	0
{20, 15, 0.8, 0.5}	158	139	139	47	0	203	165	165	4	0
{20, 15, 0.2, 0.3}	158	141	141	9	0	205	174	174	3	0
{20, 15, 0.2, 0.4}	161	141	141	45	0	203	172	172	0	0
{20, 15, 0.2, 0.6}	159	140	140	2	0	203	189	189	4	0
average				31.92					27.46	

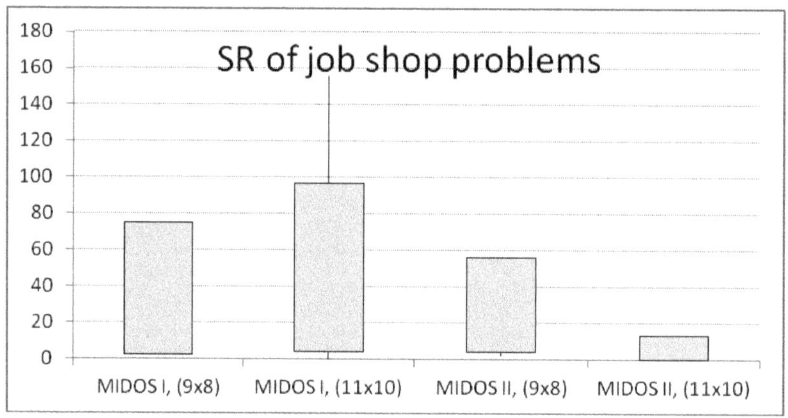

Figure 7. The SR of the predictive schedules achieved by the ACO and MIDOS rules for scheduling problems: (9 × 8) and (11 × 10).

6. Conclusions

In the paper, the predictive-reactive (proactive with prediction) method for joint scheduling of production and maintenance tasks was presented. The presented method can improve the work of maintenance team. Machine failure causes great losses as a result of downtime, the need to replace parts or even modiffication of the production plan to take into account the fact that the given machine or device need to be repaired for a longer period. The analysis of historical data on the machine uptimes allowes one to plan the replacement of elements, machine inspection and may contribute to extending the machine uptime.

The original value of the paper was the development of the method of a basic schedule generation with the application of the Ant Colony Optimisation (ACO). A predictive schedule was built by planning the technical inspection of the machine at time of the predicted failure time. Flexible operations are allocated to the machine during an increased risk of failure. Next, the influence of the disturbance on the predictive schedule using robustness measures was examined.

In the future, the presented method for generating predictive schedules will be compared with the genetic algorithm, immune and clonal selection algorithms. ACO algorithms are alternative methods of searching the solution space for scheduling problems. The presented algorithm may, however, contribute to the development of a method that reflects the operation of the production system and the nature of disturbances, and improves the system operation.

Author Contributions: Conceptualization, I.P.; Methodology, I.P.; Software, I.P.; Validation, I.P.; Formal Analysis, A.B. and D.K.; Investigation, I.P.; Resources, I.P., D.K. and A.B.; Data Curation, I.P.; Writing-Original Draft Preparation, I.P.; Writing-Review & Editing, D.K., and A.B.; Visualization, D.K., I.P. and A.B.; Supervision, I.P.; Funding Acquisition, A.B. All authors have read and agreed to the published version of the manuscript.

Funding: This research was funded from the statutory grant of the Wroclaw University of Science and Technology.

Conflicts of Interest: The authors declare no conflict of interest.

References

1. Boudjelida, A. On the robustness of joint production and maintenance scheduling in presence of uncertainties. *J. Intell. Manuf.* **2019**, *30*, 1515–1530. [CrossRef]
2. Liu, Q.M.; Dong, M.; Peng, Y. A dynamic predictive maintenance model considering spare parts inventory based on hidden semi-Markov model. *Proc. Inst. Mech. Eng. Part C* **2013**, *227*, 2090–2103. [CrossRef]
3. Mehta, S.V.; Uzsoy, R.M. Predictable scheduling of a job subject to breakdowns. *IEEE Trans. Robot. Autom.* **1998**, *14*, 365–378. [CrossRef]
4. Cui, W.W.; Lu, Z.; Li, C.; Han, X. A proactive approach to solve integrated production scheduling and maintenance planning problem in flow shops. *Comput. Ind. Eng.* **2018**, *115*, 342–353. [CrossRef]
5. Alagöz, O.; Azizoğlu, M. Rescheduling of identical parallel machines under machine eligibility constraints. *Eur. J. Oper. Res.* **2003**, *149*, 523–532. [CrossRef]
6. Bali, N.; Labdelaoui, H. Optimal Generator Maintenance Scheduling Using a Hybrid Metaheuristic Approach. *Int. J. Comput. Intell. Appl.* **2015**, *14*, 1550011. [CrossRef]
7. Cui, W.W.; Lu, Z.; Pan, E. Integrated production scheduling and maintenance policy for robustness in a single machine. *Comput. Oper. Res.* **2014**, *47*, 81–91. [CrossRef]
8. Boryczka, U. *Anthill Optimization Algorithms*; Wydawnictwo Uniwersytetu Śląskiego: Katowice, Poland, 2006. (In Polish)
9. Kamrul Hasan, S.M.; Sarker, R.; Essam, D. Genetic algorithm for job-shop scheduling with machine unavailability and breakdowns. *Int. J. Prod. Res.* **2011**, *49*, 4999–5015. [CrossRef]
10. Benbouzid-Sitayeb, F.; Ammi, I.; Varnier, C.; Zerhouni, N. Applying Ant Colony Optimization for the Joint Production and Preventive Maintenance Scheduling Problem in the Flowshop Sequencing Problem. In Proceedings of the 3rd International Conference on Information and Communication Technologies: From Theory to Applications, Damascus, Syria, 7–11 April 2008. [CrossRef]
11. Fei, Y.; Ma, H. Multi-objective joint optimization of batch-discrete hybrid flow shop scheduling integrated with machine maintenance. In Proceedings of the 2018 5th International Conference on Industrial Engineering and Applications (ICIEA), Singapore, 26–28 April 2018. [CrossRef]

12.   Nourelfath, M.; Châtelet, E. Integrating production, inventory and maintenance planning for a parallel system with dependent components. *Reliability Eng. Syst. Safety* **2012**, *101*, 59–66. [CrossRef]

13.   Berrichi, A.; Yalaoui, F.; Amodeo, L.; Mezghiche, M. Bi-Objective Ant Colony Optimization approach to optimize production and maintenance scheduling. *Comput. Oper. Res.* **2010**, *37*, 1584–1596. [CrossRef]

14.   Paprocka, I.; Grabowik, C.; Kempa, W.M.; Krenczyk, D.; Kalinowski, K. The influence of algorithms for basic-schedule generation on the performance of predictive and reactive schedules. *Conf. Ser. Mater. Sci. Eng.* **2018**, *400*, 1757–8981. [CrossRef]

15.   Paprocka, I.; Skołud, B. A Hybrid-Multi Objective Immune Algorithm for predictive and reactive scheduling. *J. Sched.* **2017**, *20*, 165–182. [CrossRef]

16.   Paprocka, I.; Kempa, W.M. Searching for a Method of Basic Schedules Generation which influences over the Performance of Predictive and Reactive Schedules. In *Advances in Intelligent Systems and Computing*; Springer: New York, NY, USA, 2016; Volume 524, pp. 233–242.

applied sciences

Article

Dynamic Analysis Method for Fault Propagation Behaviour of Machining Centres

Liming Mu [1,2], Yingzhi Zhang [1,2,*], Jintong Liu [1,2], Fenli Zhai [1,2] and Jie Song [1,2]

[1] Key Laboratory of Reliability of CNC Equipment, Ministry of Education, Changchun 130022, China; moulm19@mails.jlu.edu.cn (L.M.); jintongliu1324@jlu.edu.cn (J.L.); difl19@mails.jlu.edu.cn (F.Z.); jiesong19@mails.jlu.edu.cn (J.S.)

[2] School of Mechanical and Aerospace Engineering, Jilin University, Changchun 130022, China

[*] Correspondence: zhangyz@jlu.edu.cn

Abstract: Fault propagation behaviour analysis is the basis of fault diagnosis and health maintenance. Traditional fault propagation studies are mostly based on a priori knowledge of a causality model combined with rule-based reasoning, disregarding the limitations of experience and the dynamic characteristics of the system that cause deviations in the identification of critical fault sources. Thus, this paper proposes a dynamic analysis method for fault propagation behaviour of machining centres that combines fault propagation mechanisms with model structure characteristics. This paper uses the design structure matrix (DSM) to establish the fault propagation hierarchy structure model. Considering the correlation of fault time, the fault probability function of a component is obtained and the fault influence degree of nodes are calculated. By introducing the Copula and Coupling degree functions, the fault influence degree of the edges between the same level and different levels are calculated, respectively. This paper constructs a fault propagation intensity model by integrating the edge betweenness and uses it as an index to analyze real-time fault propagation behaviour. Finally, a certain type of machining centre is taken as an example for specific application. This study can provide as a reference for the fault maintenance and reliability growth of a machining centre.

Keywords: machining centre; DSM; Copula function; fault propagation intensity; fault propagation behaviour

Citation: Mu, L.; Zhang, Y.; Liu, J.; Zhai, F.; Song, J. Dynamic Analysis Method for Fault Propagation Behaviour of Machining Centres. *Appl. Sci.* **2021**, *11*, 6525. https://doi.org/10.3390/app11146525

Academic Editors: José Machado, Filomena Soares, Dariusz Mazurkiewicz, Dario Antonelli and Katarzyna Antosz

Received: 12 May 2021
Accepted: 14 July 2021
Published: 15 July 2021

Publisher's Note: MDPI stays neutral with regard to jurisdictional claims in published maps and institutional affiliations.

1. Introduction

CNC technology and CNC machine tools are enabling the development of technologies and basic equipment towards emerging high-technology and cutting-edge industries [1]. Numerical control technology is extensively used in many countries worldwide to improve the capacity and level of the equipment manufacturing industry, and to improve market adaptability and competitiveness [2]. Machining centres are widely used because of their strong flexible processing capabilities. They have a strong technical advantage in the manufacturing field and have become the main processing equipment for various manufacturing enterprises.

A machining centre is a complex system composed of multiple components. Given the influence of system structure, working environment, human factors, and maintenance level, the usage of a machining centre becomes more complex. The system fault is not only related to the independent fault of the component but also to the propagation faults [3]. If the fault cannot be eliminated in a timely manner, it will affect the progress of the entire production and even cause the contract to fail to be performed as scheduled, resulting in irreparable economic losses to a company [4]. Currently, many companies have low levels of fault diagnosis and maintenance in their machining centres, especially in the fault diagnosis of a machining centre [5]. The cost [6] and time [7] spent on locating a fault source of a machining centre cannot be ignored and when the cause of the fault is determined, the time for troubleshooting will be significantly shortened. Therefore, clarifying the real-time fault propagation behaviour of a machining centre is crucial for the prediction and elimination of faults.

The current fault diagnosis methods can be summarized into four categories [8,9]: knowledge-based fault diagnosis [10–12], model-based fault diagnosis [13–15], signal-based fault diagnosis [16–18], and hybrid method-based fault diagnosis (a method that combines two or more methods) [19–22]. Fault diagnosis for machining centres mainly include diagnosis methods based on fault information monitoring, training models, and fault trees.

The diagnosis method based on fault information monitoring is concerned with monitoring information characteristics of each system component of a machining centre by means of sensors [23]. Through the test analysis software and the corresponding model, the fault information feature extraction is realized and the fault location is determined [24]. However, because the sensor is highly susceptible to the influence of the surrounding environment, the collected signal is not accurate enough and leads to a certain deviation of the diagnosis results. This method cannot detect all the usage information of a system; hence, it is mostly used for the fault diagnosis of system components and cannot realize the fault diagnosis of an entire system.

The diagnosis method based on training model is concerned with training models such as support vector machines [25–27] and neural networks [28–30] on the basis of constructing a machine fault information database. Then, on the basis of the input information that represents the fault symptom, the output information that reflects the fault cause is directly derived to realize the judgment of the machine fault; However, this method cannot accurately determine the fault location of a machine. Concurrently, when the machine is very expensive, establishing test samples is difficult, hence collecting numerous samples to train a model is impossible.

The diagnosis method based on fault trees is concerned with listing all the possible causes of machine faults successively from top to bottom [31]. By establishing the fault tree of a machining centre, faults can be checked individually [32]. However, this method uses the dependency of each fault mode to construct the fault tree and realizes the fault diagnosis on the basis of the simple logic gate and the average fault rate of the bottom event. Due to the neglecting of the correlation of fault mode, a deviation in the calculation of fault rate occurs and the structural characteristics of fault trees are not considered, thus resulting in a wrong diagnosis.

Given the shortcomings of the above diagnosis methods, some scholars use a petri net, cellular automaton, and complex networks with topological characteristics such as regular networks, scale-free networks, small-world networks, and random networks to establish fault propagation models to study fault propagation [33–36]. However, given the dependence of a model on the structure and the correlation of faults, a deviation occurs in the fault mechanism analysis and fault source location of a machining centre. In addition, the fault influence degree of system components obtained by traditional methods is mostly constant; in contrast, the fault influence degree of each system component will change over time. Thus, achieving the ideal effect in the fault diagnosis strategy is difficult.

Therefore, this paper proposes a fault propagation intensity evaluation method that is used to dynamically describe the fault propagation process of machining centre system components. Compared with the existing methods, the DSM-based fault propagation hierarchy structure model of machining centres established in this paper can more clearly demonstrate the relationship between system components. The fault influence degree of system components obtained in this paper are also time-varying, thereby more in line with engineering practice. Moreover, for the calculation of the influence degree of the edge between components, most of the previous studies used the same method to calculate the fault influence degree of each level and did not consider the difference of fault influence between different levels that will inevitably lead to the deviation in the fault propagation analysis results. This paper considers the difference of fault influence degree between different levels and uses the Copula function and coupling function to calculate the fault influence degree of the edges between the same level and different levels, respectively. The structural characteristics of the model and fault mechanism of the system is considered synthetically, thus the description is more reasonable. We use the value of the fault propagation intensity as an index to study the fault propagation behaviour of a machining centre, to identify the critical fault propagation path of a machining centre, and to provide a theoretical basis and practical reference for later fault detection and maintenance.

The remainder of this paper is organized as follows. Section 2 describes the method for evaluating the fault propagation intensity of machining centres on the basis of a fault propagation model. Section 3 introduces the dynamic analysis method of fault propagation behaviour of machining centres on the basis of the fault propagation intensity. Section 4 provides a case application of a machining centre to demonstrate the effectiveness of the proposed method. Section 5 engages in a discussion. Lastly, Section 6 presents our conclusions.

2. Method for Evaluating Fault Propagation Intensity of Machining Centres on the Basis of the Fault Propagation Model

Based on the basic working process and principle of machining centres, a machining centre is divided into modules and the relevant faults are identified in combination with the field fault data. Considering the fault correlation of components, a hierarchy structure model of machining centres' fault propagation on the basis of DSM is established. The Johnson method is applied to correct the component fault order and construct a time-dependent component fault probability model. On this basis, the importance of component nodes is calculated to reflect the fault influence degree of component nodes. Considering the differences of fault influence degree of the edges at different levels, this paper uses the Copula function to calculate the influence degree of the edges at the same level and uses the coupling degree function to calculate the influence degree of the edges between different levels. Then, the component fault probability model is integrated to calculate the node's probability of fault propagation. Given that the topology of the model will also affect the propagation of the fault, the fault propagation intensity of machining centres can be evaluated by fusing the probability of the fault propagation with the edge betweenness of the structural model. The evaluation process of fault propagation intensity in machining centres is shown in Figure 1.

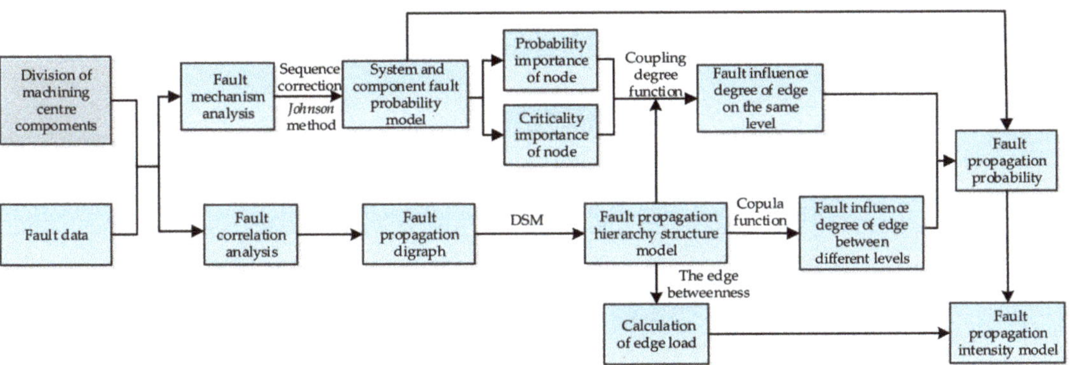

Figure 1. Evaluation process of the fault propagation intensity in machining centres.

2.1. Construction of the Hierarchy Structure Model of Fault Propagation in Machining Centres Based on DSM

2.1.1. General System Hierarchy Structure Modelling Process

The system structure model describes the relationship structure between the components of the system with the concept of set. The model is a diagram that reflects the relationship between the components of the system. For machining centres, however, the relationship diagram is difficult to utilize for clearly reflecting the relationships between the components of the system. Through matrix and hierarchy processing, the related digraph of each system component is transformed into a hierarchy structure model that can clearly understand the structure of the system and relationship between the components of the system. Its general modelling process is shown in Figure 2.

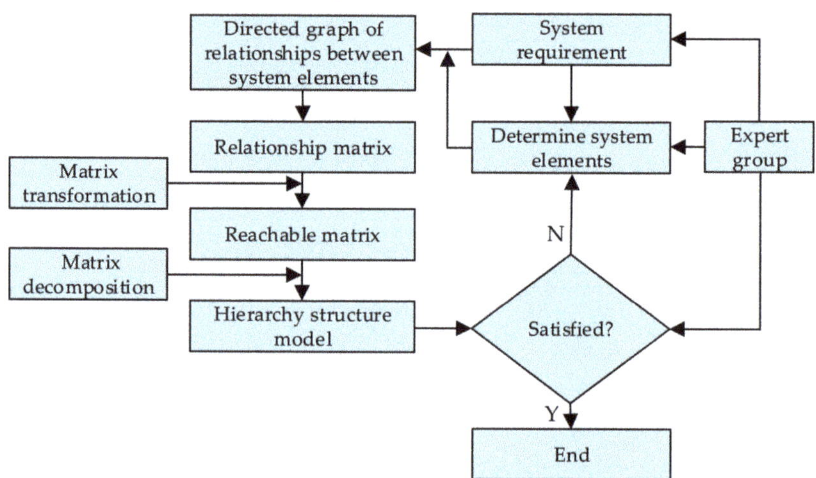

Figure 2. Schemes follow the same formatting.

2.1.2. Construction of the Hierarchy Structure Model of Fault Propagation in Machining Centres on the Basis of DSM

In accordance with the general modelling process of the hierarchy structure model, the machining centre is first divided into n system components on the basis of the working process and principles of the machining centre, and the fault correlation analysis of the collected fault data is conducted and aided by the relevant experience of the system structure function and the fault diagnosis manual. Furthermore, the fault time of each system component in the machining centre and the fault propagation relationship among the components is determined. The system component is expressed as a node set: $V = \{v_1, v_2, \cdots, v_n\}$. The fault propagation relationship and connection relationship between system component nodes are represented by directed edge set $E = \{\langle v_1, v_2 \rangle, \langle v_2, v_3 \rangle, \cdots, \langle v_m, v_n \rangle\}$. Thus, a digraph ($G = (V, E)$) of fault propagation of a machining centre can be obtained. Then, according to the fault propagation digraph, the direct fault influence between the system components is expressed with a relation matrix (A) and the reachable matrix is obtained according to the relation matrix. In addition, the reachable matrix is decomposed to realize the construction of the fault propagation hierarchy structure model.

To establish a clearer fault propagation hierarchy structure model, the design structure matrix (DSM) method is introduced [37]. In the form of binary values, '0' and '1' or '×' and a blank space represent the relationship between the row and column elements in the design structure matrix (**M**). The design structure matrix can be regarded as a reachable matrix. The elements in the design structure matrix are defined as follows:

$$m_{ij} = \begin{cases} 1, v_i \text{ has at least one dependency on } v_j \\ 0, v_i \text{ has nothing to do with } v_j \end{cases} \quad (i = 1, 2, \cdots, n; \ j = 1, 2, \cdots, n) \quad (1)$$

The fault influence relationship between the system components of a machining centre can also be expressed as the dependency relationship between the elements in the design structure matrix. At this time, the elements are defined as follows:

$$m_{ij} = \begin{cases} 1, v_i \text{ has at least one effect on } v_j \\ 0, v_i \text{ has no effect on } v_j \end{cases} \quad (i = 1, 2, \cdots, n; \ j = 1, 2, \cdots, n) \quad (2)$$

The modelling principle of the hierarchy structure model is based on the design structure matrix (**M**) and according to the knowledge of matrix, row–column transformation is conducted. In addition, the design structure matrix is converted into the lower triangular matrix as much as possible to reduce the existence of positive and negative transfer relations in the matrix. To eliminate feedback information between modules, the risk of iteration is reduced [38]. The construction process of the fault propagation hierarchy structure model based on DSM is shown in Figure 3.

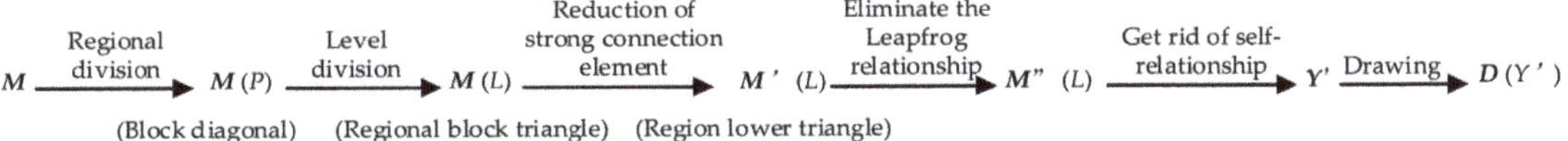

Figure 3. Flow chart of the hierarchy structure model construction based on DSM.

Based on the hierarchy sequence obtained by the DSM-based hierarchy model construction process, the appearance layer is classified as the fault absorption layer in the fault propagation model, the root layer is classified as the fault initiation layer, and the remaining layers are classified as the fault propagation layers. The number of fault propagation layers may vary according to the needs of the research object. Based on this, the hierarchy structure model of fault propagation of system components is drawn.

2.2. Calculation of Fault Propagation Probability of Machining Centre System Components

2.2.1. Calculation of Machining Centre System Components' Fault Probability

All *n* data of machining centres' fault data and right truncation data are sorted and recorded as $j(1 \leq j \leq n)$. Then, only *m* fault data of a certain component of a machining centre are sorted from small to large and are recorded as $i(1 \leq i \leq m)$. The order number *n* of the *i*-th fault data is as follows:

$$r_i = r_{i-1} + \frac{n + 1 - r_{i-1}}{n + 2 - j} \quad (3)$$

The corrected component fault time order number is substituted into the median rank equation to obtain the empirical distribution function of the component, as follows:

$$r_i = \frac{r_i - 0.3}{n + 0.4} \quad (4)$$

The Weibull distribution commonly used in engineering is taken as the hypothetical distribution of the fault interval time of machining centre components. In this paper,

Weibull distribution is used to construct the reliability model and the cumulative fault distribution function is as follows:

$$F(t) = 1 - \exp\left[-\left(\frac{t}{\beta}\right)^{\gamma}\right], t \geq 0,$$ (5)

where β represents scale parameter, $\beta > 0$, γ represents shape parameter, and $\gamma > 0$, t is a time variable, $t \geq 0$.

Equation (5) is transformed as follows:

$$\ln\left[\ln\left(1/(1 - F(t)))\right)\right] = \gamma \ln t + \gamma \ln \beta$$ (6)

The linear regression model is fitted between $\ln\left[\ln(1/(1 - F(t)))\right]$ on the left side of the equation and $\ln t$ on the right side. The Weibull model parameter value $\hat{\beta}$ and $\hat{\gamma}$ can be obtained on the basis of the fault information and empirical distribution function value. The model test value (ρ) is calculated by the linear correlation coefficient test method. Next, the minimum value (ρ_α) of the correlation coefficient is calculated on the basis of the number of fault data (n) and significance level (α). When $\rho > \rho_\alpha$, it is considered that $\ln\left[\ln(1/(1 - F(t)))\right]$ and $\ln t$ are linearly related and the fault data is subject to the assumed distribution. Otherwise, the hypothesis is rejected.

2.2.2. Determination of the Influence Degree between Components of Machining Centres

Importance refers to the contribution to the fault probability of the whole machine when a system component fails; it is a time-dependent function and can characterize both the structure of the system and a parameter of the reliability of the system components [39]. The importance of system components is quantitatively described and called the importance of system components. In this paper, it is regarded as the influence degree of a system component to reflect the influence degree of a whole machine fault caused by the fault of a certain system component.

(1)   Calculation of Probability Importance of System Component Nodes

In the late 1960s, Birnbaum put forward the Birnbaum probability importance for practical production [40]. The idea of probability importance is introduced by considering the contribution of reducing the fault probability of system components to reducing the fault probability of the whole system.

When any system component of a machining centre fails, it may cause the whole machine to fail. Hence, the relationship between the system components of a machining centre can be regarded as the series relationship. According to this structural relationship, the fault probability function of the system components can be used to describe the fault probability function of a whole machining centre. The fault probability function is a function of time and their relationships are expressed in Equation (7):

$$F_Z(t) = 1 - [(1 - F_1(t))(1 - F_2(t)) \cdots (1 - F_i(t))].$$ (7)

In addition, the equation for calculating the probability importance of system component nodes are shown in Equation (8):

$$I_P(v_i) = \frac{\partial F_Z(t)}{\partial F_i(t)} \quad i = 1, 2, \cdots, n,$$ (8)

where $I_P(v_i)$ is the probability importance of the system component node, $F_Z(t)$ is the fault probability function of the whole machine, and $F_i(t)$ is the fault probability function of the system component node. For any two-system component nodes, regarding v_i and v_j, if $I_P(v_i) > I_P(v_j)$ exists at some time then v_i is more important than v_j.

(2)   Calculation of the Criticality Importance of Nodes

The probability importance expresses that the fault probability of the system component node changes by one unit, resulting in the change of the fault probability of the whole system. Given that the probability of the fault of each system component node is different, the difficulty of generating a unit change is also different. Therefore, the criticality importance is introduced to describe the properties that the probability importance cannot reflect. Criticality importance was first proposed by Lambert [41]. In the case of a given fault of a whole system, the conditional probability that the whole machine is in a certain state when the critical system component node v_i fails at time t. The core idea is to improve the non-reliable system components rather than to further improve the reliable system components.

Therefore, the criticality importance ($I_C(v_i)$) of the system component node is defined as:

$$I_C(v_i) = \frac{F_i(t)}{F_Z(t)} \cdot \frac{\partial F_Z(t)}{\partial F_i(t)} = \frac{F_i(t)}{F_Z(t)} \cdot I_P(v_i).$$ (9)

For a machining centre, when the system component node v_i fails and the fault is passed to the component node v_j at a certain value, v_j is affected by the fault and the component node v_i is also affected by the component node v_j. There also exists a certain influence value on the directed edge between the two component nodes. This value represents the influence ability of the component node v_i fault to cause the fault of its connected component node v_j. However, the fault influence degree of the edge of the same level differs from that of the different layers, hence it should be considered separately.

2.2.3. Calculation of the Fault Influence Degree of Edges in the Same Level Based on the Copula Function

In this paper, it is assumed that a whole machine has n fault-related system components and its reliability function is $R(t_i)$, $i = 1, 2, \cdots, n$ and the joint reliability function of these system components is $R(t_1, t_2, \cdots, t_n)$. Based on the Copula theory and Sklar theorem [42–44], a Copula function can be uniquely determined. It is expressed as follows:

$$C(R_1(x), R_2(x), \cdots, R_n(x)) = \exp\left\{ -\left[\sum_{i=1}^{n} (-lnR_i(t))^{1/\theta} \right]^{\theta} \right\},$$ (10)

where θ is the parameter of the Copula function to characterize the degree of association between the system components ($\theta \in (0, 1]$).

Given that the calculation process of the Copula function parameters is relatively complex, this paper introduces a relatively intelligent artificial fish swarm algorithm [45] and uses MATLAB software to solve the parameters of the Copula function.

The following relationship exists between the fault probability function and reliability function:

$$F(t) = 1 - \exp\left[-\left(\frac{t}{\beta} \right)^{\gamma} \right] = 1 - R(t).$$ (11)

As expressed in Equation (10), the degree of correlation between components obtained from the perspective of the reliability function is represented by θ and as expressed in Equation (11), $F(t) = 1 - R(t)$. Therefore, the fault influence degree of the edge at the same level in a machining centre is $I(v_i, v_j) = 1 - \theta$ and the greater the value is, the greater the fault influence degree of the edge will be.

2.2.4. Calculation of the Fault Influence Degree of Edges at Different Levels Based on the Coupling Degree Function

In Reference [46], the effect function is used to describe the impact of a certain subsystem change on the whole machine, while the coupling degree model is used to characterize the coupling degree between subsystems. Thus, the importance of the node is taken as the effect function value and a model of the fault coupling degree is established to reflect the fault influence value between the components. The calculation equation is expressed in Equation (12):

$$I(v_i, v_j) = 2 \left[\frac{I_C(v_i) I_C(v_j)}{(I_C(v_i) + I_C(v_j))^2} \right]^{1/2}, \tag{12}$$

where $I(v_i, v_j) \in (0, 1)$ and is the fault influence degree between v_i and v_j. This value is used to characterize the effect of this edge on fault propagation. $I_C(v_i)$ and $I_C(v_j)$ are the critical importance values of v_i and v_j, respectively.

2.2.5. Construction of the Fault Propagation Probability Model for Machining Centre System Components

Based on the hierarchy model of a machining centre and the fault propagation mechanism, the state of a system component node depends on its own fault probability, which the influence of upstream component node is dependant on. The fault influence between the nodes reflects the possibility of a fault of the component node to cause another component fault to be associated with it. Therefore, by integrating the fault probability ($F_i(t)$) of the node and the fault influence degree ($I(v_i, v_j)$) between the system component nodes, the value of the fault propagation probability between system components can be obtained. The equation is as follows:

$$P(v_i, v_j)(t) = F_i(t) I(v_i, v_j) \tag{13}$$

2.3. Evaluation of Fault Propagation Intensity of a Machining Centre Based on the Fault Propagation Model

Given that the influence of the structural characteristics of the hierarchy model of fault propagation cannot be disregarded, the edge betweenness that represents the structural characteristics of the model is considered [47]. The larger the value is, the stronger its influence on the whole model will be.

In this paper, the edge betweenness is defined as the proportion of the number of times that all paths pass through the edge $E\langle v_i, v_j \rangle$ in the graph model. The edge betweenness of the directed edge $E\langle v_i, v_j \rangle$ connecting v_i and v_j is denoted by $L(v_i, v_j)$ that can be calculated by the following equation:

$$L(v_i, v_j) = \sum_{\substack{v_i, v_j, v_e, v_f \\ (e,f) \neq (i,j)}} \frac{\kappa_{ef} E\langle v_i, v_j \rangle}{\kappa_{ef}}, \tag{14}$$

where κ_{ef} is number of paths between any of the nodes, and v_e, v_f, and $\kappa_{ef} E\langle v_i, v_j \rangle$ are the number of paths between v_e and v_f passing through edge $E\langle v_i, v_j \rangle$.

In this paper, the fault propagation intensity model of a machining centre is established from two angles of the fault propagation probability and the edge betweenness of the hierarchy model. This value of the fault propagation intensity is used as an index to measure the severity of the impact of the fault on the whole machine through the path. The calculation equation for defining the fault propagation intensity $In(v_i, v_j)$ of a machining centre is as follows:

$$In(v_i, v_j) = \frac{1}{2} \left(\frac{P(v_i, v_j)}{\Sigma P(v_i, v_j)} + \frac{L(v_i, v_j)}{\Sigma L(v_i, v_j)} \right), \tag{15}$$

where $\Sigma P(v_i, v_j)$ is the sum of the fault propagation probabilities of each directed edge. $\Sigma L(v_i, v_j)$ is the sum of the edge betweenness of each directed edge.

3. Dynamic Analysis of the Fault Propagation Behaviour of a Machining Centre Based on the Fault Propagation Intensity

To analyze the real-time fault propagation behaviour of a machining centre, the propagation range of the fault should be clarified to determine whether the propagation of the fault can be achieved. As expressed by the system and product safety manual [48], the occurrence of the fault propagation is within a certain interval and when the probability of fault propagation is greater than the threshold of 10^{-8}, the case of fault propagation may occur.

Therefore, the probability of v_i fault to propagate to v_j can be obtained as follows:

$$P_E(v_i, v_j) \;=\; \sum_{b=1}^{w} P_b(v_i, v_j), \left(P_b(v_i, v_j) \;=\; \prod P(v_i, v_j)\right), \tag{16}$$

where b is one of the several paths in which a fault propagates from node v_i to v_j. $P_b(v_i, v_j)$ is the fault propagation probability of the path. w is the number of paths from node v_i to v_j.

On the basis of Equation (16), it can be clarified whether the fault can propagate in the fault propagation model and then realize the determination of critical nodes and the fault propagation path. The process of the determination of critical nodes at each level is shown in Figure 4.

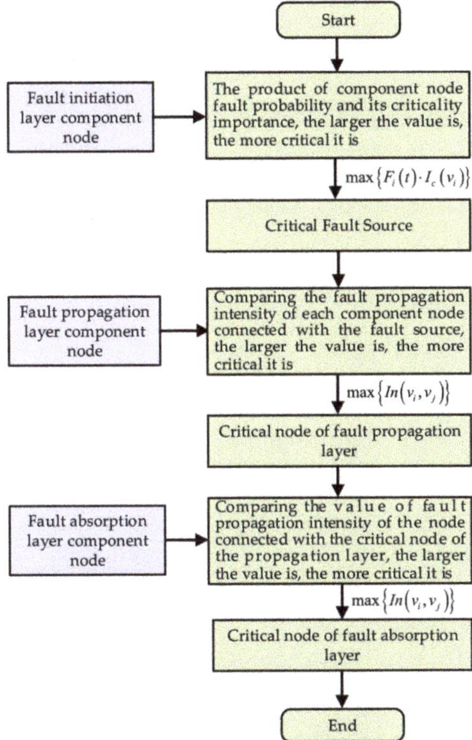

Figure 4. Flow chart expressing how to determine critical nodes at each level.

As illustrated in Figure 4, to determine the system components of the fault initiation layer, the product of component fault probability and its own criticality importance at time t is calculated; in turn, the larger the value, the more critical it is. The critical component of the fault initiation layer is considered as the fault source. To determine the system components of the fault propagation layer, according to the hierarchy model of the fault propagation intensity of a machining centre, the system components in the propagation layer connected with the fault source are determined at time *t* and the fault propagation intensity values of the system components connected with the fault source are calculated; in turn, the larger the value, the more critical it is. Similarly, the fault propagation intensity values of each system component connected with the critical nodes of the fault propagation layer in the fault absorption layer are calculated; in turn, the larger the value, the more critical it is and the critical nodes of the fault absorption layer are determined.

The path composed of critical nodes is the critical fault propagation path, according to which fault prevention and maintenance can be conducted.

4. Case Application of a Machining Centre

The fault propagation behaviour analysis method proposed in this paper has a certain general applicability and can provide guidance for the analysis of fault propagation behaviour of other complex systems. This paper selects the machining centre commonly used by various manufacturing companies as an example to illustrate the specific application.

4.1. Implementation

In this section, we take MDH series horizontal machining centres as the research object that are mainly used for processing rotary parts. We collected 108 on-site fault information details of 36 machining centres of this series during the course of one year. After fault analysis, we can determine whether each component fault is an independent fault or a related fault. If it is a related fault, the antecedent component that caused the component fault will be determined through fault analysis. For example, when the workpiece cannot be clamped, the tool magazine system (T) is the direct fault location but the root cause is that the workpiece cannot be clamped due to insufficient pressure in the hydraulic system (H). At this time, the faulty component is the tool magazine system (T) and its antecedent component is the hydraulic system (H). Considering the existence of this propagation fault, there is likely to be a directed arrow pointing from the hydraulic system (H) to the tool magazine system (T). Similarly, when the servo motor fails, the fault location is the feed system (F) but the root cause is the abnormality of the spindle system (B). Therefore, there is likely to be a directed arrow from the spindle system (B) to the feed system (F). In this manner, we can identify other related faults. The statistical analysis results of the related faults are shown in Table 1.

Table 1. Statistical analysis of the related faults in machining centre system components.

Fault Component	Antecedent Component	Frequency
T	B	3
F	B	2
B	NC	1
F	NC	1
T	H	1
B	H	1
B	D	1
NC	D	1
W	D	1
F	R	1
T	R	1
B	R	1
F	W	1
F	K	1
B	Q	2
T	Q	1

As expressed in Table 1, combined with the knowledge of graph theory, the fault propagation relationship of a machining centre is modelled. We do not consider components with uncorrelated faults such as the workbench (*U*). Thus, we can obtain the fault propagation digraph of a machining centre as shown in Figure 5.

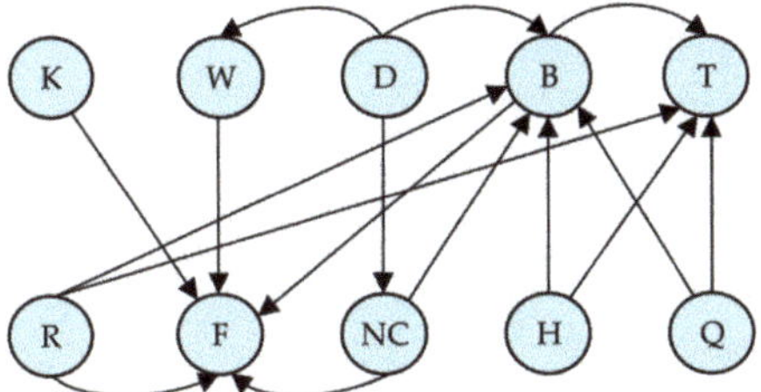

K—Chip removal system   R—Lubrication system

W—Cooling system   F—Feed system

D— Electrical system   NC—Numerical control system

B—Spindle system   H—Hydraulic system

T—Tool magazine   Q—Pneumatic system

Figure 5. Fault propagation digraph of a machining centre.

As expressed in the fault propagation digraph of a machining centre in Figure 5, the relationship matrix (*A*) and the design structure matrix (*M*) can be obtained as follows:

$$
A =
\begin{array}{c}
\ \\ B \\ T \\ F \\ NC \\ H \\ D \\ Q \\ R \\ W \\ K
\end{array}
\begin{array}{c}
\begin{array}{ccccccccccc}
B & T & F & NC & H & D & Q & R & W & K
\end{array} \\
\left[
\begin{array}{cccccccccc}
0 & 1 & 1 & 0 & 0 & 0 & 0 & 0 & 0 & 0 \\
0 & 0 & 0 & 0 & 0 & 0 & 0 & 0 & 0 & 0 \\
0 & 0 & 0 & 0 & 0 & 0 & 0 & 0 & 0 & 0 \\
1 & 0 & 1 & 0 & 0 & 0 & 0 & 0 & 0 & 0 \\
1 & 1 & 0 & 0 & 0 & 0 & 0 & 0 & 0 & 0 \\
1 & 0 & 0 & 1 & 0 & 0 & 0 & 0 & 1 & 0 \\
1 & 1 & 0 & 0 & 0 & 0 & 0 & 0 & 0 & 0 \\
1 & 1 & 1 & 0 & 0 & 0 & 0 & 0 & 0 & 0 \\
0 & 0 & 1 & 0 & 0 & 0 & 0 & 0 & 0 & 0 \\
0 & 0 & 1 & 0 & 0 & 0 & 0 & 0 & 0 & 0
\end{array}
\right]
\end{array}
$$

$$
M =
\begin{array}{c}
\ \\ B \\ T \\ F \\ NC \\ H \\ D \\ Q \\ R \\ W \\ K
\end{array}
\begin{array}{c}
\begin{array}{ccccccccccc}
B & T & F & NC & H & D & Q & R & W & K
\end{array} \\
\left[
\begin{array}{cccccccccc}
1 & 1 & 1 & 0 & 0 & 0 & 0 & 0 & 0 & 0 \\
0 & 1 & 0 & 0 & 0 & 0 & 0 & 0 & 0 & 0 \\
0 & 0 & 1 & 0 & 0 & 0 & 0 & 0 & 0 & 0 \\
1 & 1 & 1 & 1 & 0 & 0 & 0 & 0 & 0 & 0 \\
1 & 1 & 1 & 0 & 1 & 0 & 0 & 0 & 0 & 0 \\
1 & 1 & 0 & 1 & 0 & 1 & 0 & 0 & 1 & 0 \\
1 & 1 & 1 & 0 & 0 & 0 & 1 & 0 & 0 & 0 \\
1 & 1 & 1 & 0 & 0 & 0 & 0 & 1 & 0 & 0 \\
0 & 0 & 1 & 0 & 0 & 0 & 0 & 0 & 1 & 0 \\
0 & 0 & 1 & 0 & 0 & 0 & 0 & 0 & 0 & 1
\end{array}
\right]
\end{array}
$$

The element 1 in the matrix *M* indicates that the fault of the system component *i* impacts *j*. Contrarily, 0 indicates that no impact exists.

Thus, on the basis of the modelling process of the DSM fault propagation hierarchy structure model, the fault propagation DSM model of a machining centre can be obtained as shown in Figure 6. The diagonal elements in Figure 6 are represented by black squares.

	T	F	B	W	K	NC	H	Q	R	D
T	■									
F		■								
B	X	X	■							
W		X		■						
K		X			■					
NC		X	X			■				
H	X		X				■			
Q	X		X					■		
R	X	X	X						■	
D			X	X		X				■

Figure 6. The fault propagation DSM model of a machining centre.

On the basis of Figure 6, we can derive the result of module division after the DSM modeling processing: the first-layer system components are (T, F), the second-layer system components are (B, W, K), the third-layer system components are (NC, H, Q, R), and the fourth-layer system component is (D). The system components of the first layer, which is the appearance layer, are classified as the fault absorption layer in the fault propagation model. The system components of the fourth layer, the root layer, are classified as the fault initiation layer and the remaining layers are classified as the fault propagation layers. Thus, the fault propagation hierarchy structure model of a machining centre can be obtained as shown in Figure 7.

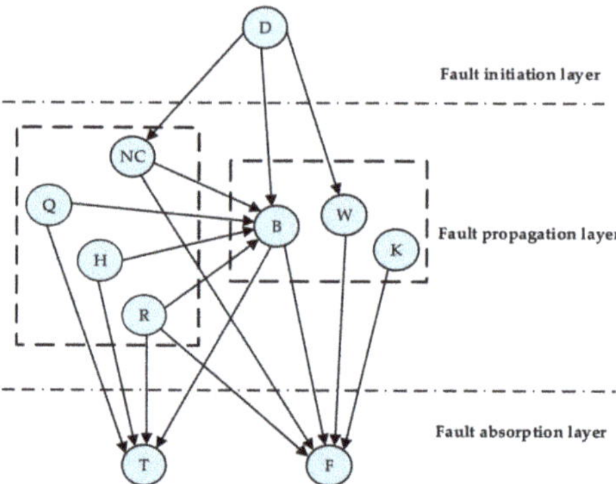

Figure 7. Hierarchy structure model of fault propagation in a machining centre.

On the basis of obtaining the fault information of a machining centre, we calculated the fault probability function of each component according to the calculation method presented in Section 2.2.1. Considering the impact of the timing truncation test and the fault time truncation on the sequence of the fault data, we use the Johnson method to modify it. The parameters of the fault probability model are then estimated and the distribution hypothesis test is passed. Finally, the fault probability function of each component can be obtained as shown in Table 2 and the function curve is illustrated in Figure 8.

Similarly, the fault probability function of a whole machining centre is $F_Z(t) = 1 - \exp\left[-(t/620.984)^{0.951}\right]$.

According to Table 2 and Equations (9) and (10), the probability importance and criticality importance of each system component node can be obtained at any time. The function curves of probability importance and criticality importance of each system component node are illustrated in Figures 9 and 10, respectively.

Table 2. Fault probability function of system components in a machining centre.

Component Code	Fault Probability Function
T	$F_T(t) = 1 - \exp\left[-(t/2791.105)^{1.015}\right]$
B	$F_B(t) = 1 - \exp\left[-(t/7832.664)^{0.641}\right]$
R	$F_R(t) = 1 - \exp\left[-(t/8161.512)^{0.791}\right]$
H	$F_H(t) = 1 - \exp\left[-(t/8905.062)^{0.656}\right]$
NC	$F_{NC}(t) = 1 - \exp\left[-(t/7912.399)^{0.542}\right]$
D	$F_D(t) = 1 - \exp\left[-(t/7935.696)^{0.771}\right]$
Q	$F_Q(t) = 1 - \exp\left[-(t/27503.400)^{0.524}\right]$
F	$F_F(t) = 1 - \exp\left[-(t/2606.440)^{1.295}\right]$
W	$F_W(t) = 1 - \exp\left[-(t/4145.720)^{0.910}\right]$
K	$F_K(t) = 1 - \exp\left[-(t/5960.776)^{0.931}\right]$
U	$F_U(t) = 1 - \exp\left[-(t/6869.766)^{0.773}\right]$

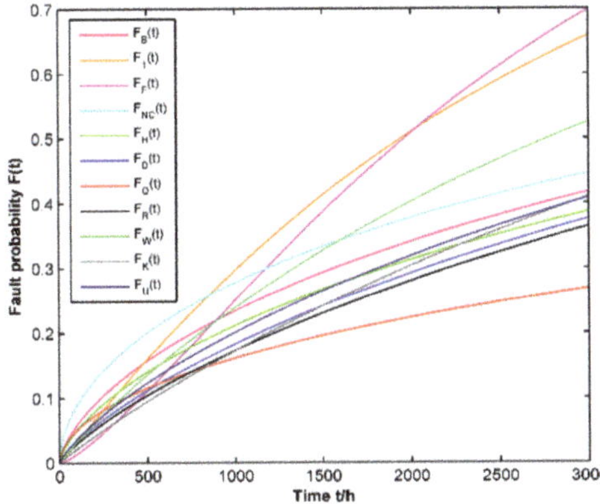

Figure 8. Fault probability function curve of machining centre system components.

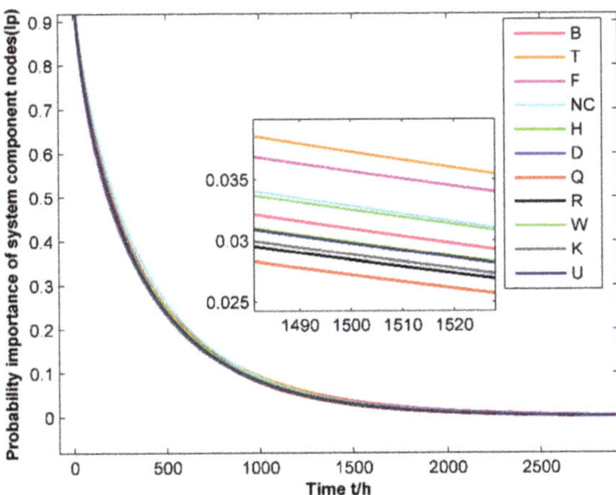

Figure 9. Probability importance function curve of system component nodes.

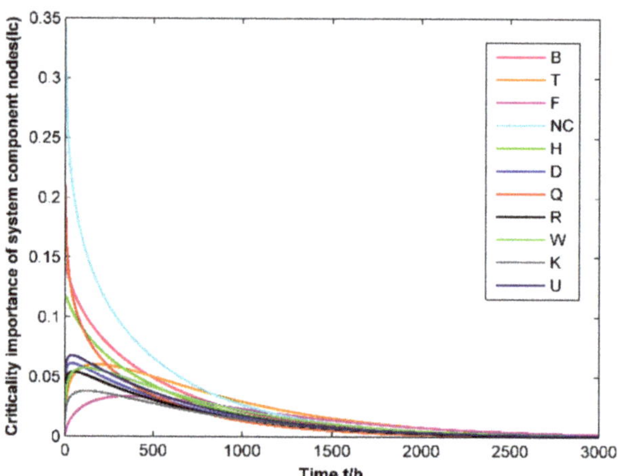

Figure 10. Criticality importance function curve of system component nodes.

Given that the values of probability and the criticality importance of system components are varied, this paper takes t = 1500 h as an example to illustrate this concept. By substituting the fault probability values of system components obtained from Table 2 into Equations (9) and (10), the probability importance and criticality importance of each system component node at t = 1500 h can be obtained as shown in Table 3.

Table 3. Probability importance and criticality importance of each system component at 1500 h.

Component Code	$I_P(v_i)$	$I_C(v_i)$
B	0.1399	0.0455
T	0.1669	0.0754
K	0.1613	0.0692
NC	0.1485	0.0550
H	0.1350	0.0400
D	0.1305	0.0350
Q	0.1230	0.0267
R	0.1285	0.0329
W	0.1471	0.0534
K	0.1305	0.0350

On the basis of Equations (10) and (11), combined with Figure 7, the edge fault influence value of the same level at 1500 h can be calculated as shown in Table 4.

Table 4. Fault influence degree of the edge at the same level at 1500 h.

Directed Edge	$I\left(v_i,v_j\right)$
$E(v_R,v_B)$	0.9870
$E(v_H,v_B)$	0.9876
$E(v_Q,v_B)$	0.9808
$E(v_{NC},v_B)$	0.9914

In reference to Equation (12), Table 3, and Figure 7, the edge fault influence value between different levels in the fault propagation hierarchy structure model of a machining centre at 1500 h is calculated. The results are detailed in Table 5.

Table 5. Fault influence degree of the edge between different levels at 1500 h.

Directed Edge	$I\left(v_i,v_j\right)$	Directed Edge	$I\left(v_i,v_j\right)$
$E(v_D,v_B)$	0.9915	$E(v_B,v_T)$	0.9660
$E(v_R,v_T)$	0.9153	$E(v_B,v_F)$	0.9783
$E(v_D,v_{NC})$	0.9751	$E(v_R,v_F)$	0.9344
$E(v_D,v_W)$	0.9781	$E(v_H,v_T)$	0.9484
$E(v_W,v_F)$	0.9917	$E(v_K,v_F)$	0.9446
$E(v_{NC},v_F)$	0.9934	$E(v_Q,v_T)$	0.8741

On the basis of Tables 4 and 5, the 1500 h fault propagation hierarchy model of a machining centre based on fault influence degree can be obtained as shown in Figure 11.

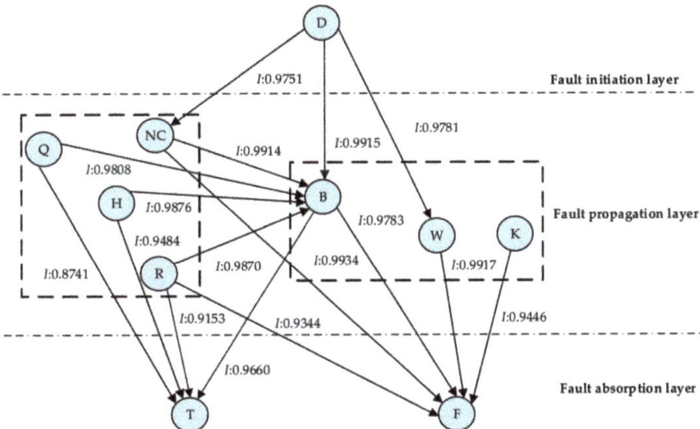

Figure 11. Hierarchy structure model of fault propagation of a machining centre based on the fault influence degree.

By substituting the data in Tables 2, 4 and 5 into Equation (13), the fault propagation probability of each system component at any time can be obtained. In addition, the calculation results of the 1500 h example are presented in Table 6.

Table 6. Fault propagation probability of the directed edge at 1500 h.

Directed Edge	$P\left(v_i,v_j\right)$	Directed Edge	$P\left(v_i,v_j\right)$
$E(v_R,v_B)$	0.3308	$E(v_Q,v_B)$	0.2397
$E(v_D,v_B)$	0.3246	$E(v_B,v_T)$	0.2639
$E(v_{NC},v_B)$	0.3315	$E(v_B,v_F)$	0.2358
$E(v_H,v_B)$	0.2830	$E(v_R,v_T)$	0.2365
$E(v_D,v_{NC})$	0.2866	$E(v_R,v_F)$	0.2534
$E(v_D,v_W)$	0.2274	$E(v_H,v_T)$	0.2284
$E(v_W,v_F)$	0.2109	$E(v_K,v_F)$	0.1919
$E(v_{NC},v_F)$	0.2153	$E(v_Q,v_T)$	0.1711

On the basis of the fault propagation hierarchy model of a machining centre in Figure 7, all the paths between any two nodes in the fault propagation digraph can be obtained and the paths are listed in Table 7. In reference to Equation (14) and Table 7, the edge betweenness of the fault propagation hierarchy model is calculated and the results are presented in Table 8.

Table 7. Statistical results of any two node paths.

Any Two Nodes	Paths
D→F	D→NC→F, D→B→F, D→W→F, D→NC→B→F
D→B	D→B, D→NC→B
NC→F	NC→F, NC→B→F
D→T	D→NC→B→T, D→B→T
Q→F	Q→B→F
Q→T	Q→T, Q→B→T
R→T	R→T, R→B→T
R→F	R→F, R→B→F
H→T	H→T, H→B→T
D→W	D→W
Q→B	Q→B
D→NC	D→NC
H→B	H→B
NC→B	NC→B
NC→T	NC→B→T
H→F	H→B→F
R→B	R→B
B→T	B→T
W→T	W→T
K→F	K→F
B→F	B→F

Table 8. The edge betweenness of the fault propagation hierarchy model of a machining centre.

Directed Edge	$L\left(v_i,v_j\right)$	Directed Edge	$L\left(v_i,v_j\right)$
$E(v_R,v_B)$	1	$E(v_Q,v_B)$	1.5
$E(v_D,v_B)$	0.75	$E(v_B,v_T)$	3.5
$E(v_{NC},v_B)$	2.75	$E(v_B,v_F)$	3.5
$E(v_H,v_B)$	1.5	$E(v_R,v_T)$	0
$E(v_D,v_{NC})$	1.5	$E(v_R,v_F)$	0
$E(v_D,v_W)$	0.25	$E(v_H,v_T)$	0
$E(v_W,v_F)$	0.25	$E(v_K,v_F)$	0
$E(v_{NC},v_F)$	0.25	$E(v_Q,v_T)$	0

Table 8 reveals that the edge betweenness of different directed edges differs. Therefore, the influence of the structural characteristics of the model on the fault propagation also must be considered. The larger the edge betweenness value is, the more important the edge is in the fault propagation of a whole machine and greater consideration is required when the probability of being selected as the fault propagation path increases.

Therefore, on the basis of Equation (15), Tables 6 and 8, the fault propagation intensity of each directed edge at 1500 h can be obtained as shown in Table 9.

Table 9. Fault propagation intensity of each directed edge of a machining centre fault propagation model at 1500 h.

Directed Edge	$In\left(v_i,v_j\right)$	Directed Edge	$In\left(v_i,v_j\right)$
$E(v_R,v_B)$	0.1396	$E(v_Q,v_B)$	0.0775
$E(v_D,v_B)$	0.1400	$E(v_B,v_T)$	0.0314
$E(v_{NC},v_B)$	0.1231	$E(v_B,v_F)$	0.0581
$E(v_H,v_B)$	0.0740	$E(v_R,v_T)$	0.0262
$E(v_D,v_{NC})$	0.0521	$E(v_R,v_F)$	0.0267
$E(v_D,v_W)$	0.0486	$E(v_H,v_T)$	0.0368
$E(v_W,v_F)$	0.0686	$E(v_K,v_F)$	0.0477
$E(v_{NC},v_F)$	0.0212	$E(v_Q,v_T)$	0.0283

In reference to Figure 11 and Table 9, a hierarchy structure model of the fault propagation intensity of a machining centre can be obtained as shown in Figure 12.

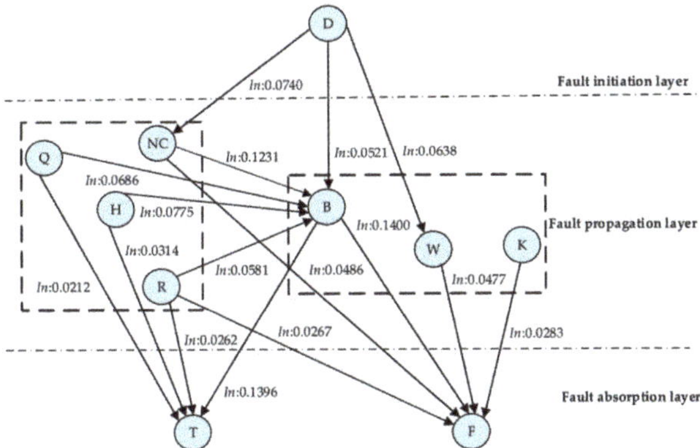

Figure 12. Hierarchy structure model of fault propagation intensity in a machining centre.

In reference to Equation (16) and Figure 11, the fault propagation probability values of each path in the fault propagation hierarchy model of a machining centre at 1500 h can be calculated as shown in Table 10.

Table 10. The fault propagation probability value of each path in the fault propagation model of a machining centre at 1500 h.

Paths	$P_E\left(v_i,v_j\right)$
D→NC→F	0.004799
D→NC→B→F	0.000342
D→NC→B→T	0.000338
D→B→F	0.004219
D→B→T	0.004178
D→W→F	0.004714
Q→T	0.042622
Q→B→F	0.003326
Q→B→T	0.003293
H→T	0.063015
H→B→T	0.004647
H→B→F	0.004692
R→T	0.052492
R→F	0.053345
R→B→T	0.003963
R→B→F	0.004001
K→F	0.056592

As expressed in Table 10, the fault propagation probability of each path is greater than the threshold value of 10^{-8}; thus, a fault propagation phenomenon exists in the model.

Based on Table 9 and the hierarchy structure model of fault propagation in a machining centre, for the fault initiation layer there is only the electrical system component D. Hence, the critical node of the fault initiation layer is D. For the fault propagation layer at 1500 h, $In(v_D, v_{NC}) = 0.0740 > In(v_D, v_B) = 0.0521 > In(v_D, v_W) = 0.0368$, thus the critical node currently is NC. This indicates that the fault is more likely to be transferred from the electrical system to the numerical control system. Given that $In(v_{NC}, v_B) = 0.1400 > In(v_{NC}, v_F) = 0.1396$, the critical node is B. Similarly, for the fault absorption layer system component node, $In(v_B, v_F) > In(v_B, v_T)$, thus the feed system F is the critical component of this layer. Therefore at 1500 h, the critical nodes are D, NC, B, and F, and the path composed of critical nodes is the critical fault propagation path. At 1500 h, the most likely propagation path of the fault is D→NC→B→F.

Similarly, the fault propagation intensity value of each directed edge of a machining centre at any time can be determined; in turn, the critical fault nodes and paths of a machining centre at any time can be obtained. This paper arbitrarily chooses the running time of 200 h and 5000 h as examples and draws the comparison diagram of the fault propagation intensity value of each directed edge as shown in Figure 13.

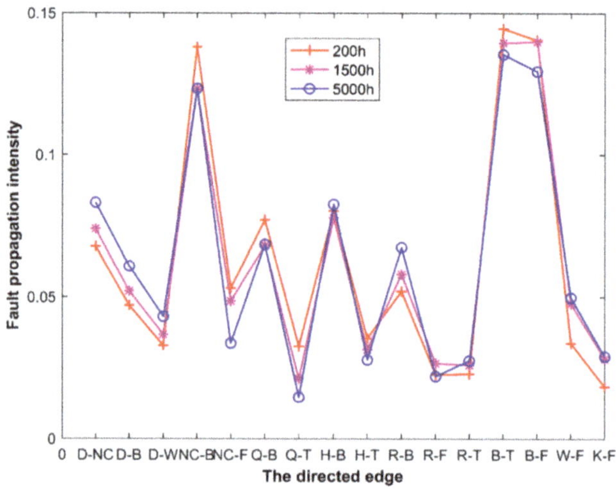

Figure 13. Comparison diagram of the fault propagation intensity of each directed edge at different times.

In reference to Figure 13, the fault propagation intensity values of each directed edge are varied at different times and the critical fault propagation paths of the machining centre at different times are distinct. The critical fault propagation path of 200 h is D→NC→B→T, at 1500 h the critical fault propagation path is D→NC→B→F, and at 5000 h the critical fault propagation path is D→NC→B→T.

4.2. Comparison Analysis

The proposed method in this paper is compared with the importance evaluation method proposed in Reference [49]. When evaluating the importance of machine tool system components, the method in Reference [49] only evaluated the importance of components from the perspective of the fault propagation mechanism and did not consider the structural characteristics of the model. Combining the application examples in this paper, when the model structure characteristics are not considered and only the fault propagation mechanism is considered, the fault propagation probability value of each directed edge at 1500 h can be calculated according to Equations (10)–(13). Combined with the fault propagation hierarchy structure model of the machining centre in Figure 6, the fault propagation hierarchy structure model of the machining centre based on the fault propagation probability is drawn as shown in Figure 14. The model only considered the fault propagation mechanism of the machining centre.

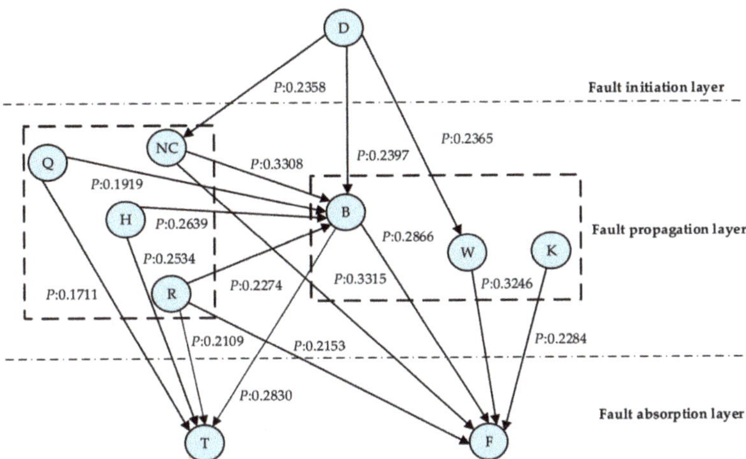

Figure 14. Hierarchy structure model of the machining centre based on the fault propagation probability.

As expressed in Figure 14 at 1500 h, the fault source of the machining centre is component D. The fault will propagate along the components with a high probability of fault propagation. As $P(v_D, v_B) = 0.2397 > P(v_D, v_W) = 0.2365 > P(v_D, v_{NC}) = 0.2358$, the fault will preferentially propagate along D→B and because $P(v_B, v_F) = 0.2866 > P(v_B, v_T) = 0.2830$, the fault is more likely to propagate along B→F. The critical fault propagation path at this time is D→B→F and the critical nodes are D, B, and F. That is, at 1500 h, the fault is most likely to be transmitted as such: electrical system→spindle system→feeding system. However, according to the method proposed in this paper, the result of calculation and analysis is D→NC→B→F. There is a certain difference in the critical fault propagation path obtained by the two methods that is mainly because the method based on the fault propagation probability does not consider the structural characteristics of the model. The component NC plays an important role in the fault propagation structure model; when it fails, it will have a greater impact on the entire system, thus requiring attention. The influence of the structural characteristics of the model on the propagation of faults cannot be ignored. Therefore, the method proposed in this paper is more conducive to the analysis of the fault propagation behaviour of a machining centre and the analysis result is more reasonable.

5. Discussion

In reference to Tables 6 and 8, the edge betweenness of the edge with a high probability of fault propagation is not necessarily large. For example, the fault propagation probability of the directed edge $E(v_{NC}, v_B)$ is the largest but its edge betweenness is not the largest. The edges with the largest betweenness are $E(v_B, v_T)$ and $E(v_B, v_F)$. Through comparative analysis of these examples, we can determine there is likely to be a certain deviation in the analysis of the fault propagation behaviour of a machining centre based on a single index. Therefore, the fault propagation mechanism and the structural characteristics of the model should be integrated to identify the critical fault nodes and critical fault propagation paths.

This paper takes the machining centre running for 1500 h as an example to explain the proposed method. We determine that the critical fault propagation path of a machining centre at 1500 h is D→NC→B→F. Therefore, for a machining centre of the same model, when the running time is 1500 h, the electrical system (D), numerical control system (NC), spindle system (B), and feed system (F) must be considered. These system components are more likely to fail and measures can be taken in advance to avoid faults.

Figure 13 illustrates that the critical system components and critical fault propagation paths are different at different stages of a machining centre's operation. Therefore, according to the fault propagation and evolution law of each system component of a machining centre, staff can make corresponding adjustments to the degree of attention necessary for the components of a machining centre during different stages of operation and can formulate appropriate fault prevention strategies.

6. Conclusions

This paper presents a dynamic analysis method of fault propagation behaviour of machining centres that can identify the critical fault propagation paths and nodes of a machining centre at any time. On this basis, fault warning and preventive maintenance can be conducted in a targeted manner, thereby reducing the economic loss and safety hazards of manufacturing enterprises due to equipment fault.

The method proposed in this paper mainly embodies the following advantages:

(1)   The DSM-based fault propagation hierarchy structure model of a machining centre established in this paper can more clearly demonstrate the relationship between the system components in the form of a design structure matrix.

(2)   There are certain differences in the influence degree of the edge between the components at the same level and the different levels. By introducing the Copula and the Coupling degree functions, the fault influence degree of edges between the same level and different levels are calculated, respectively. In this way, it is possible to more accurately measure the impact of faults between components.

(3)   Considering the structural characteristics of the model and the fault mechanism of the system, a fault propagation intensity model of a machining centre is constructed and a quantitative description of the severity of fault propagation on the components is realized. According to the fault propagation intensity of the components, the critical fault propagation paths and nodes of a machining centre can be identified, can provide a reference for the fault maintenance, and encourage reliability growth of machining centres.

This paper demonstrates the effectiveness and practicability of the proposed method through the application of the specific case. In the future, the proposed fault propagation behaviour analysis method can be extended to other complex electromechanical products through sensor technology, rather than remaining at the level of theoretical guidance.

Author Contributions: Conceptualization, L.M. and Y.Z.; methodology, L.M.; software, J.L.; validation, L.M. and Y.Z.; formal analysis, J.L.; investigation, L.M. and J.S.; resources, Y.Z.; data curation, Y.Z.; writing—original draft preparation, L.M.; writing—review and editing, L.M.; visualization, Y.Z. and F.Z.; supervision, J.L.; project administration, Y.Z.; funding acquisition, Y.Z. and L.M. All authors have read and agreed to the published version of the manuscript.

Funding: This research was funded by the Jilin Province Science and Technology Development Plan Project (Grant No. 20190302104GX), the Jilin Provincial Natural Science Foundation (Grant No. 20170101212JC), and the project 101832020DJX037 was supported by the Ph.D. Graduate Interdisciplinary Research of Jilin University.

Institutional Review Board Statement: Not applicable.

Informed Consent Statement: Not applicable.

Data Availability Statement: Not applicable.

Conflicts of Interest: The authors declare no conflict of interest.

References

1.   Li, X.F. Research on the status QUO and development trend of numerical control technology. In Proceedings of the 2018 International Conference on Computational Science and Engineering (ICCSE 2018), Qingdao, China, 3–4 November 2018.
2.   Martinova, L.I.; Martinov, G.M. Prospects for CNC machine tools. *Russ. Eng. Res.* **2019**, *39*, 1080–1083. [CrossRef]
3.   Xing, L.D.; Zhao, G.L.; Wang, Y.J.; Xiang, Y.S. Reliability modelling of correlated competitions and dependent components with random failure propagation time. *Qual. Reliab. Eng. Int.* **2020**, *36*, 947–964. [CrossRef]

4. Peukert, B.W.; Archenti, A. Dynamic interaction between precision machine tools and their foundations. *Int. J. Autom. Technol.* **2020**, *14*, 386–398. [CrossRef]

5. Gao, J.Z.; Yin, Y.F.; Fiondella, L.; Liu, L.J. Recovery of coupled networks after cascading failures. *J. Syst. Eng. Electron.* **2018**, *29*, 650–657.

6. Zhang, Y.Z.; Mu, L.M.; Shen, G.X.; Yu, Y.; Han, C.Y. Fault diagnosis strategy of CNC machine tools based on cascading failure. *J. Intell. Manuf.* **2019**, *30*, 2193–2202. [CrossRef]

7. Sheng, B.; Deng, C.; Xiong, Y.; Luo, Z.J.; Wang, Y.H. Fault diagnosis for CNC machine tool based on mapping model. *Appl. Mech. Mater.* **2014**, *607*, 739–742. [CrossRef]

8. Gao, Z.W.; Cecati, C.; Ding, S.X. A survey of fault diagnosis and fault-tolerant techniques-part I: Fault diagnosis with model-based and signal-based approaches. *IEEE Trans. Ind. Electron.* **2015**, *62*, 3757–3767. [CrossRef]

9. Gao, Z.W.; Cecati, C.; Ding, S.X. A survey of fault diagnosis and fault-tolerant techniques—part II: Fault diagnosis with knowledge-based and hybrid/active approaches. *IEEE Trans. Ind. Electron.* **2015**, *62*, 3768–3774. [CrossRef]

10. Feng, J.; Yao, Y.; Lu, S.; Liu, Y. Domain knowledge-based deep-broad learning framework for fault diagnosis. *IEEE Trans. Ind. Electron.* **2021**, *68*, 3454–3464. [CrossRef]

11. Kaluder, S.; Fekete, K.; Jozsa, L.; Klaic, Z. Fault diagnosis and identification in the distribution network using the fuzzy expert system. *Eksploat. Niezawodn. Maint. Reliab.* **2018**, *20*, 621–629. [CrossRef]

12. Zhou, Q.; Yan, P.; Liu, H.; Xin, Y.; Chen, Y. Research on a configurable method for fault diagnosis knowledge of machine tools and its application. *Int. J. Adv. Manuf. Technol.* **2018**, *95*, 937–960. [CrossRef]

13. Elimelech, O.; Stern, R.; Kalech, M. Structural abstraction for model-based diagnosis with a strong fault model. *Knowl. Based Syst.* **2018**, *161*, 357–374. [CrossRef]

14. Kouadri, A.; Hajji, M.; Harkat, M.-F.; Abodayeh, K.; Mansouri, M.; Nounou, H.; Nounou, M. Hidden markov model based principal component analysis for intelligent fault diagnosis of wind energy converter systems. *Renew. Energy* **2020**, *150*, 598–606. [CrossRef]

15. Song, W.L.; Xiang, J.W.; Zhong, Y.T. A simulation model based fault diagnosis method for bearings. *J. Intell. Fuzzy Syst.* **2018**, *34*, 3857–3867. [CrossRef]

16. Karabacak, Y.E.; Ozmen, N.G.; Gumusel, L. Worm gear condition monitoring and fault detection from thermal images via deep learning method. *Eksploat. Niezawodn. Maint. Reliab.* **2020**, *22*, 544–556.

17. Liang, L.; Wen, H.B.; Liu, F.; Li, G.; Li, M.L. Feature extraction of impulse faults for vibration signals based on sparse non-negative tensor factorization. *Appl. Sci.* **2019**, *9*, 3642. [CrossRef]

18. Nayana, B.R.; Geethanjali, P. Analysis of statistical time-domain features effectiveness in identification of bearing faults from vibration signal. *IEEE Sens. J.* **2017**, *17*, 5618–5625. [CrossRef]

19. Jung, D.; Ng, K.Y.; Frisk, E.; Krysander, M. Combining model-based diagnosis and data-driven anomaly classifiers for fault isolation. *Control. Eng. Pract.* **2018**, *80*, 146–156. [CrossRef]

20. Slimani, A.; Ribot, P.; Chanthery, E.; Rachedi, N. Fusion of model-based and data-based fault diagnosis approaches. *IFAC-PapersOnLine* **2018**, *51*, 1205–1211. [CrossRef]

21. Tabaszewski, M.; Szymanski, G.M. Engine valve clearance diagnostics based on vibration signals and machine learning methods. *Eksploat. Niezawodn. Maint. Reliab.* **2020**, *22*, 331–339. [CrossRef]

22. Zuber, N.; Bajric, R. Gearbox faults feature selection and severity classification using machine learning. *Eksploat. Niezawodn. Maint. Reliab.* **2020**, *22*, 748–756. [CrossRef]

23. Wu, Z.; Zhang, Q.; Cheng, L.F.; Tan, S.Y. A new method of two-stage planetary gearbox fault detection based on multi-sensor information fusion. *Appl. Sci.* **2019**, *9*, 5443. [CrossRef]

24. Gangsar, P.; Tiwari, R. Signal based condition monitoring techniques for fault detection and diagnosis of induction motors: A state-of-the-art review. *Mech. Syst. Signal. Process.* **2020**, *144*, 106908. [CrossRef]

25. Cho, K.-H.; Jo, H.-C.; Kim, E.-S.; Park, H.-A.; Park, J.H. Failure diagnosis method of photovoltaic generator using support vector machine. *J. Electr. Eng. Technol.* **2020**, *15*, 1669–1680. [CrossRef]

26. Goyal, D.; Choudhary, A.; Pabla, B.S.; Dhami, S.S. Support vector machines based non-contact fault diagnosis system for bearings. *J. Intell. Manuf.* **2020**, *31*, 1275–1289. [CrossRef]

27. Piltan, F.; Prosvirin, A.E.; Sohaib, M.; Saldivar, B.; Kim, J.M. An SVM-based neural adaptive variable structure observer for fault diagnosis and fault-tolerant control of a robot manipulator. *Appl. Sci.* **2020**, *10*, 1344. [CrossRef]

28. Gunerkar, R.S.; Jalan, A.K.; Belgamwar, S.U. Fault diagnosis of rolling element bearing based on artificial neural network. *J. Mech. Sci. Technol.* **2019**, *33*, 505–511. [CrossRef]

29. Marmouch, S.; Aroui, T.; Koubaa, Y. Statistical neural networks for induction machine fault diagnosis and features processing based on principal component analysis. *IEEJ Trans. Electr. Electron. Eng.* **2021**, *16*, 307–314. [CrossRef]

30. Wang, B.; Ke, H.W.; Ma, X.D.; Yu, B. Fault diagnosis method for engine control system based on probabilistic neural network and support vector machine. *Appl. Sci.* **2019**, *9*, 4122. [CrossRef]

31. Ruijters, E.; Stoelinga, M. Fault tree analysis: A survey of the state-of-the-art in modelling, analysis and tools. *Comput. Sci. Rev.* **2015**, *15–16*, 29–62. [CrossRef]

32. Chen, X. Research on fault early warning and the diagnosis of machine tools based on energy fault tree analysis. *Proc. Inst. Mech. Eng. Part B J. Eng. Manuf.* **2019**, *233*, 2147–2159. [CrossRef]

33. Adler, C.O.; Dagli, C.H. A study of the effect of basic network characteristics on system-of-system failure propagation. *Procedia Comput. Sci.* **2014**, *36*, 345–352. [CrossRef]
34. Jin, L.; Wang, X.J.; Zhang, Y.; You, J.W. Cascading failure in multilayer networks with dynamic dependency groups. *Chin. Phys. B* **2018**, *27*, 737–744. [CrossRef]
35. Maza, S. Diagnosis modelling for dependability assessment of fault-tolerant systems based on stochastic activity networks. *Qual. Reliab. Eng. Int.* **2015**, *31*, 963–976. [CrossRef]
36. Zakari, A.; Lee, S.P.; Chong, C.Y. Simultaneous localization of software faults based on complex network theory. *IEEE Access* **2018**, *6*, 23990–24002. [CrossRef]
37. Browning, T.R. Design structure matrix extensions and innovations: A survey and new opportunities. *IEEE Trans. Eng. Manag.* **2016**, *63*, 27–52. [CrossRef]
38. Son, H.; Kwon, Y.; Park, S.C.; Lee, S. Using a design structure matrix to support technology roadmapping for product-service systems. *Technol. Anal. Strateg. Manag.* **2018**, *30*, 337–350. [CrossRef]
39. Li, Y.F.; Zhang, Y.J.; Dai, B.C.; Zhang, L. Dynamic importance analysis of components with known failure contribution of complex systems. *Math. Probl. Eng.* **2018**, *2018*, 8534065. [CrossRef]
40. Miziuła, P.; Navarro, J. Birnbaum importance measure for reliability systems with dependent components. *IEEE Trans. Reliab.* **2019**, *68*, 439–450. [CrossRef]
41. Peng, H.; Coit, D.W.; Feng, Q.M. Component reliability criticality or importance measures for systems with degrading components. *IEEE Trans. Reliab.* **2012**, *61*, 4–12. [CrossRef]
42. Hosseini Nodeh, Z.; Babapour Azar, A.; Khanjani Shiraz, R.; Khodayifar, S.; Pardalos, P.M. Joint chance constrained shortest path problem with Copula theory. *J. Comb. Optim.* **2020**, *40*, 110–140. [CrossRef]
43. Montes, I.; Miranda, E.; Palessoni, R.; Vicig, P. Sklar's theorem in an imprecise setting. *Fuzzy Sets Syst.* **2015**, *278*, 48–66. [CrossRef]
44. Peng, W.; Zhang, X.L.; Huang, H.Z. A failure rate interaction model for two-component systems based on copula function. *Proc. Inst. Mech. Eng. Part O J. Risk Reliab.* **2016**, *230*, 278–284. [CrossRef]
45. Neshat, M.; Sepidnam, G.; Sargolzaei, M.; Toosi, A.N. Artificial fish swarm algorithm: A survey of the state-of-the-art, hybridization, combinatorial and indicative applications. *Artif. Intell. Rev.* **2014**, *42*, 965–997. [CrossRef]
46. Fang, C.L.; Liu, H.M.; Li, G.D. International progress and evaluation on interactive coupling effects between urbanization and the eco-environment. *J. Geogr. Sci.* **2016**, *26*, 1081–1116. [CrossRef]
47. Shahrivari Joghan, H.; Bagheri, A.; Azad, M. Weighted label propagation based on local edge betweenness. *J. Supercomput.* **2019**, *75*, 8094–8114. [CrossRef]
48. Guo, L. Research on Failure Propagation Mechanism for Machine Integration System. Master's Thesis, Beijing Jiaotong University, Beijing, China, 2015.
49. Zhang, Y.Z.; Liang, S.B.; Liu, J.L.; Cao, P.L.; Luan, L. Evaluation for machine tool components importance based on improved leaderRank. *Proc. Inst. Mech. Eng. Part O J. Risk Reliab.* **2021**, *235*, 331–337.

applied sciences

Article

A Hybrid Multi-Objective Evolutionary Algorithm-Based Semantic Foundation for Sustainable Distributed Manufacturing Systems

Veera Babu Ramakurthi [1], V. K. Manupati [1], José Machado [2,*] and Leonilde Varela [3]

[1] Department of Mechanical Engineering, National Institute of Technology, Warangal 506004, Telangana, India; rvbabu.nitw@gmail.com (V.B.R.); manupativijay@gmail.com (V.K.M.)
[2] MEtRICs Research Center, Mechanical Engineering Department, Campus of Azurém, University of Minho, 4800-058 Guimarães, Portugal
[3] Algoritmi Research Centre, Department of Production and Systems, Campus of Azurém, University of Minho, 4800-058 Guimarães, Portugal; leonilde@dps.uminho.pt
* Correspondence: jmachado@dem.uminho.pt

Abstract: Rising energy prices, increasing maintenance costs, and strict environmental regimes have augmented the already existing pressure on the contemporary manufacturing environment. Although the decentralization of supply chain has led to rapid advancements in manufacturing systems, finding an efficient supplier simultaneously from the pool of available ones as per customer requirement and enhancing the process planning and scheduling functions are the predominant approaches still needed to be addressed. Therefore, this paper aims to address this issue by considering a set of gear manufacturing industries located across India as a case study. An integrated classifier-assisted evolutionary multi-objective evolutionary approach is proposed for solving the objectives of makespan, energy consumption, and increased service utilization rate, interoperability, and reliability. To execute the approach initially, text-mining-based supervised machine-learning models, namely Decision Tree, Naïve Bayes, Random Forest, and Support Vector Machines (SVM) were adopted for the classification of suppliers into task-specific suppliers. Following this, with the identified suppliers as input, the problem was formulated as a multi-objective Mixed-Integer Linear Programming (MILP) model. We then proposed a Hybrid Multi-Objective Moth Flame Optimization algorithm (HMFO) to optimize process planning and scheduling functions. Numerical experiments have been carried out with the formulated problem for 10 different instances, along with a comparison of the results with a Non-Dominated Sorting Genetic Algorithm (NSGA-II) to illustrate the feasibility of the approach.

Keywords: text mining; network-based distributed manufacturing systems; moth flame optimization algorithm; support vector machines; Naive Bayes; random forest; decision trees; supplier classification

Citation: Ramakurthi, V.B.; Manupati, V.K.; Machado, J.; Varela, L. A Hybrid Multi-Objective Evolutionary Algorithm-Based Semantic Foundation for Sustainable Distributed Manufacturing Systems. *Appl. Sci.* **2021**, *11*, 6314. https://doi.org/10.3390/app11146314

Academic Editor: Luis M. Camarinha-Matos

Received: 19 May 2021
Accepted: 29 June 2021
Published: 8 July 2021

1. Introduction

Increasing competition, coupled with advancing computing technologies and the advent of decentralization in the supply chain, has led to the attainment of a shorter product life cycle, reducing production costs and responding to customer demands with greater flexibility. Thus, manufacturing units are now leaning toward a distributed manufacturing environment far from the traditional approach of promptly manufacturing products [1]. This involves multiple processes consisting of classification of manufacturing units, assignment of tasks as per product category on the basis of requirements, and information exchange within various units of an enterprise and between firms. All these together represent parameters of a compound scenario needed to be refined. In this paper, the implications of the proposed classification and optimization-/simulation-based integrated approach for the considered system is presented.

Managing supplier relationships and estimating the level of risk involved with various categories of suppliers, their capabilities, core services, constraints, target industries, and customers are some of the parameters for classification and selection. Dealing with selection of suppliers on such a large scale becomes tedious; thus, it is required to subsume advanced techniques for supplier classification. This research incorporates text mining based on supervised machine-learning models as one of the approaches for a better resolution of the aforementioned supplier classification problem. Text mining is the operation through which information from unstructured text documents is extracted by devising non-trivial patterns and trends through statistical pattern learning. It contains various steps, viz., pre-processing, structuring of the input text data, extracting patterns, and classification of information from various sources into a pre-defined genre [2]. Irrespective of any kind of manufacturing system, the prominent functions are process planning and scheduling. The former includes raw materials, semi-finished products, machine tools, and process information. The latter includes the allocation of resources over a set of constraints for manufacturing various entities [3]. The delay of time between planning and execution phase demands the modification of process plans. In other words, it is estimated that 20–30 percent of the already-generated process plans has to be modified in a given life cycle as a result of sequential processing of planning and scheduling in the existing systems [4].

This study seeks to address the following questions:

- How can supervised machine-learning models be employed to find an efficient method for supplier classification, categorizing suppliers based on specific tasks?
- What type of mathematical model can be developed considering the optimization of various conflicting objectives such as completion time, energy consumption, interoperability, machine utilization rate, service utilization, and utility?
- In what way can evolutionary algorithms be utilized to optimize scheduling and planning?
- What are the benefits of the proposed approach on the considered problem, and how would these effects influence the manufacturing system in a real-time environment?
- How can the effectiveness of the proposed Hybridized evolutionary (HMFO) algorithm be validated?

In this paper, the context of the problem considered is a distributed manufacturing environment where different enterprises are geographically distributed, in which coordination and collaboration among such enterprises for mutual exchange of information without any significant loss of data represent a challenge. Motivated by the nature of the problem and the factors considered above, this study, in a distributed network manufacturing system, pursues the classification, coordination, and communication to optimize scheduling and planning. This paper introduces three different steps which have been duly scrutinized to achieve the desired optimization of parameters, i.e., makespan, energy consumption, machine utilization rate, and reliability of services. Initially, a supplier classification problem is explored under a manufacturing setup in which the suppliers are classified into task specific suppliers using text mining. Thereafter, a many-objective mathematical model is developed to achieve the above-mentioned competing objectives. We then compared the effectiveness of the above propounded approach by equating the outcomes with the proposed multi-objective Hybrid Moth Flame Optimization (HMFO) algorithm with another benchmark algorithm such as NSGA-II.

The research contributions of this paper can be outlined as given below:

- Proposing an integrated text-mining assisted process planning framework for distributed manufacturing systems;
- Employing a machine-learning-based text-mining method to identify the potential enterprises and sharing of resources effectively across the network and tested its feasibility with various other machine-learning algorithms;
- A multi-objective evolutionary algorithm-based Hybridized Moth Flame Optimization Algorithm (HMFO) is used to solve the considered problem in the scenario of distributed gear manufacturing industries;

The results of the proposed HMFO method are validated with the Non-Dominated Sorting Genetic Algorithm (NSGA-II) to evaluate their usefulness by both experimental and practical instances. Finally, the superiority of the proposed HMFO is confirmed with the help of various performance indicators.

In this paper, Section 2 deals with the literature, Section 3 describes the problem and the developed mathematical model developed. Section 4 explains the proposed framework and algorithm for the text-mining approach. In Section 5, experimentation with a case study of the gear manufacturing industry is presented, and the corresponding outcomes are explained in Section 6. The paper is concluded in Section 7 by providing scope for future work.

2. Literature Review

The section discusses a review of text mining, Integration of Process Planning and Scheduling (IPPS), interoperability, and evolutionary algorithm-based approaches implemented in the proposed methodology.

A review has been conducted for knowledge discovery and text-mining techniques with data-mining attributes, namely depiction, explanation, classification, estimation, grouping, and evolution in the domain of manufacturing [5]. An algorithm developed with K-means and support vector machine (SVM) clustering algorithms to examine the polarity of text and group the online hotspot detection forums into clusters depending upon their similarities [6]. An ontology was implemented based on text-mining strategy to extract the fault system information from the unstructured natural language text. All the information regarding the manufacture of the product was represented as a knowledge with the help of an ontology naming product to share this knowledge in any platform by achieving interoperability [7]. At the same time, the Naive Bayes algorithm approach was adopted for classification of manufacturing supplier. Later, Decision Trees and Random Forests were utilized for supervised machine-learning model-based digit classification using Waikato Environment for Knowledge Analysis (WEKA) and have performed comparison on multiple performance parameters such as Kappa Statistic, Precision, Recall, and F-measure [8]. A resource allocation strategy with help of big data and Machine-Learning (ML) techniques is proposed to find the perfect forecasting of energy consumption patterns [9]. A text-mining technique was proposed for classification of sustainable environmental indices for service and manufacturing system. They also established relationships between indicator utility levels and company characteristics [10]. A support SVM classification algorithm was presented to classify the supplier's text data of various web pages into manufacturing and non-manufacturing suppliers [11]. An e-commerce strategy was adopted for monitoring specific features of the enterprises. It describes how the records are obtained automatically from a corporate website using supervised classification algorithms [12].

Process Planning and Scheduling (PPS) mentions the requirement of manufacturing resources, operations, and routes that are possible to manufacture a product and allocate the operations of all the jobs on machines, without disturbing the actual precedence relationships in the process plans [13]. In traditional manufacturing system, process planning and scheduling were carried out in a step-by-step process. To overcome the adverse effects by a conventional way of PPS, researchers have identified the need to integrate both PPS and have found the benefits of it in case of networked manufacturing environment. Manufacturing decision making (MADEMA) approach was proposed for the assignment of work center resources with multiple decision-making criteria in order to have an effective utilization for IPPS problem [14]. Later, the modified above-proposed MADEMA model consists of five basic steps and mainly focuses on finding alternative machines and solving the Integration of Process Planning and Scheduling (IPPS) problem [15]. Furthermore, a net-man strategic framework proposed where an operational mechanism is introduced for manufacturing organizations that helps to change their operations on a timely basis with the help of forming distributed manufacturing networks that help as a performance enabler for manufacturers in the Networked Manufacturing System (NMS) [16]. The agent-based system has been proposed with the distributed ruler method for distributed manufacturing systems as a function-based decomposition method to accomplish the process planning and scheduling, and the feasibility is illustrated through different case studies [17]. A Disruptive Innovation-Like Algorithm (DILA) is presented to minimize the tardiness for the job and to obtain the optimal schedule for one machine by varying setup times at irregular interval of time [18]. A novel way for formalizing datasets and concepts utilized for ontology were embedded in the product itself and hence made it interoperable in NMS. Furthermore, a two-level nested solution algorithm was implemented by developing a hybrid adaptive genetic algorithm (HAGA) to achieve optimal process plans for multiple jobs in NMS. The feasibility of the approach is investigated through numerical experiments [19]. The Binary Spring Search Algorithm (BSSA) based on simulation of Hookes law is employed to solve various optimization problems. Moreover, the results obtained by BSSA are compared with other standard binary algorithms, namely the grasshopper mechanism, bat algorithm, etc. [20]. The mobile-agent-based system, introduced for IPPS in NMS to prove the consistency of the proposed model comparison, was made with the Controlled Elitist Non-dominated sorting GA [21].

The effectiveness of the integration of production planning and scheduling approach is compared and proved with conventional sequential scheduling approach. An evolutionary algorithm-based GA approach was adopted for the scheduling of integrated manufacturing and distribution systems [22]. A two-loop algorithm consists of the longest processing-time rule-based tabu search, which is proposed to obtain the optimal schedule in IPPS by minimizing the total cost for manufacturing and maintenance of machines arranged in series in case of NMS [23]. A hybrid dynamic-DNA assisted an evolutionary algorithm proposed to solve N-person non-co-operative game in context to produce various optimal schedule for several jobs in a NMS [24]. A Chaotic Particle Swarm optimization (C-PSO) algorithm for IPPS problem in network manufacturing and compared it with other benchmark algorithms like Genetic Algorithm (GA), Simulated Annealing (SA), and hybrid algorithm to prove its superiority [25]. A Hybrid Particle Swarm Optimization (H-PSO) was described for IPPS and delivery route planning. This was implemented utilizing multi-purpose machines to minimize the cost and earliness and tardiness of the jobs [26]. A logic-based Benders decomposition (LBBD) algorithm mainly separates the decision variables into two sub-categories, i.e., master problem deals with the process plan and sub-problem deals with the sequencing has been emphasized that can solve the IPPS problem for finding an exact solution [27]. A MILP formulation was proposed to minimize the storage cost and workforce cost in the airline industry. This problem mainly deals with the maintenance and repair operations allocation, and it is an NP-hard problem solved by evolutionary algorithms [28]. A combination of algorithms, namely H-PSO and GA with special operators, has been presented to deal with the existed uncertainty in IPPS problem. Later, standard problems are considered to validate the effectiveness of the presented hybrid approach [29]. To overcome the trapping of solutions to a local optimum while solving the multi-modal functions by using several heuristic algorithms, a Comprehensive Learning Particle Swarm Optimizer (CLPSO) has been proposed that combines the advantage of Local Search (LS) strategy along with the excellent global search ability of Particle Swarm Optimizer (PSO). In this work, several multi-modal benchmark functions such as CEC2013 are tested for determining the effectiveness of CLPSO-LS algorithm [30].

A dynamic scheduling method based on event-triggered dynamic task scheduling (EDS) proposed to get the optimal services times for cloud manufacturing systems. A case study of numerical control machines has been considered to prove the effectiveness of the proposed methodology. [31]. Furthermore, generic mediator architecture for effective coordination and task planning in a Distributed Manufacturing Environment (DME) were developed [32]. Moreover, a review has been presented on the articles related to game theory and optimization methods for several applications of problems. Furthermore, there was a classification into various categories where game theory is useful for increasing the effectiveness of optimization; optimization methods are useful for solving the game theory problems; and a combination of game theory and optimization also may be useful for efficient solving of other classes of problems. The proposed classification was based on four criteria: mainly based on nature of optimization (classic or modern), based on the number of objectives (single or multi), and based on the type of game theory [33]. Subsequently, the reduction of maximum completion time, tardiness, and production cost along with the optimal schedule are generated with the help of GA integrated with Gantt chart (GC) methodology for DMS. A case study of manufacturing scenario with six jobs and twelve machines considered and solved with the proposed GA-GC method [34]. Likewise, resource, management and part agents are considered in a multi-agent-based system to make decisions in a timely manner with proper co-ordination to generate optimal process plans in a distributed scheduling environment [35].

The question lies in how to make the proper choice of action without deviating from the optimal strategy in an uncertain environment that exists in distributional robust optimization problems where the decision maker is not sure about the distribution of uncertainty that exists in the problem. In such a situation, this work gives insight into exploring the alternative ways that help find the distribution of uncertainty by the decision maker, based on the observations found from the experiments. Algorithms are proposed to find the local optima and are derived from a common evolutionary stable strategy to explore their convergence rate by using a mean estimate [36]. A memetic algorithm was discussed for the minimization of makespan for a distributed assembly permutation flow-shop scheduling problem to obtain accurate results [37]. A multi-objective-based mixed-integer programming model was implemented with the consideration of makespan and total traveling distance as objective functions, and a GA-based heuristic approach was proposed to obtain the optimum results in virtual manufacturing cells (VMC) [38]. To solve the scheduling and maintenance planning simultaneously in DMS, an evolutionary-based GA is proposed, and the performance was validated by comparing with the other algorithm [39]. An IPPS problem was solved with the Simulated Annealing (SA) approach that contains the added flexibilities of process, operation, scheduling to optimize the utilization of machines, and production cost in a DMS [40]. A GA was introduced to identify a near-optimal configuration in a manufacturing network. The performance of GA-derived alternate designs is held in comparison with the output of an intelligent search algorithm. A new approach named hybrid of Estimation of Distribution Algorithm (EDA) was employed to increase the profit in forwarding supply chain and to reduce the carbon footprints in a closed-loop supply chain network system [41]. A framework model was presented for the feasibility and merits and their applications of complex networks in advanced manufacturing systems [42]. A particle swarm optimization with a hill-climbing approach was proposed for minimizing the functions that make span and energy consumption in distributed manufacturing systems [43].

To avoid the difficulty of applying the heuristics for the combinatorial nature of problems like crude oil operation scheduling problem, initially, the problem was converted to an assignment problem-related to tanks and distillers, and later a chromosome was implemented, which helps the further application of meta-heuristics like NSGA-II for optimization of refinery schedule. A case study of a china refinery with three distillers and ten charging tankers with multiple objectives was considered and tested successfully [44]. A modified particle swarm optimization was developed to generate an optimal process plan, and its performance was verified through five independent experiments and a comparison with other meta-heuristic algorithms in the domain of flexible process planning research [45]. Recent advances of multi-objective genetic algorithms (MoGA) were described with differential evolution (HSS-MoEA-DE) to solve several multi-objective scheduling problems in manufacturing systems [46]. A hybrid harmony search and genetic algorithm (HSGA) was proposed for an integrated job maintenance scheduling problem in NMS [47]. A hybrid algorithm was proposed to solve the lot sizing and IPPS problem in the case of plastic molding industry. The proposed approach was compared with the simulated annealing for test its effectiveness [48].

3. Problem Description

Here, thirty-six medium-scale gear manufacturing industries located in a distributed manner across the southern part of India were considered as a case study for investigation for the prospect of providing an optimum solution with a composite multi-objective evolutionary algorithm approach assisted by a classifier. The systematic diagram of the gearbox that is being manufactured and its major parts, i.e., gear, shaft, coupling flanges, key, bearing inner-outer race, and bearing ball in a gearbox is mainly considered in this study and, as shown in Figure 1 has been considered for further investigation. Our research addresses areas of great concern related to finding a suitable supplier according to product consignment, interoperability, lack of efficient techniques, tools, and methods for enhancing the productivity of the system. The process usually begins with the customers in a network-based manufacturing service requesting resources through a particular supplier. However, the search and selection of an appropriate supplier are time-consuming, with the consumer having little or no information about the respective capability narratives. We intend to categorize suppliers that manufacture gearbox related products in the market from their capability narratives and textual information collected via multiple product sourcing and supplier discovery platforms through text mining. Further, efficient supplier classification through the use of supervised machine-learning algorithms is implemented.

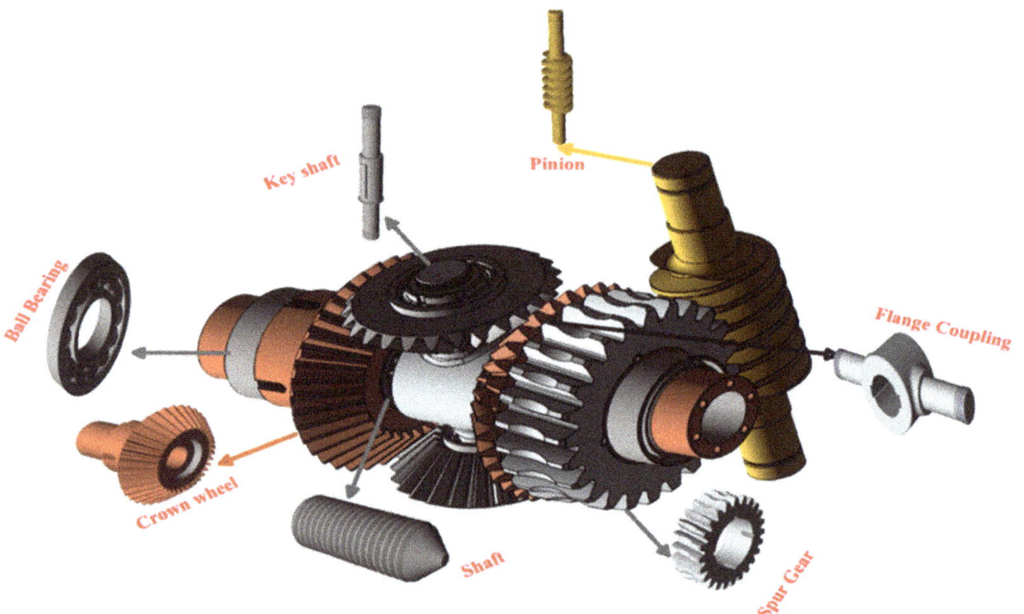

Figure 1. Various components in the gear box.

The output obtained in the form of task-specific suppliers from the proposed supervised learning algorithm is fed as input for the considered network-based manufacturing system which consists of a set of job orders given by the customers denoted by n.

Each job has numerous process plans by which it can be implemented. A set of available machines is distributed geographically to perform necessary operations in a process plan for the completion of the job. Considering the scenario of the current network-based manufacturing environment, the above background setting presents a challenge in terms of optimizing the objectives, i.e., completion time, energy consumption, machine utilization rate, and service utilization. As the problem is computationally complex and NP-hard in nature, it becomes tedious to solve the above scenario. Thus, there is a necessity for an efficient and effective approach for an optimal process plan of the considered jobs. This research aims to provide a solution to approach an optimal process plan of the considered jobs by fulfilling the above-mentioned objectives.

Hence, an integrated machine-learning-based evolutionary algorithmic approach where the outcome of a supervised algorithm is used as an input to an evolutionary algorithm that was considered. The mathematical model involved is presented in the below section, and its respective notations are explained in Table 1. The present problem requires a few assumptions that are very important to mention below.

1. Job pre-emption is prohibited.
2. Until the previous job is completed, the successive job cannot be processed.
3. Only one job can be processed in an enterprise at a time.
4. The reliability of a machine with respect to time is constant, and its value is the same for that particular machine for every operation while processing a job.
5. At time $t = 0$, all machines and jobs are concurrently available.
6. The operations of every job and its respective sequence consisting of future processing tasks need to be pre-defined.

Table 1. Notations used in mathematical model.

Notation	Description
E	The number of all the available jobs
G	The number of all the available machines
H_v	The number of all the available alternative process plans of job v.
Q_{vpk}	pth alternative process plan for kth operation of job v
S_{vp}	The number of all the available operations in the pth alternative process plan of the job v
L	Maximum completion time of vth job from the all the available process plans
D_{vkpr}	For operation Q_{vkp} corresponding processing time of the on machine r
B	An arbitrary Integer which is a very large positive integer.
C_v	The completion time till the processing of job v
C_{vkpr}	The earliest completion time till the operation Q_{vkp} on machine r
E_{vkr}	Indicates energy consumption for processing kth operation of job v on machine r
Rel_{vk}	Indicates reliability of the kth operation of job v

Decision Variables:

X_{vp} 1 The pth alternative process plan of job v is selected

0 Under other conditions

$Y_{vkpwtur}$ 1 The operation Q_{vkp} preceding over the operation Q_{wtu} on given machine r

0 Under other conditions

Z_{vkpr} 1 If given machine r is selected for Q_{vkp}

0 Under other conditions

3.1. Objectives

$$\text{Minimization of makespan } (L_{\min}) = \text{Max} C_{vkpr} \tag{1}$$

$$\text{Minimization of Machine Utilization } (U_v) = \frac{\sum_{v=1}^{E} D_{rv}}{\sum_{r=1}^{G} (mct_r - mst_r)} \tag{2}$$

$$\text{Minimization of energy consumption } (E) = \sum_{v=1}^{E} \sum_{k=1}^{S_{vp}} \sum_{r=1}^{G} E_{vkr} \tag{3}$$

$$\text{Minimization of Reliability (R)} = \prod_{k=1}^{S_{vp}} \text{Rel}_{vk} \tag{4}$$

where D_{rv} represents processing time of job v on the rth machine, and mct_r indicates finishing time of rth machine, i.e., the time taken to finish the final operation on rth machine. mst_r is the start time of rth machine.

3.2. Subject to Constraints

The initial operation ($k = 1$) in the possible process plan p of job v is mentioned as

$$C_{vp1r} + B(1 - X_{vp}) \geq D_{vp1r}$$
$$v \in [1, E], \ p \in [1, H_v], r \in [1, G] \tag{5}$$

The final operation for the possible process plan p of job v is mentioned below

$$C_{vpS_{vp}r} - B(1 - X_{vp}) \leq C_{vpkr}$$
$$v \in [1, E], \ p \in [1, H_v], r \in [1, G] \tag{6}$$

Different operations for the same job having precedence constraints are unable to be processed simultaneously.

$$C_{vpkr} - C_{vp(k-1)r_1} + B(1 - X_{vp}) \geq D_{vkpr}$$

$$v \in [1, E], \ p \in [1, H_v], k \in [1, S_{vp}], r, r_1 \in [1, G] \tag{7}$$

Every machine is able to process only one operation at a time and is expressed as

$$C_{vpkr} - C_{wutr} + BY_{vpkwtur} \geq D_{vkpr}$$

$$v, w \in [1, E], \ p, u \in [1, H_v], k, t \in [1, S_{vp}], r \in [1, G] \tag{8}$$

Among the available process plans, there is the possibility to choose only one alternative process plan

$$\sum_{v=1}^{E} X_{vp} = \begin{cases} 1 \ \text{if processplan } /p/ \text{ is selected from job } /v/ \\ 0 \ \text{otherwise zero} \end{cases} \tag{9}$$

$$p \in [1, H_V]$$

One machine only must be chosen for each operation.

$$\sum_{r=1}^{G} Z_{vkpr} = \begin{cases} 1 \ \text{if processplan } /p/ \text{ is selected from job } /v/ \\ 0 \ \text{otherwise zero} \end{cases} \tag{10}$$

$$v \in [1, E], \ p \in [1, H_v], k \in [1, S_{vp}]$$

Table 1 presents the notation used in the mathematical model. Equations (1)–(4) represent an optimization of process parameters like minimizing makespan, maximization of machine utilization, minimization of energy consumption, and maximization of reliability, respectively. Precedence constraints of the operations are represented by Equations (5) and (6); more specifically after the finishing of operation of a particular job, only the next operation must start. Equation (7) represents different operations for the same job having precedence constraints that are unable to be processed simultaneously. Equation (8) represents that each machine is able to process only one operation at a time and is expressed as a constraint for the machine. Equation (9) indicates that among the available process plans there is the possibility to choose only one alternative process plan, and Equation (10) represents that one machine only must be chosen for each operation.

4. A Framework of the Proposed Classifier-Assisted Evolutionary Algorithm Approach

In this section, the proposed classifier-assisted evolutionary algorithm approach as a framework is explained. A distributed manufacturing network environment is considered. Figure 2 represents the proposed approach. In this model, the process gets initiated with the customer requests for a specific product. These requests are handled by enterprise user (EU) and customer user (CU), which are service providers in network-based manufacturing service. CU is an organization that accepts requests of different products from varied customers to complete the consignment agreement. To complete the accepted tasks, the available potential suppliers are evaluated from their database for assigning the task. Here, the main role of CU is to assign the tasks to the appropriate suppliers/manufacturers/distributors, etc., and to monitor their activities regularly to finish the task effectively and efficiently. On the other side, EU also accepts multiple requests from customers; unlike CU, it has the capacity to provide some of the services on its own due to its own manufacturing unit. The remaining services are fulfilled by assigning the task through potential enterprises as sub-contracting. In this study, we consider the EU service path where some of the services are fulfilled on their own. The next step would be the selection of the most appropriate supplier from the list of potential enterprises to whom the customer request must be forwarded. These suppliers are either maintained in the knowledge base or available as text corpus in the form of capability narratives. Initially, categorization into manufacturing and non-manufacturing units is carried out. Following this, text mining is implemented to finally perform supervised machine-learning models-based classification to differentiate the above dataset of suppliers into task-specific suppliers.

The above outcome of task-specific suppliers is further considered in a networked manufacturing environment to undergo three different stages, i.e., order pool, task pool, and service pool for processing requirements of the jobs to its final outcome. To execute the process of the desired product, requests are sent to the order pool where the information of product specifications and their requirements are sorted and stored. These stored orders in the order pool have classified the tasks into individual tasks by taking into consideration factors such as reliability, task priority, processing time, and serviceability identified in this work. They contain a wide variety of tasks initiating from the procurement of raw goods to dispatching the processed product to customers. Among the group of enterprises, the enterprises that are needed to meet the product and process requirements are chosen to carry out the manufacturing functions, i.e., process planning and scheduling for optimal solutions.

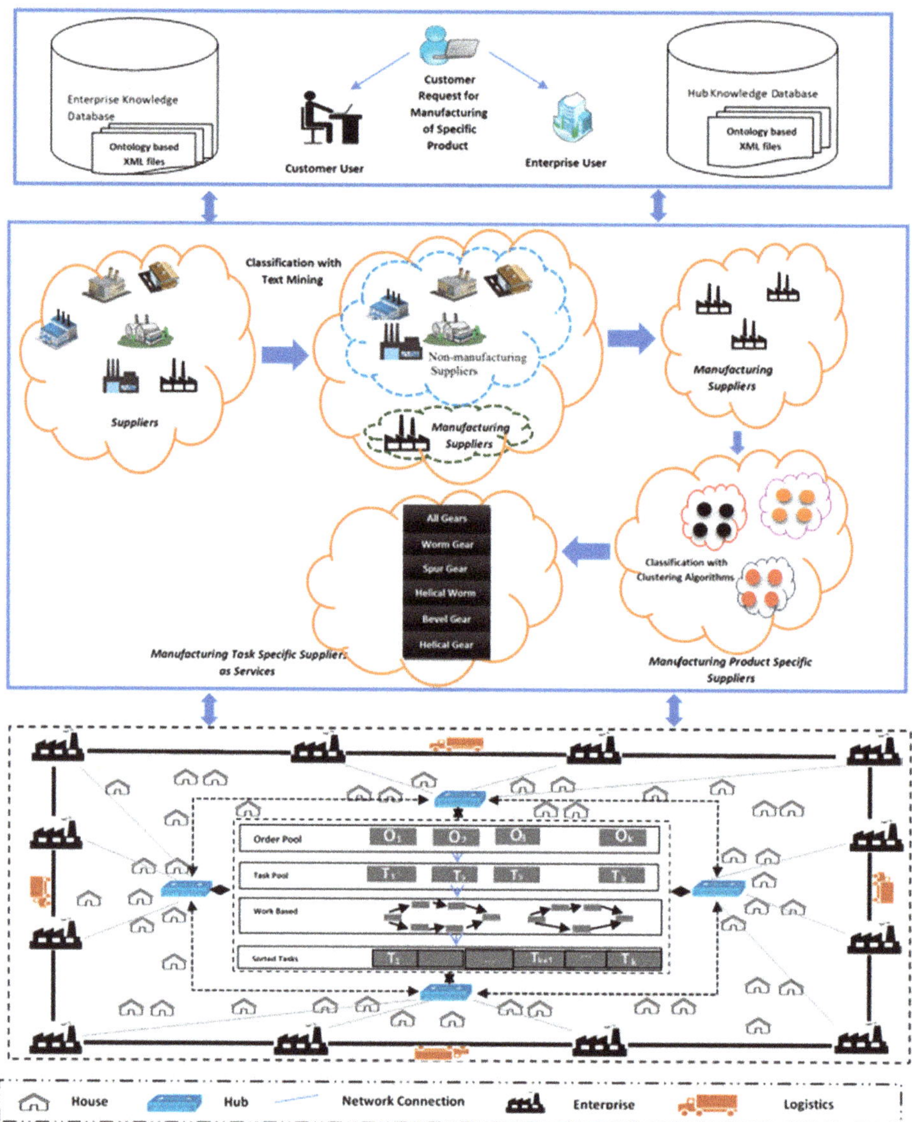

Figure 2. Framework of the proposed network manufacturing approach.

5. Experimentation Part Text-Mining

5.1. Task-Specific Supplier Classification through Supervised Machine-Learning Algorithms Based on Text Mining

For the purpose of supplier classification, numerous suppliers representing the gear manufacturing industry in India are taken as a case study. A flowchart explaining the methodology for supplier classification is shown in Figure 3. After pre-processing and mining are applied to the above dataset, the suppliers are classified into manufacturing and non-manufacturing suppliers. Later, the manufacturing suppliers are further classified into task-specific suppliers with the help of various supervised machine-learning algorithms. The performance of these algorithms is validated with different performance measures. The above approach is implemented using R and WEKA.

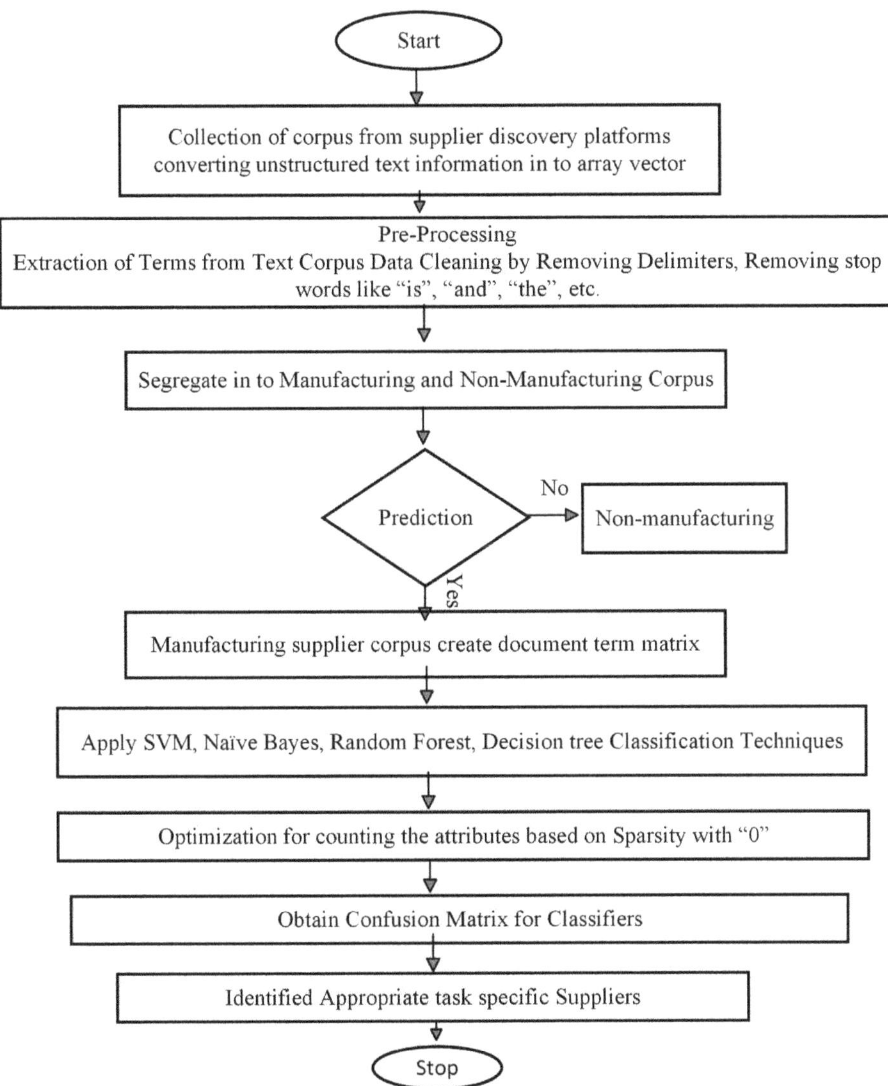

Figure 3. Flowchart for the proposed text-mining approach.

Step 1. *Creation of Supplier Corpus.*

For the purpose of text mining, a corpus of suppliers representing the gear manufacturing industry was created. This corpus was constructed with the help of capability narratives and textual portfolios accumulated via multiple product sourcing and supplier discovery platforms such as Thomas Net, Procure Search, and Supply and Demand Chain Executive, among several others. The enterprises fall into five different categories of gear manufacturing as shown in Table 2. To accumulate any gear not falling into one of the above categories, a miscellaneous type 'All types of gear' was created. A test corpus of 40 different gear firms was also created to later validate our approach and classification performance. Out of these four, data are removed from corpus due to inadequate information on several parameters, such as Types of Gears, Types of Machines, Industries Served, etc. This unstructured textual information is then read and converted into vectors in R.

Table 2. Various types of gear manufacturing.

Category 1	Bevel gear
Category 2	Helical and worm
Category 3	Helical gear
Category 4	Spur gear
Category 5	Worm gear
Category 6	All types of gear

Step 2.   *Pre-Processing of Text Corpus and Creation of Document Term Matrix.*

The prepared text corpus usually consists of delimiters, blank spaces, punctuation mark, and stop words. These need to be removed before the application of machine-learning models to remove any unnecessary bias during training. The corpus is thus subjected to data cleaning in this stage to eliminate the above entities. To proceed to the later stages, two separate corpuses for manufacturing and non-manufacturing were created. The created manufacturing and non-manufacturing corpus are subjected to training and testing with varied weightages depending upon the frequency of their occurrence as explained [10]. The document term matrix was also created by selecting the features which are critical in the classification. The features can be represented in the form of a word cloud on the basis of varying sparsity measure which indicates the numeric occurrence estimate of the feature in the overall dataset. This measure can then be used to eliminate those features which are not distributed entirely over the dataset and cannot be used to accurately classify the dataset. This matrix is then converted into a comma-separated value CSV file to be later used in training and testing for machine-learning models. Figure 4 represents the word clouds which were formed for the six different gear classes with a varying sparsity measure. Namely, Figure 4a represents word Cloud for Worm at 0.77 sparsity, Figure 4b Word Cloud for Spur 0.90 sparsity, Figure 4c Word Cloud for Helical at 0.77 sparsity Figure 4c Word Cloud for Worm and Helical at 0.90, Figure 4c Word Cloud for Bevel at 0.90, and Figure 4f Word Cloud for All Types of gears at 0.77 sparsity are represented. Through a hit-and-trial approach, 0.77 was selected as the optimum. Several industry-specific information is also extracted through the use of regular expressions as shown in Table 3.

Table 3. Information to be extracted from mining and gear classification categories.

Types of machines (CNC, LATHE)
Types of operations (milling, drilling, and grinding)
Types of gears (spur, helical, and bevel)
Types of materials (steel, aluminum, bronze, and brass)
Types of certifications (ISO 9000, ISO 14000
Types of manufacturing process (casting, forging, and extrusion)

Step 3.   *Classification into Task-Specific Suppliers.*

The manufacturing corpus represented through document term matrix in the format of comma-separated values is subjected to classification algorithms such as Support Vector Machines, Decision Tree, Naïve Bayes, and Random Forests to classify them into task-specific suppliers. This is implemented with WEKA [49]. Training is performed on a dataset comprising numerous random capability narratives and testing on the above 36 capability narratives of gear manufacturing industries to classify them in one of the industrial categories. The performance of various classification algorithms is validated through the confusion matrix and other performance measures such as Kappa statistic, precision, recall, F-measure, etc., as shown in Table 4 obtained through WEKA. Figure 5 shows the confusion matrix which gives an idea about the number of instances that are classified in various categories, leading the classification to be either False Positive (FP), True Positive (TP), False Negative (FN), or True Negative (TN). The Decision Tree was found to be the best among all models with 0.932 precision the least relative absolute error at 9.6%, followed by Naïve Bayes at 0.73. SVM and Random Forest, having performed below par. Figure 6 shows the results of the text mining depicting the enterprise information. It gives the detailed information of an enterprise such as types of gears manufactured, types of machines used, types of industries served, and major clients of that particular enterprise.

a) Word Cloud for Worm at 0.77　　b) Word Cloud for spur at 0.90　　c) Word Cloud for helical at 0.77

d) Worm and Helical at 0.90　　e) Word Cloud for Bevel at 0.90　　f) Word Cloud for All types at 0.77

Figure 4. (**a**) Word Cloud for Worm at 0.77, (**b**) Word Cloud for Spur 0.90, (**c**) Word Cloud for Helical (**d**) Word Cloud for Worm and Helical at 0.90, (**e**) Word Cloud for Bevel at 0.90, and (**f**) Word Cloud for All Types of gears at 0.77 sparsity.

Naïve Bayes							Random forest							SVM							Decision tree						classified as
a	b	c	d	e	f		a	b	c	d	e	f		a	b	c	d	e	f		a	b	c	d	e	f	
1	1	2	0	0	1	a	0	2	2	0	1	0	a	0	0	0	0	3	2	a	2	1	0	0	2	0	a—all types
0	5	1	0	0	0	b	0	6	0	0	0	0	b	0	4	0	0	0	2	b	0	6	0	0	0	0	b – bevel gear
1	0	2	0	0	1	c	0	0	2	1	0	1	c	0	1	0	0	1	2	c	0	0	4	0	0	0	c –helical and worm
0	0	0	6	0	0	d	0	0	0	6	0	0	d	0	0	0	3	3	0	d	0	0	0	6	0	0	d – helical gear
1	0	0	0	7	0	e	0	0	0	0	7	1	e	0	0	0	0	7	1	e	0	0	0	0	8	0	e – spur gear
1	0	1	0	0	5	f	0	0	0	0	0	7	f	0	0	0	0	1	6	f	0	0	0	0	0	7	f – worm gear

Figure 5. Confusion matrices of Naïve Bayes, Random Forest, SVM, and Decision Trees.

Table 4. Various performance measures for machine-learning algorithms Decision Tree (J48), Naïve Bayes, Random Forest, and Support Vector Machines.

TP Rate	FP Rate	Precession	Recall	F-Measure	MCC	ROC Area	PRC Area	Class
Decision Tree (J48)								
0.400	0.000	1.000	0.400	0.571	0.604	0.881	0.694	All types
1.000	0.033	0.857	1.000	0.923	0.910	0.983	0.857	Bevel gear
1.000	0.000	1.000	1.000	1.000	1.000	1.000	1.000	Helical and worm gear
1.000	0.000	1.000	1.000	1.000	1.000	1.000	1.000	Helical gear
1.000	0.071	0.800	1.000	0.889	0.862	0.991	0.950	Spur gear
1.000	0.000	1.000	1.000	1.000	1.000	1.000	1.000	Worm gear
0.917	0.021	0.932	0.917	0.903	0.899	0.979	0.923	Weighted Avg.
Naïve Bayes								
0.200	0.097	0.250	0.200	0.222	0.114	0.490	0.181	All types
0.833	0.033	0.833	0.833	0.833	0.800	0.978	0.897	Bevel gear
0.500	0.125	0.333	0.500	0.400	0.316	0.672	0.573	Helical and worm gear
1.000	0.000	1.000	1.000	1.000	1.000	1.000	1.000	Helical gear
0.875	0.000	1.000	0.875	0.933	0.919	0.938	0.920	Spur gear
0.714	0.069	0.714	0.714	0.714	0.645	0.808	0.723	Worm gear
0.722	0.046	0.738	0.722	0.727	0.681	0.838	0.750	Weighted Avg.
Random Forest								
0.000	0.000	0.000	0.000	0.000	0.000	0.771	0.435	All types
1.000	0.067	0.750	1.000	0.857	0.837	1.000	1.000	Bevel gear
0.500	0.063	0.500	0.500	0.500	0.438	0.969	0.817	Helical and worm gear
1.000	0.033	0.857	1.000	0.923	0.910	1.000	1.000	Helical gear
0.875	0.036	0.875	0.875	0.875	0.839	0.996	0.986	Spur gear
1.000	0.069	0.778	1.000	0.875	0.851	1.000	1.000	Worm gear
0.778	0.045	0.669	0.778	0.717	0.692	0.964	0.898	Weighted Avg.
Support Vector Machines								
0.000	0.000	0.000	0.000	0.000	0.000	0.500	0.139	All types
0.667	0.033	0.800	0.667	0.727	0.683	0.817	0.589	Bevel gear
0.000	0.000	0.000	0.000	0.000	0.000	0.500	0.111	Helical and worm gear
0.500	0.000	1.000	0.500	0.667	0.674	0.750	0.583	Helical gear
0.875	0.286	0.467	0.875	0.609	0.497	0.795	0.436	Spur gear
0.857	0.241	0.462	0.857	0.600	0.507	0.808	0.423	Worm gear
0.483	0.093333	0.454833	0.483	0.433	0.3935	0.695	0.380	Weighted Avg.

5.2. Proposed Multi-Objective Evolutionary Algorithms

A nature-inspired population-based algorithm called Moth Flame Evolutionary Optimization (MFEO) algorithm introduced bfy MirjalIli [50] that works on the consideration of the natural transverse movement of moths in nature. Moths can travel very long distances on a straight-line path. However, interestingly, along with the straight-line, moths travel spirally near to the light sources which converge into an optimized path for them to reach their destination. Based on this phenomenon, the Moth Flame Evolutionary Optimizing algorithm was developed. The effectiveness of this algorithm over other algorithms (GA, PSO, ACO) is clearly shown by Mirjalili by considering several benchmark functions and case studies [50]. In this work, a hybridized form of the Moth Flame Evolutionary Optimizing algorithm (HMFEO) is presented. We have considered non-dominated sorting Pareto approach for the proposed algorithm hybridization. This has been implemented to the moth flame optimization with non-dominated sorting and crowding distance operators, and a flowchart for the same is presented in Figure 7.

The parameters for the HMFO technique are specified for the implementation of algorithms shown in Table 5 with the number of moths at 200 and the maximum number of iterations at 1500. The upper boundary and lower boundary values are specified based on the test data input.

```
Source

Console    Terminal

D:/Species Distribution/Gear Manufacturing classification/Data/

[1] "Sakthi gears "
Types of gears -> Sakthi gears --- Spur gear

Types of Machines -> uses CNC turning, vertical machining and gear cutting machines.

Types of Industries Served -> we are able to employ skilled manpower and develop customized solution for diverse sectors like
  Textile, Compressor, Transmission, automotive industries, special valves, Gears & Gear Boxes and critical components.integra
ted with allied industries like reputed Iron & Steel Foundries, Fabrication and Finishing Industries and NABL accredited labo
ratories

Major clients -> integrated with allied industries like reputed Iron & Steel Foundries, Fabrication and Finishing Industries
and NABL accredited laboratories


[1] "Shrivik "
Types of gears -> Shrivik --- spur gear

Types of Machines -> manufactures Lub-oil-pumps, water pumps, fuel feed pumps, governors, connecting rods, marine engine gear
s etc.uses CNC, Lathe, milling, grinding and lapping machines.

Types of Industries Served -> entered into Automotive Components manufacturing field with an idea of doing things differently
.manufactures Lub-oil-pumps, water pumps, fuel feed pumps, governors, connecting rods, marine engine gears etc.

Major clients -> major customers are Greaves cotton Ltd, Continental engines Ltd, Simpson & co Ltd, Delphi tvs Diesel systems
.


[1] "Nikaki gears "
Types of gears -> Nikaki gears ---  spur gears

Types of Machines -> NIKAKI Gears has today made a prominent name in the gear industry as a leading supplier, manufacturer of
  industrial gears and custom made gears.we have high accuracy machine for gear cutting and profile grinding.uses cnc, lathe,
milling machines.

Types of Industries Served -> NIKAKI Gears has today made a prominent name in the gear industry as a leading supplier, manufa
cturer of industrial gears and custom made gears.serves fertilizer, tyre, sugar, chemical, pharmaceuticals, paper, wind mill,
  food industries.

Major clients -> Data category not found


[1] "Karthic gears "
Types of gears -> Karthic gears -- Spur gears

Types of Machines -> uses lathe, milling and grinding machines.

Types of Industries Served -> supplies gears for the automobile industry.

Major clients -> Data category not found
```

Figure 6. Screen shot of extracted enterprise information and classification into task-specific supplier with text mining.

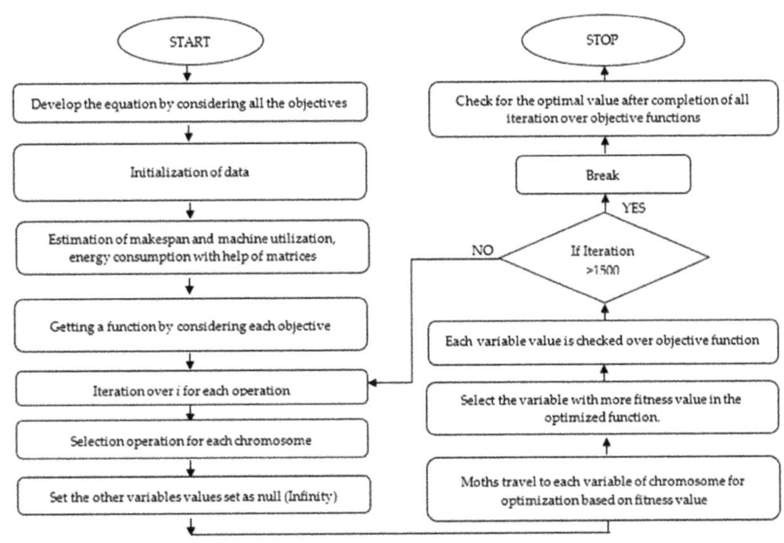

Figure 7. Flowchart of the proposed HMFO.

Table 5. Initialization of parameters for proposed solution algorithm.

Process Parameters	HMFEO	NSGA II
Population Size/No. of Moths	200	200
Number of generations	1500	1500
Mutation Probability	-	0.07
Cross-Over Probability	-	0.76

Step 1. In HMFO, potential solutions are represented as moths and variables are represented as position in the moth space. A matrix consists of all the moths (n), and their dimension is d.

$$L = \begin{bmatrix} L_{11} & L_{12} & \cdots & L_{1d} \\ L_{21} & L_{22} & \cdots & L_{2d} \\ L_{31} & L_{32} & \cdots & L_{3d} \\ L_{41} & L_{42} & \cdots & L_{4d} \end{bmatrix}$$

Initialization of moth population and their spaces are defined with the time matrices and their corresponding inputs. In this proposed HMFEO, a new type of encoding schema was presented to suit the problem nature. The encoding scheme for makespan is presented in Figure 8.

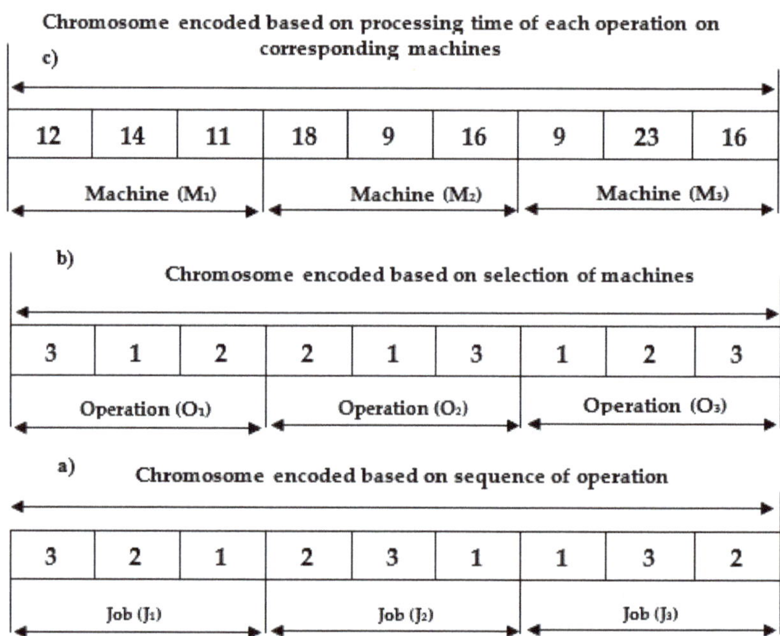

Figure 8. Representation of chromosome initialization for make span.

The example of encoding schema is represented in Figure 8, and the encoding consists of three parts. If we observe from bottom to top, Figure 8a is the encoding based on a sequence of operations of each job, which can determine the sequence of operations needed to produce a job. Figure 8b is the encoding based on machines, which can choose the machine for each operation. Figure 8c is the encoding based on the processing times of each machine for the corresponding operation. Therefore, a chromosome in Figure 8 shows three jobs, which consist of nine operations, and will be processed on three different machines. The processing sequence of this chromosome can be represented as $Q_{ik}{}^{j}$ that the jth operation of the ith job that will be processed on the kth machine. Where $Q_{23}{}^{1}$ is the first operation of the second job that will be processed on the third machine. Based on the encoding, schema time matrices were obtained, and the time matrices for makespan can be represented as follows. Makespan = zeros(mach, opns, pp, jobs); makespan(:,:,2, 3) matrix below indicating the processing time values of machines for the corresponding operation for the third job and second process plan. The remaining values in the matrix are kept as zeros.

	M1	M2	M3	M4	M5	M6	M7	M8	M9	M10	M11	M12
O_1	[12	18	9	0	0	0	0	0	0	0	0	0;
O_2	14	9	23	0	0	0	0	0	0	0	0	0;
O_3	11	16	16	0	0	0	0	0	0	0	0	0;
O_4	0	0	0	0	0	0	0	0	0	0	0	0;
O_5	0	0	0	0	0	0	0	0	0	0	0	0];

Step 2. Based on the above information, the energy consumption matrix (Equation (11)) is obtained through multiplication with the corresponding time matrix with specified energy consumption input mentioned in Table 6. The encoding schema is specified in Figure 9.

Table 6. Energy and reliability data.

Machine	M1	M2	M3	M4	M5	M6	M7	M8	M9	M10	M11	M12
Energy consumption	15	29	32	14	11	12	19	24	14	16	22	11
Reliability	0.76	0.82	0.78	0.84	0.84	0.92	0.89	0.94	0.88	0.95	0.84	0.92

$$\text{EC matrix } (o, m, p, j) = L \text{ matrix } (o, m, p, j) * E \text{ (Rated energy matrix)}; \tag{11}$$

The encoding schema is represented as shown in Figure 9, and the encoding consists of three parts. Figure 9a is the encoding based on a sequence of operations of each job, which can determine the sequence of operations needed to produce a job. Figure 9b shows the encoding based on machines, which can choose the machine for each operation. Figure 9c is the encoding based on the energy consumption of each machine for the corresponding operation of a particular job. Based on the encoding schema time, matrices were obtained, and the time matrices for energy consumption can be represented as follows. Energy-consumption(:,:,1,2) indicates energy consumption matrix for second job first process plan.

	M1	M2	M3	M4	M5	M6	M7	M8	M9	M10	M11	M12
O_1	[175	312	198	0	0	0	0	0	0	0	0	0;
O_2	400	234	330	0	0	0	0	0	0	0	0	0;
O_3	325	429	352	0	0	0	0	0	0	0	0	0;
O_4	0	0	0	0	0	0	0	0	0	0	0	0;
O_5	0	0	0	0	0	0	0	0	0	0	0	0];

The above matrix indicating the energy consumption values of machines for the corresponding operation for the third job and second process plan. The remaining values in the matrix are kept as zeros.

Step 3. A score function is defined that helps to select a suitable process plan, and it is shown in Equation (12), where a higher score value indicates that the probability of selecting the process plan is lesser. The lower the score value, the better the process plan. The formula for the score function is shown below.

$$\text{Score} = (L * E_{vk})/(R_{vk}) \tag{12}$$

where L represents makespan, E_{vk} indicates energy consumption for job v on machine k, and R_{vk} indicates reliability for job v on machine k.

Step 4. A matrix CL is formed considering all the moths into the objective functions that are stored in FK represented below.

$$C = \begin{bmatrix} CL_1 & CL_2 & CL_3 & CL_4 \end{bmatrix}^{T}$$

A flame matrix with a similar size to the moth's matrix is considered that stores the fitness values. Even though the moth (L) matrix and flame (C) matrix consist of solutions, the difference is that moths are search agents, whereas the flame indicates the best position of moths.

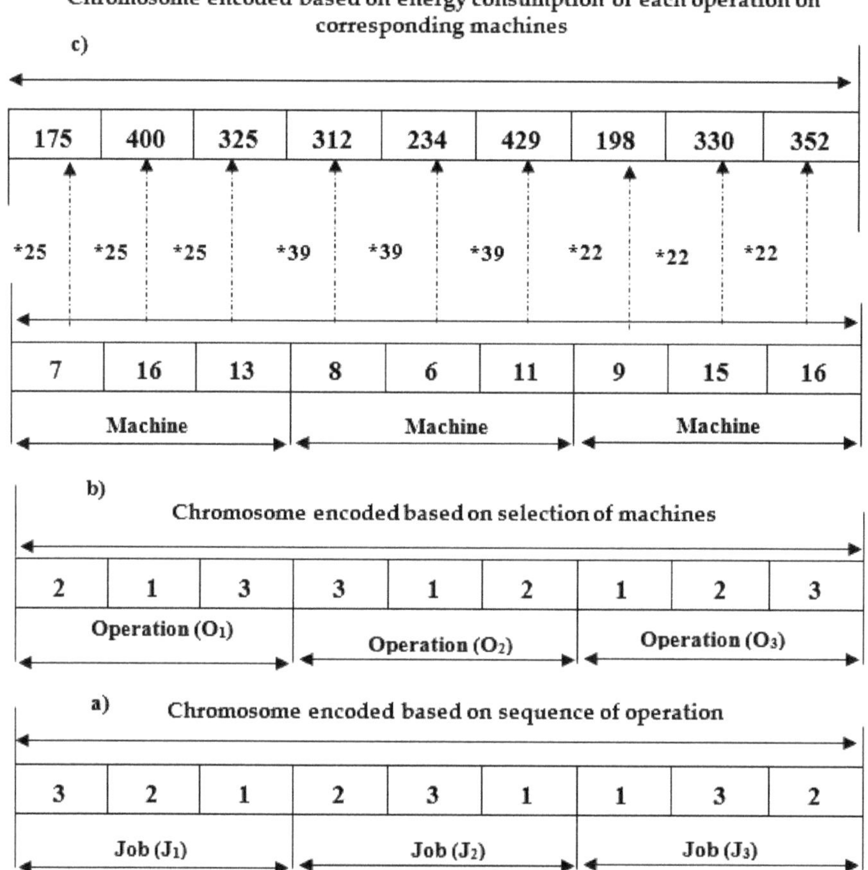

Figure 9. Representation of chromosome initialization for energy consumption.

Step 5. After an appropriate process plan is selected, we consider the rows as individual light sources through which we find the minimum values.

Step 6. The matrices are explored row-wise to find a minimum entry in their respective rows once the required inputs are received and the search space is clearly initialized.

Step 7. Moths maintain the best solution by updating its position Equation (13) by moving around the flag that is dropped by themselves during the search process. Update the position of the moth with respect to one flame. The spiral motion follows the Equation (13) represented as

$$Z(L_x, C_y) = S_x \cdot e^{at} \cdot \cos(2 \prod t) + L_y \tag{13}$$

L_x indicates the xth moth, C_y indicates the yth flame, and Z indicates spiral function. S_x is the distance of xth moth for yth flame, $S_x = \|C_y - L_x\|$, which is a constant defining shape of spiral motion. Where t $[-1, 1]$,

Step 8. After finding the minimum entry in the summed matrix and converting all ∞'s to 0s, the sum of all the values is found in their respective objective function matrices.

Step 9. Finally, we solve the function to generate optimal values for all objectives.

The most likely solution to our problem is selected based on the most appropriate fitness value. First, the data from various gear manufacturing enterprises are collected. The collected data contain the information regarding make span of the jobs, energy consumption, reliability of machines, and service utilization rate. The proposed HMFO algorithm is run on the Mathworks on Lenovo with an Intel processor with Windows 10 as the OS with 64 Gigabyte of RAM.

In this study, 10 different instances which are real-life case data have been considered by setting the numerical quantity of jobs to be 6 and machines to be 12, where each job has a varied number of substitutable process plans. Every process plan has a different set of operations. For each operation, there is a different set of machines which are capable enough to process the required task. For example, the instance 6 shown in the below Table 7 has 6 jobs and 12 machines in total. Job 1 has two alternative process plans and these process plans have three and three operations, respectively. One operation processed on a machine at a time.

Table 7. Results of the experimental instances with makespan and energy consumption values.

	Jobs	Machines	Processing Time Range	GA-SA (Instance 1 to 32) GA-MA (Instance 33 to 35)		Proposed HMFO	
				Makespan	Energy Consumption	Makespan	Energy Consumption
Instance 1	3	5	[1, 10]	41	138.1	30.8	26.3
Instance 2	3	7	[1, 10]	54.1	205.4	43	190.5
Instance 3	3	10	[1, 10]	61.2	229.1	49.7	204
Instance 4	3	5	[1, 50]	190.3	708.7	171	617
Instance 5	3	7	[1, 50]	252.8	960.6	226.7	834
Instance6	3	10	[1, 50]	333.8	1273.3	301.7	1110
Instance7	3	5	[1, 100]	375.4	1307.1	319.3	1134
Instance8	3	7	[1, 100]	531.9	1895.4	516.7	1644.9
Instance9	3	10	[1, 100]	729.1	2830.5	698	2467.5
Instance10	5	5	[1, 10]	35	140.4	24	126.1
Instance11	5	7	[1, 10]	46	186.3	30	169
Instance12	5	10	[1, 10]	51.5	199.9	43.7	177
Instance13	5	5	[1, 50]	165.5	671.5	149.8	583.9
Instance14	5	7	[1, 50]	225.2	951.2	201.7	828
Instance15	5	10	[1, 50]	317	1303.6	306.8	1139
Instance16	5	5	[1, 100]	325.5	1253.2	311.7	1098
Instance17	5	7	[1, 100]	436.9	1909	410	1663
Instance18	5	10	[1, 100]	610.3	2587.5	598	2257
Instance19	7	5	[1, 10]	28.7	110.9	19.6	96.7
Instance20	7	7	[1, 10]	39.3	162.2	24	146
Instance21	7	10	[1, 10]	56.5	241.6	49	218
Instance22	7	5	[1, 50]	159.7	607	143	524
Instance23	7	7	[1, 50]	220.8	919.1	206	846
Instance24	7	10	[1, 50]	304.6	1310.5	265.002	1150.9
Instance25	7	5	[1, 100]	351	1422.9	305.37	1287.9
Instance26	7	7	[1, 100]	426.1	1978.3	370.7	1725
Instance27	7	10	[1, 100]	625.9	2664.1	544.5	2319.7
Instance28	10	10	[1, 200]	939.04	9873.2	816.96	8597.6
Instance29	15	15	[1, 200]	1554.12	22,505.2	1352.08	19,579.5
Instance30	20	20	[1, 200]	4778.07	80,577.2	4156	70,102.16
Instance31	20	20	[1, 200]	7753.04	100,073.4	6749	87,263.8
Instance32	20	20	[1, 200]	15,062.5	197,787.5	13,115	172,975.7
Instance 33	18	15	[1, 200]	531	13,340.3	502	12,986
Instance 34	18	15	[1, 200]	810	2036.32	739	1956
Instance 35	18	15	[1, 200]	680	2267.88	593	1837.6

The processing time of various operations of jobs on several machines in different process plans, the energy consumption, and reliability of all the machines are known earlier. Table 6 provides the energy consumed per unit time by each machine and the reliability of the machine. To solve the considered problem, a newly established bio-inspired Moth Flame Optimization algorithm (HMFO) was adopted, and further, it was mapped according to problem nature. The operations were assigned to the machines in such a way that the considered objective functions are satisfied and an optimal sequence is obtained. The above-discussed approach is implemented for all formulated instances to find the robustness of the algorithm.

6. Discussion and Results

6.1. Validation of Proposed HMFO Algorithm with the Experimental Instances

To validate our approach toward optimization of makespan and energy consumption, we consider some experimental instances from the literature. Table 7 shows the results of the experimental instances with makespan and energy consumption values. We have calculated makespan and energy consumption for around a total of 35 experiments with the data available from (instances 1–32) mentioned in [51] and (instances 33–35) mentioned in [52]. A comparison of our proposed HMFO results with results carried out with a Simulated Annealing Genetic Algorithm (SA-GA) for instances 1–32. For most of the instances, the proposed HMFO gives better makespan and energy consumption values when compared with existing SA-GA makespan and energy consumption values. We also compared our proposed HMFO results with results carried out by [52], a Genetic Algorithm-based Memetic Algorithm (GA-MA) for instances 33–35. For most of the instances, the proposed HMFO gives better makespan and energy consumption values when compared with existing GA-MA makespan and energy consumption values. All the above results indicate the better performance of proposed HMFO with some of the experimental instances which were already proposed in the literature.

6.2. Evolution of Proposed HMFO with Practical Instances

After proving the effectiveness of the proposed HMFO with practical instances in Table 8. Furthermore, the effectiveness of the proposed algorithm is tested on different problems of practical instances with the aim to minimize makespan and energy consumption, maximizing machine utilization and reliability. Table 8 describes the optimal process plans chosen for each job in all the instances. Out of the various alternative process plans, only one process plan per job is chosen depending upon the score value. The lower the score value, the better the process plan. Thus, whichever process plan gives the lowest score value, that process plan is selected. For example, in instance 6, the process plans selected for the jobs 1–6 are 1, 1, 1, 1, 2, and 1, respectively.

Table 8. Optimal process plans selected for each job for all practical instances 1–10.

	Case		Process Plans Selected							
	Jobs	Machines	J1	J2	J3	J4	J5	J6	J7	J8
Instance1	6	6	2	2	2	2	2	2	-	-
Instance2	6	6	1	1	2	2	1	2	-	-
Instance3	6	8	3	1	2	3	3	4	-	-
Instance4	8	8	2	2	2	2	1	2	1	3
Instance5	8	8	2	1	2	1	1	1	2	2
Instance6	6	12	1	1	1	1	2	1	-	-
Instance7	6	12	2	2	2	3	2	2	-	-
Instance8	6	12	2	2	2	2	2	2	-	-
Instance9	6	12	2	1	1	2	2	1	-	-
Instance10	6	12	2	1	1	3	2	1	-	-

Table 9 shows the optimal values of makespan and energy consumption for all different problem instances (1–10) of HMFO and NSGA-II. These are the Pareto optimal values obtained by the simultaneous optimization of all the objectives. For example, in instance1, for proposed HMFO, the Pareto optimal value of makespan is 30, and energy consumption is 8906 and with NSGA-II the values of makespan and energy consumption as 43 and 9083, respectively. For instance6, for proposed HMFO, the value of makespan is 986, and energy consumption is 15,553. With NSGA-II, we obtained the values of makespan and energy consumption as 1083 and 17,694, respectively. To validate the performance of the proposed algorithm, 10 different instances were considered, and their considered objectives makespan and energy consumption values are shown in Table 9. In Table 9, clear observation indicates Pareto optimal values of makespan and energy consumption values are dependent on the number of jobs and machines. Interestingly makespan and energy consumption values for instance 3, i.e., six jobs and eight machines (6 × 8) case is more when compared to instances other instances 1, 2, i.e., six jobs, six machine cases (6 × 6), and instances 4 and 5, i.e., eight jobs, eight machine case (8 × 8). A similar trend is also shown in the literature for 6 × 8 cases for makespan and energy consumption values. In almost all instances, the makespan and energy consumption values of HMFO are lesser and far better than those given by NSGA-II. Thus, upon comparing both the algorithms we can conclude that the optimum values are attained in the case of HMFO, which proves the effectiveness of the proposed multi-objective evolutionary algorithm.

Table 9. Results of the practical instances with makespan and energy consumption values.

	Jobs	Machines	Proposed HMFO		NSGA-II	
			Makespan	Energy Consumption	Makespan	Energy Consumption
Instance 1	6	6	30	8906	35	9083
Instance 2	6	6	27	8325	38	8700
Instance 3	6	8	179	39,658	187	44,920
Instance 4	8	8	42	13,089	50	14,040
Instance 5	8	8	50	10,129	62	11,189
Instance6	6	12	986	15,553	1083	17,694
Instance7	6	12	1179	15,224.5	1256	16,785
Instance8	6	12	1026	13,881	1094	14,205.5
Instance9	6	12	669	14,298	756	14,898.5
Instance10	6	12	814	14,143.5	884	14,678

For instances 6–10, even though there are equal instances of jobs and machines, the makespan and energy consumption values are different and depend only on the operations and types of machines used.

For a better portrayal of the optimized HMFO algorithm's results, Gantt charts have been utilized for all 10 instances. For better understanding, Gantt charts for instances 1–5 are shown separately in Figure 10, and Gantt charts for instances 6–10 are shown separately in Figure 11. Here, Figures 10 and 11 illustrate the maximum completion time for the problem with respect to instances 1–5 and instances 6–10, respectively. The X-axis of the Gantt chart indicates the average time of completion of the job (makespan), and the Y-axis denotes the machines. As stipulated in Figure 10, it is clearly observable that the makespan for all the five instances is 30, 27, 179, 42, and 50, respectively, which replicates the results that are previously mentioned in Table 9. In Figure 11, it is clearly observable that the makespan for all the five instances is 986, 1179, 1026, 669, and 814, respectively, which replicates the results that are previously mentioned in Table 9.

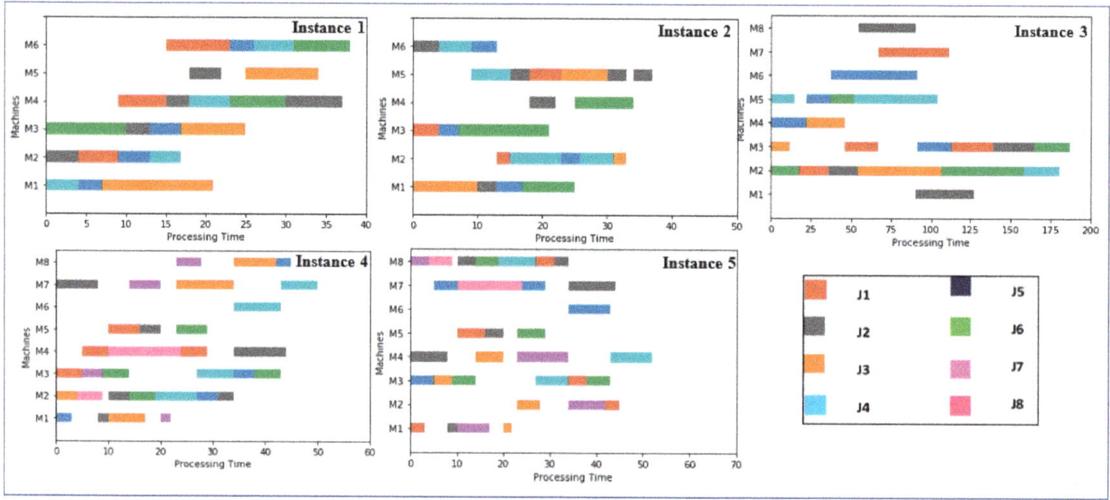

Figure 10. Gantt chart showing the makespan of instances from 1–5.

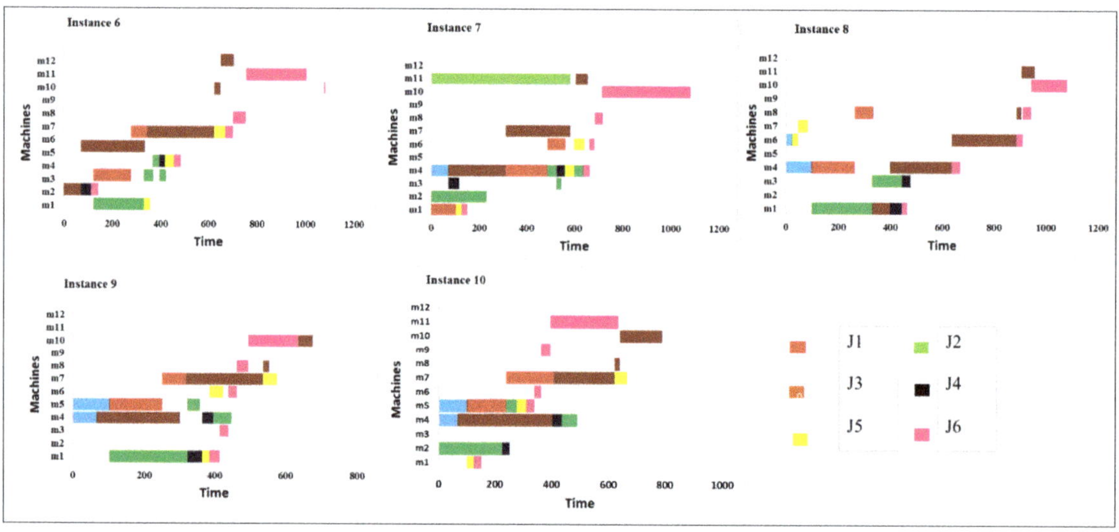

Figure 11. Gantt chart showing the make span of instances from 6–10.

In addition to comparing the performance of both the algorithms, i.e., HMFO and NSGA-II, a comparative study of the machine utilization rate of different machines for all the 10 instances is illustrated in Figures 12 and 13. It can be inferred from Figure 12 that in all instances 1–5, machine utilization rates are far better for the results that are obtained with HMFO when compared with results that are obtained with NSGA II. In Figure 13 for instance 6 in case of HMFO, machine 6 has the maximum utilization rate, and machine 9 has the minimum utilization rate. In case of NSGA-II, machine 6 has the highest, and machine 12 has the least utilization rate. In addition, it is evident from the bar graph that in the case of HMFO, machine 8 and machine 12 have zero utilization rate. This means that these machines are not at all utilized in the manufacturing process and in the case of NSGA-II, all the machines are completely utilized. Since our objective is the maximization of the service utilization rate, the machines that are not utilized can be completely removed from

being ignored and can be removed from the workspace. Since HMFO reduces the initial cost involved in installing the machines by eliminating the machines, we can conclude that HMFO gives the optimum results of the machine utilization rates.

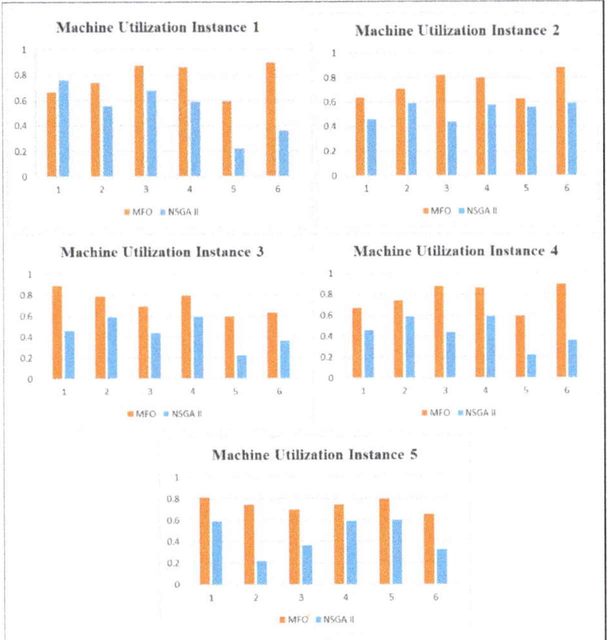

Figure 12. Utilization rate of different machines of instance 1–5.

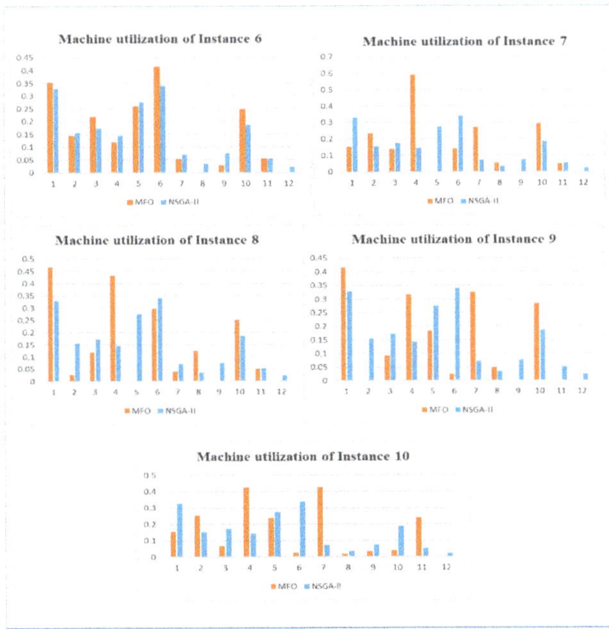

Figure 13. Utilization rate of different machines of instance 6–10.

In Table 9, the energy consumption values for all instances are plotted and shown in the Figure 14. It can be inferred that energy consumption values for all 10 instances are lower for HMFO when compared to NSGA II. Figure 15 articulates the reliability of the jobs/services provided by the selected enterprise with respect to each instance. In Figure 15, it can be inferred that for all five instances, there is not much change in the reliability. In all the instances, i.e., instances 1–10, job 3 has the highest reliability, which means machines performing job 3 have the highest probability of surviving their expected lifetime. Job 4 has the least reliability, which indicates that machines processing job 4 fail at a very early stage and cannot survive at least half of their intended lifetime. Hence, the machines involved in the manufacture of job 4 need to be improved/changed.

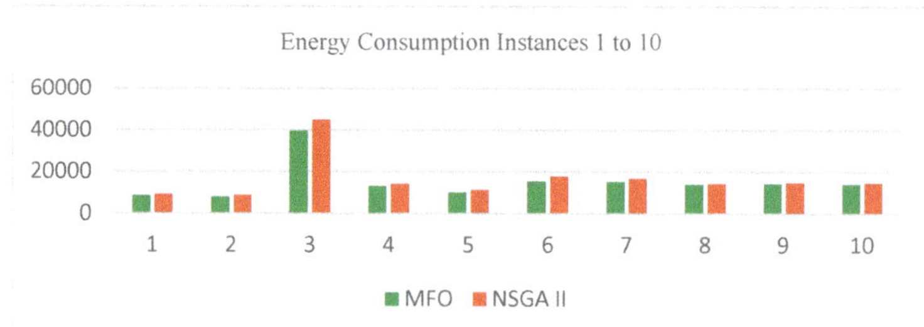

Figure 14. Energy consumption values of all instances 1–10.

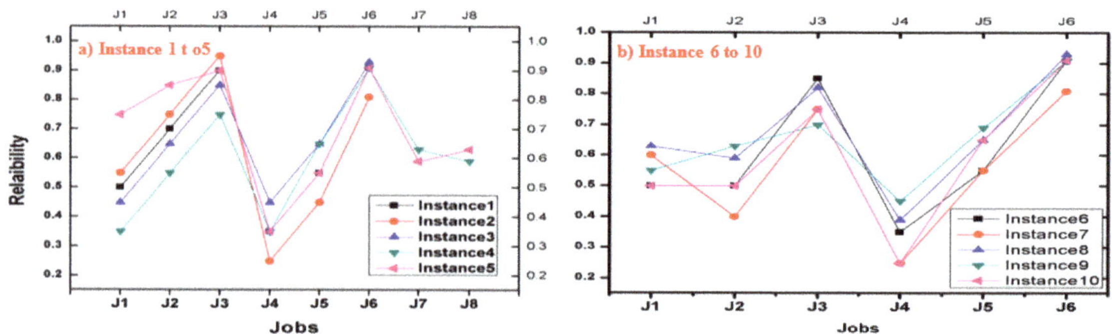

Figure 15. Reliability of different jobs of instance from 6–10.

To make the comparison of proposed algorithm with the benchmark algorithms, several performance measures were reported [53–56], and these measures were mainly useful for multi-/many-objective optimization problems. To make the process simple and effective, Pareto optimal graphs were plotted for three objectives shown in Figures 16 and 17. Figures 16 and 17 give various Pareto graphs generated by HMFO and NSGA –II, respectively. In Figures 16 and 17, the black color solutions indicate the non-dominated solutions, and blue color solutions indicate the remaining solutions. Considering effective evolution of all of the algorithms' performances gives a better picture.

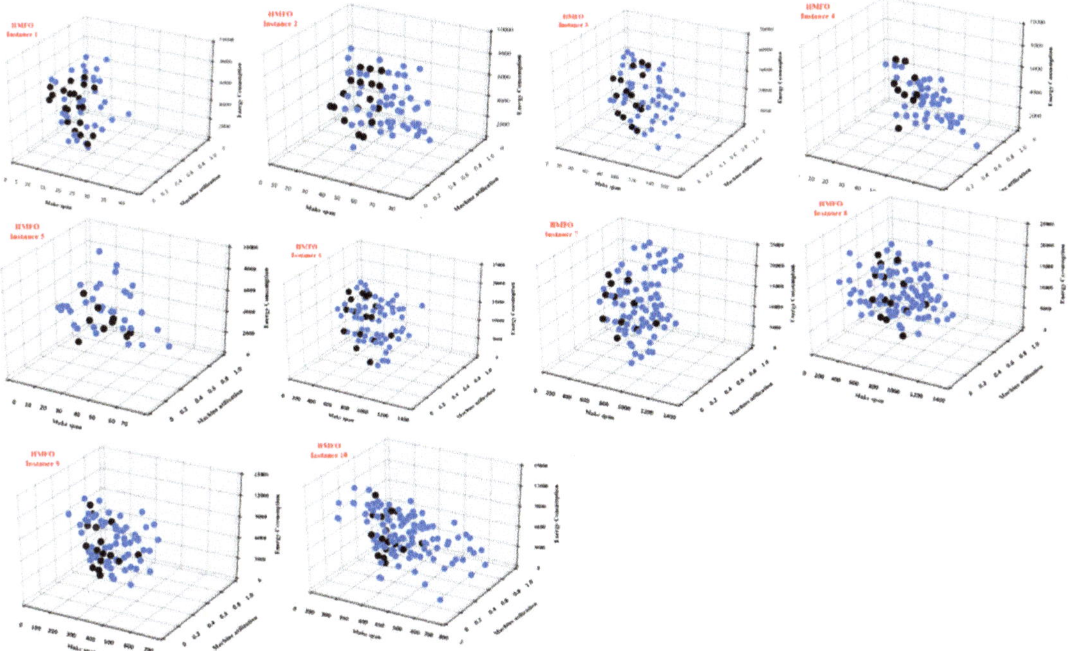

Figure 16. Pareto optimal graphs showing various solutions for three objectives makespan, energy consumption, and machine utilization for HMFO algorithm.

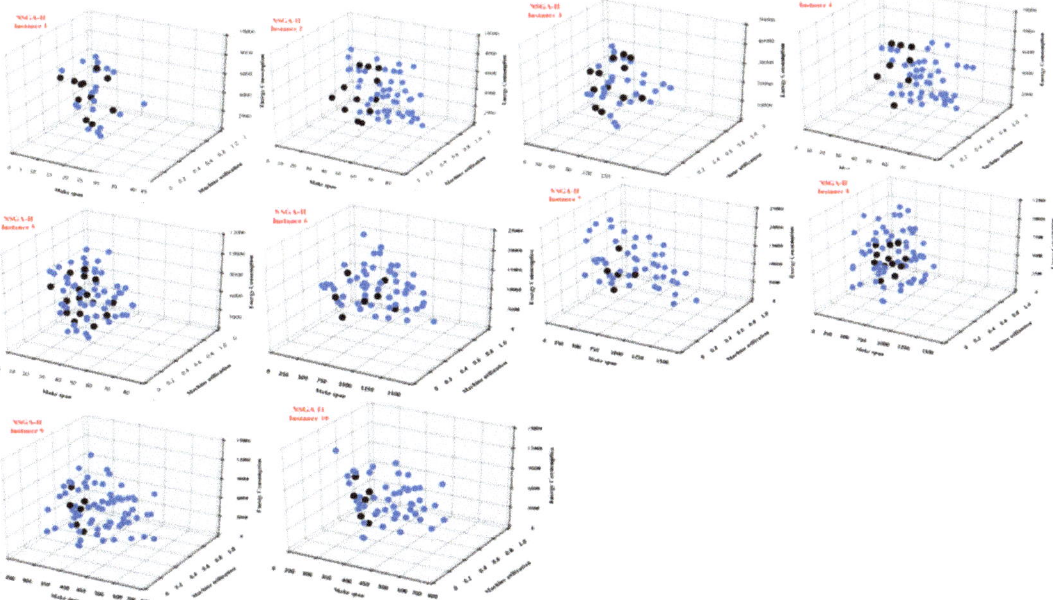

Figure 17. Pareto optimal graphs showing various solutions for three objectives makespan, energy consumption, and Machine utilization for NSGA-II algorithm.

The hypervolume (HV) is most used Performance Indicator (PI) among all for the comparison multi-objective algorithms [56]. HV is the volume occupied by the dominated Pareto front approximation '*R*' drawn from a reference point o £ O ^{L}, such that *Y* £ *R*, *R* < o. The *HV* is given by Equation (14). Here, η_L represents *L* dimensional lebesgue measure.

$$HV(R, o) = \eta_L \left(\bigcup_{Y \in R} [Y, o] \right) \tag{14}$$

The *HV* describes the region of the objective space which is weakly dominated by the approximation set. Until now, there are no particular guidelines for selecting the reference point; however, the worst possible point (i.e., dominated by all points) then the nadir point (1, 1, ... 1) is considered in most of the studies [57]. Here, the *HV* is calculated by specifying the reference point as (1.1, 1.1, ... 1.1). Out of all the available *HV* values, the best, median, and least values of hypervolume (*HV*) results are indicated with help of box-plot in Figure 18. The *HV* is calculated for various instances of problems of proposed HMFO and NSGA-II algorithms represented in Figure 18. For instance 1, in the case of HMFO, the best, median, and least values are 0.6997, 0.6297, and 0.5497, and similarly for instance 1, in the case of NSGA II, the best, median, and least *HV* values are 0.5653, 0.4853, and 0.3653, respectively. In a similar manner, the box plots were obtained for all 10 instances that are represented in Figure 18. It is well known from the literature the higher the *HV* value, the better the performance of the algorithm. In the figure, after thorough observation, the *HV* values indicate the higher values for all the instances (i.e., from instance 1 to instance 10) in case of proposed HMFO when compared to the NSGA –II algorithm. This demonstrates the superiority of the proposed HMFO over the NSGA II algorithm for better approximation. Moreover, *HV* results for the first five instances (instances 1–5) fall in a lower range when compared to the other instances (instances 6–10). This may be due to different problem scenarios that were considered in all 10 instances.

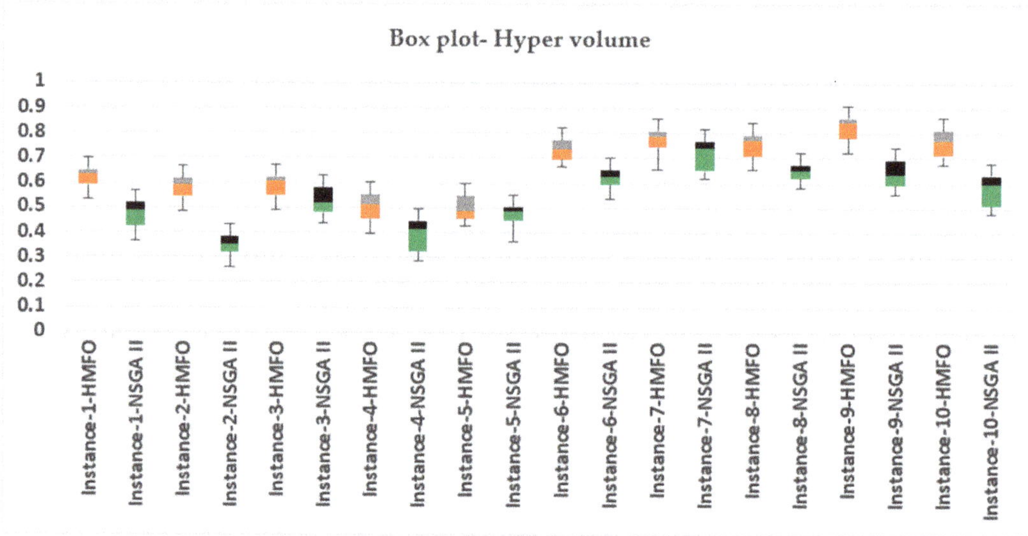

Figure 18. Comparison of HMFO and NSGA-II with Hyper-Volume (HV) results for all the 10 instances of problems.

Apart from HV indicator, the results of various other performance measures for first five instances (1–5) and last five instances (6–10) are shown in Tables 10 and 11, respectively. Out of all the available solutions, the number of Non-Dominated (ND) solutions obtained by a proposed algorithm is denoted as α, and the number of ND solutions that are not identified by the benchmark algorithm is denoted by β. For better results, it is suggested to have larger values of α and β and the ratio of β/α ratio near to one determines the effectiveness of the algorithm's strength. The percentage of ND solutions provided by a certain algorithm is expressed in terms of dominance ratio $\Upsilon\Upsilon$ in Equation (15). If larger, the dominance ratio indicates the superiority of the given algorithm.

$$\Upsilon = \frac{|B(\cup_i M_i) \setminus B(\cup_{i \neq k} M_i)|}{B(\cup_i M_i)} \tag{15}$$

where $|B(\cup_i M_i) \setminus B(\cup_{i \neq k} M_i)|$ indicates the non-dominated solutions found by the algorithm M that are unable to identify by the other standard algorithms.

Table 10. Results of performance indicators for comparison of HMFEO and NSGA II instance 1–5.

Indicator	Algorithm	Instance				
		1	2	3	4	5
α	HMFEO	**8.5**	**10.8**	**9.7**	**10.6**	**8.8**
	NSGA II	8.0	10.5	9.4	10.6	8.6
β	HMFEO	**8.1**	**10.7**	**9.5**	**10.6**	**8.6**
	NSGA II	7.0	9.5	9.3	9.3	8.3
β/α	HMFEO	**0.9529**	**0.9900**	**0.9700**	**1.0000**	**0.9700**
	NSGA II	0.8750	0.9047	0.9893	0.8773	0.9651
Υ	HMFEO	**0.5264**	**0.5079**	**0.6012**	0.5000	**0.5264**
	NSGA II	0.4736	0.4921	0.3988	0.5000	0.4736
K	HMFEO	**0.0800**	**0.0100**	**0.0300**	**0.0000**	**0.0300**
	NSGA II	0.1250	0.0953	0.0107	0.1227	0.0349
λ	HMFEO	**0.3693**	**0.0037**	**0.6289**	**0.0000**	**0.0042**
	NSGA II	12.123	9.236	8.265	8.6321	9.5632
π	HMFEO	0.4236	**0.4856**	**0.4982**	**0.4932**	**0.4495**
	NSGA II	**0.4125**	0.5563	0.6029	0.6765	0.5988
HR	HMFEO	**0.8526**	**0.7445**	0.9538	**0.6984**	**0.4495**
	NSGA II	1.2536	1.1456	**0.9469**	0.9548	0.9854

Table 11. Results of performance indicators for comparison of HMFEO and NSGA II instance 6–10.

Indicator	Algorithm	Instance				
		6	7	8	9	10
α	HMFEO	**12.5**	**13.8**	**9.7**	**16.6**	**9.8**
	NSGA II	12.0	10.5	9.4	16.6	8.8
β	HMFEO	**12.1**	**12.7**	**9.5**	**16.6**	**9.6**
	NSGA II	11.0	9.2	9.3	15.0	8.3
β/α	HMFEO	**0.9682**	**0.9202**	**0.9700**	**1.0000**	**0.9795**
	NSGA II	0.9166	0.8761	0.9893	0.9036	0.9431
Ω	HMFEO	**0.5151**	**0.5070**	**0.5078**	0.5000	**0.5057**
	NSGA II	0.4848	0.4929	0.4922	0.5000	0.4942
K	HMFEO	**0.0800**	**0.0104**	**0.0310**	**0.0000**	**0.0400**
	NSGA II	0.1250	0.0835	0.0307	0.1027	0.0349
λ	HMFEO	**0.4693**	**0.0089**	**0.5259**	**0.0000**	**0.0063**
	NSGA II	12.123	9.236	8.265	8.6321	10.5632
π	HMFEO	**0.5136**	**0.4856**	**0.4982**	**0.4932**	**0.4445**
	NSGA II	0.4125	0.4563	0.5129	0.5365	0.5488
HR	HMFEO	**0.7415**	**0.8556**	**0.8629**	**0.7875**	**0.7589**
	NSGA II	0.9438	1.2465	1.1803	1.9888	0.8765

$K(a,b)$ in Equation (16) is useful for comparison of Pareto fronts, particularly to identify the weak solutions of algorithm b with an algorithm ($a > b$), which ultimately gives the correctness of the particular algorithm. Furthermore, lesser K species than the ND solutions identified by a particular algorithm are considered weaker.

$$K(a,b) = \frac{|b \in B, \exists a \in A : a > b|}{|B|} \tag{16}$$

Lesser π is necessary, and the values which are very nearer indicate the highly distributed uniformly over the Pareto front. Equation (17) values of π which is the Euclidean length between end points of the identified ND Pareto set by an algorithm is compared to the net ND Pareto front.

$$\Pi = \frac{F_f + F_l + \sum_{i=1}^{J-1} \left| F_i - \bar{F} \right|}{F_f + F_l + (J-1)\bar{F}} \tag{17}$$

λ indicates convergence power if smaller λ values are useful for identified ND solutions by the algorithm and fall very close range in the vicinity of net ND solutions for Euclidean lengths.

Hyper-volume ratio or Hyper-area Ratio (HR) is the ratio of HV ratio of the algorithms. $HR_{(R,h,o,)} = HV(R, o)/HV(h, o)$ from this lower HR is required to obtain better approximation [58]. In Tables 10 and 11, it can be confirmed that the values marked in bold are the best values for showing the superiority of the proposed HMFO algorithm over the NSGA II algorithm.

7. Conclusions

Advancements in technology, such as information and communication technologies (ICT), have changed the traditional manufacturing systems practices. This is especially true for a distributed manufacturing system due to its ability to cater to its needs such as Big Data, interoperability, timely delivery, etc. In this research, the authors have considered a case study of automotive industries which are small and medium scale in nature and are geographically distributed with the objectives such as the selection of appropriate suppliers according to product type and enhancing the system functions such as makespan, energy consumption, and increased service utilization rate, interoperability, and reliability. To execute the first objective: supplier discovery is implemented through text mining based on supervised machine-learning models. The results of classification Decision Tree (J48), Naïve Bayes, Random Forest, and Support Vector Machines are validated through various performance measures, mainly Precision, Recall, and F-Measures. Decision trees have been found to be best with a precision of 0.93 for the purpose. These selected potential suppliers and their related information have been transferred as input data to the next phase.

The flexibility and complexity of a distributed manufacturing environment create the need for investigating the multiple process plans and multiple performance measures. Hence, the research paper also investigated alternative process plans to the objective functions makespan, energy consumption, service utilization, and reliability of services. We developed a MINLP model, and by acknowledging the NP-hard nature of the above scenario, a multi-objective evolutionary algorithm was decided to be utilized for which the input of task-specific suppliers is the outcome of supervised algorithmic models. As a result, we have used a bio-inspired Moth Flame Optimization evolutionary algorithm and tuned the algorithm to fit our problem objectives.

The results demonstrate that the use of evolutionary HMFO reduces the number of machine when compared to NSGA-II proving the effectiveness of the methodology used in this research. It also provides similar results with respect to the survivability of jobs as compared to NSGA-II. Out of all the considered objective functions, energy consumption is of utmost importance because of its effect on the current manufacturing environment. An experimental comparison also reveals the effectiveness of the proposed HMFO. Various performance indicators are used to compare the superiority of the proposed HMFO over the benchmark NSGA II algorithm. Thus, the results obtained showcase the superiority of the approach mentioned in this research. Future work requires adopting an application of the proposed methodology on a wider dataset using various other many-objective evolutionary algorithms. It is important to mention here that the comparison of proposed algorithms must be performed by using the state-of-the-art many-objective optimization algorithms like Non-Dominated Sorting Genetic algorithm (NSGA) –III, S-Metric Selection Evolutionary Multi-Objective Optimization Algorithm (SMS-EMOA), EvolutionaryA Algorithm-based on Decomposition (MOEA/D), etc., for better understanding of the performance of the algorithm [59].

Author Contributions: All authors contributed equally to this work and approved the final manuscript. All authors have read and agreed to the published version of the manuscript.

Funding: The project is funded by Department of Science and Technology, Science and Engineering Research Board (DST-SERB), Statutory Body Established through an Act of Parliament: SERB Act 2008, Government of India with Sanction Order No ECR/2016/001808, and also by FCT–Portuguese Foundation for Science and Technology within the R&D Units Projects Scopes: UIDB/00319/2020, UIDP/04077/2020, and UIDB/04077/2020.

Institutional Review Board Statement: Not applicable.

Informed Consent Statement: Not applicable.

Data Availability Statement: Not applicable.

Conflicts of Interest: The authors declare no conflict of interest.

References

1. Srai, J.S.; Graham, G.; Hennelly, P.; Phillips, W.; Kapletia, D.; Lorentz, H. Distributed manufacturing: A new form of localised production? *Int. J. Oper. Prod. Manag.* **2020**, *40*, 697–727. [CrossRef]
2. Ur-Rahman, N.; Harding, J.A. Textual data mining for industrial knowledge management and text classification: A business-oriented approach. *Expert Syst. Appl.* **2012**, *39*, 4729–4739. [CrossRef]
3. Zhang, J.; Gao, L.; Chan, F.T.; Li, P. A holonic architecture of the concurrent integrated process planning system. *J. Mater. Process. Technol.* **2003**, *139*, 267–272. [CrossRef]
4. Kumar, M.; Rajotia, S. Integration of scheduling with computer aided process planning. *J. Mater. Process. Technol.* **2003**, *138*, 297–300. [CrossRef]
5. Choudhary, A.; Oluikpe, P.; Harding, J.; Carrillo, P. The needs and benefits of text mining applications on post-project reviews. *Comput. Ind.* **2009**, *60*, 728–740. [CrossRef]
6. Li, N.; Wu, D. Using text mining and sentiment analysis for online forums hotspot detection and forecast. *Decis. Support. Syst.* **2010**, *48*, 354–368. [CrossRef]
7. Kucheria, P.B.; Shaikh, I.R. Ontology Based Text Mining. Corpus ID: 212474163. Available online: https://www.semanticscholar.org/paper/Ontology-based-Text-Mining-Kucheria-Shaikh/ec8e69e2250eb5916e7aa1c782a9689bce985876 (accessed on 19 May 2021).
8. Yazdizadeh, P.; Ameri, F. A text mining technique for manufacturing supplier classification. In Proceedings of the ASME 2015 International Design Engineering Technical Conferences and Computers and Information in Engineering Conference, Boston, MA, USA, 2–5 August 2015.
9. Morariu, C.; Morariu, O.; Răileanu, S.; Borangiu, T. Machine learning for predictive scheduling and resource allocation in large scale manufacturing systems. *Comput. Ind.* **2020**, *120*, 103244. [CrossRef]
10. Park, K.; Kremer, G.O. Text mining-based categorization and user perspective analysis of environmental sustainability indicators for manufacturing and service systems. *Ecol. Indic.* **2017**, *72*, 803–820. [CrossRef]
11. Akhtar, M.D.; Manupati, V.K.; Varela, M.L.R.; Putnik, G.D.; Madureira, A.M.; Abraham, A. Manufacturing services classification in a decentralized supply chain using text mining. In *Advances in Intelligent Systems and Computing*; Kacprzyk, J., Ed.; Springer Science and Business Media LLC: Warangal, India, 2018; pp. 186–193.
12. Bianchi, G.; Bruni, R.; Scalfati, F. Identifying e-commerce in enterprises by means of text mining and classification algorithms. *Math. Probl. Eng.* **2018**, *2018*, 7231920. [CrossRef]
13. Shao, X.; Li, X.; Gao, L.; Zhang, C. Integration of process planning and scheduling—A modified genetic algorithm-based approach. *Comput. Oper. Res.* **2009**, *36*, 2082–2096. [CrossRef]
14. Chryssolouris, G.; Chan, S.; Cobb, W. Decision making on the factory floor: An integrated approach to process planning and scheduling. *Robot. Comput. Manuf.* **1984**, *1*, 315–319. [CrossRef]

15. Chryssolouris, G.; Chan, S.; Suh, N. An integrated approach to process planning and scheduling. *CIRP Ann.* **1985**, *34*, 413–417. [CrossRef]

16. Montreuil, B.; Frayret, J.-M.; D'Amours, S. A strategic framework for networked manufacturing. *Comput. Ind.* **2000**, *42*, 299–317. [CrossRef]

17. Wang, Y.H.; Yin, C.W.; Zhang, Y. A multi-agent and distributed ruler based approach to production scheduling of agile manufacturing systems. *Int. J. Comput. Integr. Manuf.* **2003**, *16*, 81–92. [CrossRef]

18. Chen, C.-L. A novel disruptive innovation-like algorithm for single-machine scheduling with sequence-dependent family setup times. *Appl. Sci.* **2021**, *11*, 986. [CrossRef]

19. Tursi, A.; Panetto, H.; Morel, G.; Dassisti, M. Ontological approach for products-centric information system interoperability in networked manufacturing enterprises. *Annu. Rev. Control.* **2009**, *33*, 238–245. [CrossRef]

20. Dehghani, M.; Montazeri, Z.; Dehghani, A.; Malik, O.; Morales-Menendez, R.; Dhiman, G.; Nouri, N.; Ehsanifar, A.; Guerrero, J.M.; Ramirez-Mendoza, R.A. Binary spring search algorithm for solving various optimization problems. *Appl. Sci.* **2021**, *11*, 1286. [CrossRef]

21. Manupati, V.; Thakkar, J.; Wong, K.; Tiwari, M. Near optimal process plan selection for multiple jobs in networked based manufacturing using multi-objective evolutionary algorithms. *Comput. Ind. Eng.* **2013**, *66*, 63–76. [CrossRef]

22. Ehm, J.; Freitag, M.; Frazzon, E.M. A heuristic optimisation approach for the scheduling of integrated manufacturing and distribution systems. *Procedia CIRP* **2016**, *57*, 357–361. [CrossRef]

23. Al-Shayea, A.; Fararah, E.; Nasr, E.A.; Mahmoud, H.A. Model for integrating production scheduling and maintenance planning of flow shop production system. *IEEE Access* **2020**, *8*, 208826–208835. [CrossRef]

24. Manupati, V.; Deo, S.; Cheikhrouhou, N.; Tiwari, M. Optimal process plan selection in networked based manufacturing using game-theoretic approach. *Int. J. Prod. Res.* **2012**, *50*, 5239–5258. [CrossRef]

25. Petrović, M.; Mitić, M.; Vuković, N.; Miljković, Z. Chaotic particle swarm optimization algorithm for flexible process planning. *Int. J. Adv. Manuf. Technol.* **2016**, *85*, 2535–2555. [CrossRef]

26. Mohammadi, S.; Al-E-Hashem, S.M.; Rekik, Y. An integrated production scheduling and delivery route planning with multi-purpose machines: A case study from a furniture manufacturing company. *Int. J. Prod. Econ.* **2020**, *219*, 347–359. [CrossRef]

27. Barzanji, R.; Naderi, B.; Begen, M.A. Decomposition algorithms for the integrated process planning and scheduling problem. *Omega* **2020**, *93*, 102025. [CrossRef]

28. Ertogral, K.; Öztürk, F.S. An integrated production scheduling and workforce capacity planning model for the maintenance and repair operations in airline industry. *Comput. Ind. Eng.* **2019**, *127*, 832–840. [CrossRef]

29. Li, X.; Gao, L.; Wang, W.; Wang, C.; Wen, L. Particle swarm optimization hybridized with genetic algorithm for uncertain integrated process planning and scheduling with interval processing time. *Comput. Ind. Eng.* **2019**, *135*, 1036–1046. [CrossRef]

30. Cao, Y.; Zhang, H.; Li, W.; Zhou, M.; Zhang, Y.; Chaovalitwongse, W.A. Comprehensive learning particle swarm optimization algorithm with local search for multimodal functions. *IEEE Trans. Evol. Comput.* **2019**, *23*, 718–731. [CrossRef]

31. Zhou, L.; Zhang, L.; Sarker, B.R.; Laili, Y.; Ren, L. An event-triggered dynamic scheduling method for randomly arriving tasks in cloud manufacturing. *Int. J. Comput. Integr. Manuf.* **2018**, *31*, 318–333. [CrossRef]

32. Maturana, F.P.; Norrie, D.H. A generic mediator for multi-agent coordination in a distributed manufacturing system. In Proceedings of the 1995 IEEE International Conference on Systems, Man and Cybernetics. Intelligent Systems for the 21st Century, Vancouver, BC, Canada, 22–25 October 2002; Institute of Electrical and Electronics Engineers (IEEE): Piscataway, NJ, USA, 2002; pp. 952–957.

33. Sohrabi, M.K.; Azgomi, H. A Survey on the combined use of optimization methods and game theory. *Arch. Comput. Methods Eng.* **2018**, *27*, 59–80. [CrossRef]

34. Jia, H.; Fuh, J.; Nee, A.; Zhang, Y. Integration of genetic algorithm and Gantt chart for job shop scheduling in distributed manufacturing systems. *Comput. Ind. Eng.* **2007**, *53*, 313–320. [CrossRef]

35. Shen, W.; Wang, L.; Hao, Q. Agent-based distributed manufacturing process planning and scheduling: A state-of-the-art survey. *IEEE Trans. Syst. Man Cybern. Part C Appl. Rev.* **2006**, *36*, 563–577. [CrossRef]

36. Gao, J.; Smyrnakis, M.; Gao, H.T.; Xu, Y.; Barreiro-Gomez, J.; Ndong, M.; Smyrnakis, M.; Tembine, H. Distributionally Robust Optimization. *Optim. Algorithms Ex.* **2018**. [CrossRef]

37. Liu, B.; Wang, K.; Zhang, R. Variable neighborhood based memetic algorithm for distributed assembly permutation flowshop. In Proceedings of the 2016 IEEE Congress on Evolutionary Computation (CEC), Vancouver, BC, Canada, 24–29 July 2016; pp. 1682–1686. [CrossRef]

38. Kesen, S.E.; Das, S.K.; Güngör, Z. A genetic algorithm based heuristic for scheduling of virtual manufacturing cells (VMCs). *Comput. Oper. Res.* **2010**, *37*, 1148–1156. [CrossRef]

39. Sortrakul, N.; Nachtmann, H.; Cassady, C. Genetic algorithms for integrated preventive maintenance planning and production scheduling for a single machine. *Comput. Ind.* **2005**, *56*, 161–168. [CrossRef]

40. Li, W.D.; McMahon, C.A. A simulated annealing-based optimization approach for integrated process planning and scheduling. *Int. J. Comput. Integr. Manuf.* **2007**, *20*, 80–95. [CrossRef]

41. Tiwari, A.; Chang, P.-C.; Tiwari, M.K.; Kandhway, R. A Hybrid Territory Defined evolutionary algorithm approach for closed loop green supply chain network design. *Comput. Ind. Eng.* **2016**, *99*, 432–447. [CrossRef]

42. Li, Y.; Tao, F.; Cheng, Y.; Zhang, X.; Nee, A. Complex networks in advanced manufacturing systems. *J. Manuf. Syst.* **2017**, *43*, 409–421. [CrossRef]
43. Tang, D.; Dai, M.; Salido, M.A.; Giret, A. Energy-efficient dynamic scheduling for a flexible flow shop using an improved particle swarm optimization. *Comput. Ind.* **2016**, *81*, 82–95. [CrossRef]
44. Hou, Y.; Wu, N.; Zhou, M.; Li, Z. Pareto-optimization for scheduling of crude oil operations in refinery via genetic algorithm. *IEEE Trans. Syst. Man, Cybern. Syst.* **2015**, *47*, 517–530. [CrossRef]
45. Wang, X.; Xing, K.; Yan, C.B.; Zhou, M. A novel MOEA/D for multiobjective scheduling of flexible manufacturing systems. *Complexity* **2019**, *2019*, 5734149. [CrossRef]
46. Lin, L.; Hao, X.-C.; Gen, M.; Jo, J.-B. Network modeling and evolutionary optimization for scheduling in manufacturing. *J. Intell. Manuf.* **2011**, *23*, 2237–2253. [CrossRef]
47. Piroozfard, H.; Wong, K.Y.; Hassan, A. A hybrid genetic algorithm with a knowledge-based operator for solving the job shop scheduling problems. *J. Optim.* **2016**, *2016*, 7319036. [CrossRef]
48. Klement, N.; Abdeljaouad, M.A.; Porto, L.; Silva, C. Lot-sizing and scheduling for the plastic injection molding industry—A hybrid optimization approach. *Appl. Sci.* **2021**, *11*, 1202. [CrossRef]
49. Beckham, C.; Hall, M.; Frank, E. WekaPyScript: Classification, regression, and filter schemes for WEKA implemented in Python. *J. Open Res. Softw.* **2016**, *4*, e33. [CrossRef]
50. Mirjalili, S. Moth-flame optimization algorithm: A novel nature-inspired heuristic paradigm. *Knowl. Based Syst.* **2015**, *89*, 228–249. [CrossRef]
51. Tang, D.; Dai, M. Energy-efficient approach to minimizing the energy consumption in an extended job-shop scheduling problem. *Chin. J. Mech. Eng.* **2015**, *28*, 1048–1055. [CrossRef]
52. Jin, L.; Zhang, C.; Fei, X. Realizing energy savings in integrated process planning and scheduling. *Processes* **2019**, *7*, 120. [CrossRef]
53. Cheng, C.-Y.; Li, S.-F.; Ying, K.-C.; Liu, Y.-H. Scheduling jobs of two competing agents on a single machine. *IEEE Access* **2019**, *7*, 98702–98714. [CrossRef]
54. Ying, K.-C.; Lin, S.-W.; Wan, S.-Y. Bi-objective reentrant hybrid flowshop scheduling: An iterated Pareto greedy algorithm. *Int. J. Prod. Res.* **2014**, *52*, 5735–5747. [CrossRef]
55. Lin, S.-W.; Ying, K.-C. A multi-point simulated annealing heuristic for solving multiple objective unrelated parallel machine scheduling problems. *Int. J. Prod. Res.* **2014**, *53*, 1065–1076. [CrossRef]
56. Audet, C.; Bigeon, J.; Cartier, D.; Le Digabel, S.; Salomon, L. Performance indicators in multiobjective optimization. *Eur. J. Oper. Res.* **2021**, *292*, 397–422. [CrossRef]
57. Ishibuchi, H.; Imada, R.; Setoguchi, Y.; Nojima, Y. How to specify a reference point in hypervolume calculation for fair performance comparison. *Evol. Comput.* **2018**, *26*, 411–440. [CrossRef] [PubMed]
58. Haber, R.E.; Beruvides, G.; Quiza, R.; Hernandez, A. A Simple multi-objective optimization based on the cross-entropy method. *IEEE Access* **2017**, *5*, 22272–22281. [CrossRef]
59. Yang, Z.; Wang, H.; Yang, K.; Bäck, T.; Emmerich, M. SMS-EMOA with multiple dynamic reference points. In Proceedings of the 12th International Conference on Natural Computation, Fuzzy Systems and Knowledge Discovery (ICNC-FSKD), Changsha, China, 13–15 August 2016; Institute of Electrical and Electronics Engineers (IEEE): Piscataway, NJ, USA, 2016; pp. 282–288.

Article

Proposal of a Monitoring System for Collaborative Robots to Predict Outages and to Assess Reliability Factors Exploiting Machine Learning

Khurshid Aliev [†] and Dario Antonelli [*,†]

Department of Management and Production Engineering, Polytechnic University of Turin, Corso Duca degli Abruzzi 24, 10138 Torino, Italy; khurshid.aliev@polito.it
* Correspondence: dario.antonelli@polito.it; Tel.: +39-01-1090-7288
† These authors contributed equally to this work.

Abstract: Industry standards pertaining to Human-Robot Collaboration (HRC) impose strict safety requirements to protect human operators from danger. When a robot is equipped with dangerous tools, moves at a high speed or carries heavy loads, the current safety legislation requires the continuous on-line monitoring of the robot's speed and a suitable separation distance from human workers. The present paper proposes to make a virtue out of necessity by extending the scope of on-line monitoring to predicting failures and safe stops. This has been done by implementing a platform, based on open access tools and technologies, to monitor the parameters of a robot during the execution of collaborative tasks. An automatic machine learning (ML) tool on the edge of the network can help to perform the on-line predictions of possible outages of collaborative robots, especially as a consequence of human-robot interactions. By exploiting the on-line monitoring system, it is possible to increase the reliability of collaborative work, by eliminating any unplanned downtimes during execution of the tasks, by maximising trust in safe interactions and by increasing the robot's lifetime. The proposed framework demonstrates a data management technique in industrial robots considered as a physical cyber-system. Using an assembly case study, the parameters of a robot have been collected and fed to an automatic ML model in order to identify the most significant reliability factors and to predict the necessity of safe stops of the robot. Moreover, the data acquired from the case study have been used to monitor the manipulator' joints; to predict cobot autonomy and to provide predictive maintenance notifications and alerts to the end-users and vendors.

Keywords: on-line monitoring; collaborative robots; human robot collaboration; machine learning

Citation: Aliev, K.; Antonelli, D. Proposal of a Monitoring System for Collaborative Robots to Predict Outages and to Assess Reliability Factors Exploiting Machine Learning. *Appl. Sci.* **2021**, *11*, 1621. https://doi.org/10.3390/app11041621

Academic Editor: Emanuele Carpanzano

Received: 11 January 2021
Accepted: 4 February 2021
Published: 10 February 2021

1. Introduction

Revolution of Industry 4.0 (I4.0) introduces new tools and technologies that can be integrated with the ones that are already exploited by factories. Several of them have already been deployed in different manufacturing sectors to improve productivity and to satisfy the expectations of consumers expectations for customisation. One such I4.0 enabling technology is the collaborative robot (cobot) which is widely deployed in industry [1,2]. A cobot allows the skills of a robot, such as precision and strength, to be combined with human dexterity and problem solving abilities [3] on a human-robot collaborative (HRC) workstation. Cobots are designed to interact with humans directly and physically within a shared workspace [4]. HRC applications that are designed on the basis of reliability and safety standards increase human trust in collaboration and improve the quality and working conditions of employees. In HRC, humans and robots share the same workspace. Cobots are specifically made to halt before any involuntary contact may harm the human coworker can cause harm. However, frequent halts induce accelerated wear and tear of the robot and increase the probability of mechanical failures. Furthermore, cobots should be light weight in order to minimise their inertia and allow them to stop suddenly. Therefore, joints have lower strength compared to standard industrial robots. Accordingly, scheduled maintenance is no longer an appropriate strategy as it cannot account for the number and severity of the forced halts of a robot. Continuous monitoring of the state of health of such robots would be preferable. During monitoring, the system generates useful knowledge and information, such as data about the robot's sensors and event logs, which are stored in historic logging databases and can be recalled to perform analytics. The smart data analytics of collected data, using machine learning techniques, offers an opportunity to monitor the health condition of industrial robots, to predict the mobile autonomy and to perform predictive maintenance, if necessary. The Internet of Things (IoT) is another emerging tool that is used for I4.0: connected devices with embedded systems that are able to interact and communicate with each other or with centralised devices. The integration of the IoT in a decision making system could improve the performance of the human-robot interaction. This is the reason why the industrial internet of things (IIoT) extension refers to its industrial application, that is to interconnect industrial machines and devices, robots, sensors and instruments that are centralised to collect, exchange and analyze data. The IIoT offers the possibility of achieving the complete design of physical cyber-systems through the integration of data processing technologies, intelligent software and sensors. On-line monitoring systems, and predictive maintenance models can be built on the basis of a large set of historical data. Several steps are involved in such a process: preprocessing of the collected data; extracting features from sensor data or feeding sensor data directly to machine learning models; training the predictive models; generating decision support models that will be able to evaluate a new data sent to the system; deploying developed models and integrating them with the system. The on-line visualisation of the health status of a robot and alerts about predicted failures will improve the human robot interaction. However, the applications of such models can go far beyond HRC. The concepts of machine learning (ML) tools used for predictive maintenance applications utilizing data available data on the internet have been discussed in recent studies, such as in [5], and a condition-based monitoring system, using ML tools, has been successfully deployed for a smart railway applications [6].

Hence, this paper has focused on developing a framework using I4.0 enabling technologies to improve reliability and safety in HRC applications. The proposed framework allows a cobot's condition to be monitored continuously during HRC. The monitoring deploys IoT connectivity, a data acquisition system, physical cyber-systems and ML tools to perform analytics. The paper is divided as follows: the relevant equipment parametersare first identified, and a description of the data acquisition framework is then given, an application to an assembly case study in which all the necessary data are collected is presented, and finally the analysis results of the considered case study are presented and discussed.

2. Research Hypothesis

In order to determine the relevant parameters that have to be monitored, it is worth analysing the most common industrial cobot applications at present in use. TTraditional robotic applications in fact exclude the access of humans to the work area and therefore limit the range of applications to production processes [7]. On the other hand, as cobots are designed to work with humans in the same shared workspace, several new applications are emerging [8].

The general requirements for collaborative robot system applications, based on ISO 10218-1:2011 [9] and ISO 10218-2:2011 [10] are described in ISO 15066:2016 Robots and robotic devices—Collaborative robots [11].

According to the ISO technical specifications, reliable safety, control and monitoring is whenever HRC processes involve heavy loads, high speeds, forces or temperatures, in a hazardous environment.

The different papers published over the last decade related to human-robot applications in the assembly, handling and welding domains as taken from the Scopus database, are listed in Figure 1. Following keywords are utilized: human robot together with collaboration assembly, welding and/or handling on the search engine of the Scopus database [12].

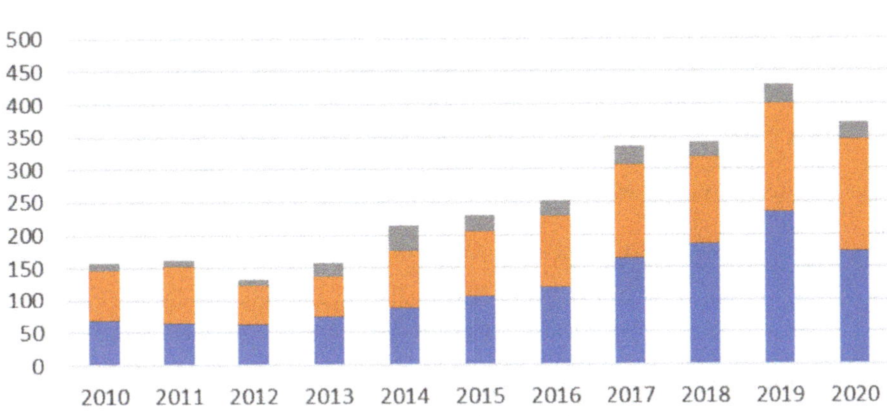

Figure 1. Recent publications of human-robot collaborative (HRC) applications in the three aforementioned domains.

Despite not claiming to be exhaustive, the chart clearly shows that most HRC applications are deployed in assembly fields [13]. Handling and process production is the second most important field that deploys collaborative robots, followed at a distance by welding. Welding applications are more complex, as they require more physical parameters in order to be precise, accurate and monitored. Table 1 offers an overview of the more studied industrial applications of cobots.

The table also shows issues that could occur in HRC applications and identifies important measuring parameters to build more reliable collaborative applications.

2.1. HRC Assembly

Human robot collaborative assembly is the action of joining two or more components together. Numerous HRC applications are already present in industry, and new solutions are continuously being proposed. Some HRC applications, with parameters being measured during collaboration in different fields, are reported in Table 1. The table also highlights the corresponding parameters necessary to monitor certain tasks. In HRC assembly, a cobot and a human can help each other during the execution of tasks. The monitoring of the physical, state, and process parameters of a cobot plays an important role in obtaining safer and more reliable collaboration. The authors of research papers [14,15] discussed the implementation of HRC assembly in manufacturing, and proposed industrial-like solutions. The importance of such parameters and variables as the load, end-effector force/torque, payload, robot and temperature of the robot and joints and robot speed were identified and classified in these researches according to the tasks of the cobots in the HRC assembly. The measurment of the performance, monitoring and prediction of the above mentioned parameters are employed in the computation of the key performance indicators (KPI)s of cobots [16–18].

2.2. HRC Handling

Handling is another widely used process in industry, for example in food manufacturing and logistics material handling. Handling involves different processes, such as grasping, packaging, glueing, palletising, surface polishing, and so forth. Cobots are used in collaborative handling applications are used for such processes as picking and placing, product testing, assembly, loading/unloading, injection and moulding as supportive devices to increase the safety of human operators and to reduce repetitive strain and accidental injuries. For example, the integration of cobots in a plastic polymer production line that produces noxious gases protects the employees in the production line from toxic gases and sharp profiles. In such processes, the physical parameters of the cobots such as the accuracy of the end-effector and the temperature ranges of the robot are monitored to provide safer collaboration [19,20].

2.3. HRC Welding

Welding is another widely used process in industry. In this direction, walk through programming has been proposed [21] for welding robots. Vision system interaction [22] in welding and augmented reality-based approaches have also been proposed [23]. Most of the proposed approaches implement cobots as assistant devices. As a result of the complexity and uncertainty of the welding process, effective practical applications, using collaborative robots, are still limited. Welding robots are currently programmed by means of lead-through or offline. Intelligent technologies, such as vision sensing, automatic programming, guiding and tracking, and real-time monitoring of the welding process, were adopted in [24] to cope with geometrical uncertainties in the weld trajectory. Thus, as shown in Table 1 such parameters as end-effector force, payload, robot temperature, joint speed, joint orientation and position are significant for the success of the welding operation and are likely to be monitored.

Table 1. Human-robot interaction (HRC) applications.

HRC Applications	Deployed Tasks	Important Parameters for HRC
Assembly	HRC assembly in a shared workspace [25]	End-effector force;
	Manual guidance collaborative assembly [14]	Payload monitoring;
	HRC integrated automotive assembly [15]	Robot Temperature; Joint Speed;
Handling	Hazardous material handling [26]	End-effector force;
	Aseptic bottling using AR [27]	Joint position and orientation;
	Collaborative surface polishing, sanding [28,29]	Robot Force; Joint Speed;
	Collaborative robot injection and moulding [30]	Speed and separation monitoring
Welding	Virtual reality HRC welding [21]	Torque/force sensors; Temperatures;
	HRC Welding Cell [22]	Position and orientation accuracy;
	Spot welding manual guidance using AR [23]	Robot Temperature; Joint Speed;

3. Methods

The proposed on-line monitoring system tracks the physical conditions of the cobot while performing HRC processes. The framework of the online monitoring system is outlined in Figure 2. Basically, the system is composed of several integrated parts: collaborative robot, that communicates with gateway using real time data exchange (RTDE) and MODBUS protocols; data acquisition, which is the gateway to cloud communication; database server, which stores the data necessary for prediction purposes and to feed the on-line monitoring dashboard; data preprocessing which extracts meaningful features from the dataset and transfers them to the ML models; machine learning models, which are exploited to predict the future behaviour of any parts subject to failure; application layer, which is deployed to allow the interactions with human operators under safe conditions. Overall, the monitoring system alerts a human operator whenever a cobot displays improper or erratic behaviour. The operator can access the dashboard remotely. For example, if the temperature of one of the joints is higher than expected, the operator can access the dashboard of the cobot to find out which working situation has led to the anomaly. Moreover, a cobot system can be integrated with additional sensors to detect gas or ambient pollution and then inform the human operator about the hazard. The present paper focuses on the integration of data acquisition and machine learning in a cobot monitoring system; networking communication and the application layer with management indicators have already been discussed in [18,31,32].

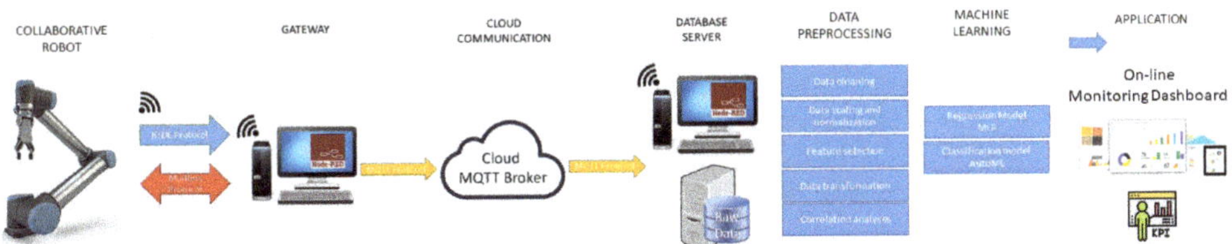

Figure 2. On-line monitoring framework.

3.1. Data Acquisition

The development of a data acquisition system for robots is an important part of a monitoring system. Data collection and data storage offer the possibility of executing deeper analyses about the connected devices and of assessing the status of the connected devices. The data collected from robots can be either physical data or event data. The former parameters that are measured by installed sensors, such as temperature, speed, vibrations, force, voltage, and current. Event data refer to the working status of a robot, and to hardware or software failures, breakdowns and so forth.

The architecture of the here presented system is composed of a collaborative robotic manipulator (Universal Robots UR3), provided with a wireless TCP/IP connection to a gateway, in order to access an Internet network over a range of around 100 m. The communication with a robotic manipulator and gateway is established using RTDE that uses TCP/IP communication on the port 30004 and robot generates output messages on 125 Hz and Modbus TCP protocols in port 502. In the system, The RTDE protocol acquires UR3 status data such as POWER OFF/ON, Emergency Stop, Protective Stop, status of programme, that is, running, paused or stopped, and other parameters necessary for monitoring [33]. In Modbus communication, the robot controller acts as a server (Slave), gateway is client (Master) that can establish connections to the robot and send standard Modbus requests to it. The server is available at the IP address of the robot controller [34]. The robot Modbus communication interface can be used to communicate to other robots, programmable logic controller (PLC)'s, Human-Machine Interface (HMI)'s or inputs and outputs (IO) devices (when the IO device is functioning as a Modbus server). In our system, the client sends a request to read specific registers that are available in the internal memory of the robot, and the robot responds by providing the requested value. The general purpose 16-bit registers present in the robot controller can contain certain discrete variables such as the tool state, tool centre point (TCP) state, joint angle, joint velocity, current, voltage, joint temperature. The system supports the Message Queuing Telemetry Transport (MQTT) protocol. MQTT is Publish/Subscribe Model which consists of three main components: publishers, subscribers, and a broker. Publishers are the lightweight sensors and devices that connect to the broker to send their data and go back to sleep whenever possible. Subscribers are applications or devices that are interested in a certain topic, or sensory data, so they connect to brokers to be informed whenever new data is received. The brokers classify sensory data in topics and send them to subscribers interested in those topics only. A device can behave as a publisher and a subscriber at the same time by publishing to specific topics and subscribing to others, the term MQTT client is used to distinguish publishers/subscribers from brokers. Node-Red is a flow based open source programming tool built upon Node.js that is used to connect hardware devices, API's and other online services belonging to the realm of Internet of Things (IoT). Node-Red provides a browser-based flow editor which can be used to create JavaScript functions in the form of interconnected blocks that together construct a flow. One of the biggest advantages of Node-Red is its ability to run at the edge of the network in the cloud and locally on a standard personal computer (PC). In the proposed framework Node-Red is ran on the standard PC and the editor is accessible via any web browser on the local network. The Node-Red dashboard is an add on module of Node-Red that is used to create and present on-line data graphical user interface(UI) on a web browser. The dashboard package allows the addition of many UI components such as buttons, sliders, leds and gauges. After representing all communication tools and protocols, data acquisition and transmission occur as follows: the robotic arm is connected to the PC(gateway) with RTDE and MODBUS protocols over Wi-Fi. On the PC, there are Node-Red nodes that read and transmit all necessary data to the server using MQTT broker. On the same PC, the Node-Red Dashboard Nodes offers the data in a graphical user interface accessible through a web browser in a real-time manner. Node-red dashboard allows the data to be presented in various forms such as charts, text fields and gauges, commands are also triggered from the GUI using sliders, switches, text fields and buttons. Additional details about communication protocols and KPI computations of the cobot on the dashboard are given in the aforementioned papers [18,31].

3.2. Data Preprocessing

The preprocessing of data helps enhanc the quality of the data and to extract meaningful insights. The data acquired from machines are normally fuzzy, biased and noisy. The preprocessing of raw data can improve the efficiency and accuracy of the ML workflows. This is why data cleaning, data integration, and feature transformation and selection are required before data can be used.

Data cleaning involves such operations as improving bad data, reducing the unnecessary elements of data, and filtering out some incorrect data that do not belong to the data set. The authors of [35] proposed different techniques such as the classic maximum likelihood procedures, like Expectation-Maximization or Multiple-Imputation for the treatment of missing and noisy data. Other authors proposed advanced ensemble missing data techniques (MDTs) [36] to improve prediction model and authors of [37] evaluate four MDTs techniques: listwise deletion (LD), mean imputation (MI), similar response pattern imputation (SRPI) and full information maximum likelihood (FMIL). The majority of authors agree to suggest using FMIL if there is enough data to afford; MI and SRPI when there is a scarcity of data. Not to use LD if data is suspect missing completely at random (MCAR).

Feature scaling which is also known as data standardisation, is another pre-processing step. It refers to the standardisation of the range of features in a data set, which means adjusting the values of numerical columns measured on different scales to a formal common scale, without changing the ranges of the values or losing information. Data normalisation means re-scaling the dimensions of data and avoiding over-weighting values. It helps to improve the overall quality of a data set [38]. Scaling intervals of [0, 1] and [−1, 1] are normally used, as shown in Equations (1) and (2)

$$[0,1] interval = \frac{actualValue - min(allValues)}{max(allValues) - min(allvalues)))} \tag{1}$$

$$[-1,1] interval = \frac{actualValue - (max(allValues) + min(allValues))/2}{(max(allValues) - min(allvalues))/2}. \tag{2}$$

Feature selection considers data composed of irrelevant and/or redundant features that could influence the performance of the trading activity to a great extent. Different feature selection such as multicollinearity, correlation coefficients and Variance Inflation Factors (VIF) are proposed by authors [39,40] to improve performance of the ML model outputs. According to the authors [39] the most commonly used techniques for numerical input and output models are correlation coefficients, such as Pearson's for a linear correlation, or rank-based methods for a nonlinear correlation. For data with numerical input and categorical output, the most common used techniques are correlation based multicollinearity coefficients and ANOVA correlation coefficients. The techniques adopted in the present framework are correlation coefficients if data coming from robot and predicting variables is numerical.To predict categorical variables of the robot multicollinearity feature selection technique is used. Data variables-collected from a cobot during human-robot collaborative applications have been evaluated using a correlation matrix and the simplest way to detect collinearity is to look at the correlation matrix of the predictors. An element of this matrix that is large in absolute value indicates a pair of highly correlated variables, and therefore a collinearity problem in the data.

Data transformation is the process where format is conveniently converted from numerical to categorical and redundant data are removed [41]. The above mentioned data pre-processing techniques have been implemented in this study to extract significant features from a cobot dataset. The collected features and variables have then been passed onto the next steps of the framework.

Correlation analyses of cobot variables during HRC.

In order t o evaluate the significance relation between predictors Xp and response Y, correlation analysis was performed. Correlation analyses provide an idea of the linearity between paired variables. The correlation coefficients between two random variables, are calculated for all the model variables as a parameter of the linear dependence [42]. The sample estimate of the correlation coefficient r_{xy} is computed for two variables, X and Y, dataset as:

$$r_{XY} = \frac{cov(XY)}{\sigma_X \sigma_Y}. \tag{3}$$

In Equation (3), σ_X and σ_Y represent the standard deviations of X and Y.

The correlation coefficients in the correlation matrix are then presented with values in the $[-1, 1]$ interval that have the following meanings:

$$r = \begin{cases} 0, \text{ means that there is no linear relationship (X and Y are linearly uncorrelated);} \\ 1, \text{ indicates a perfect positive linear relationship with X and Y varying in the same direction;} \\ -1, \text{ indicates a perfect negative linear relationship, with X and Y varying in the opposite direction;} \end{cases} \tag{4}$$

The correlation coefficients can be symmetrically arranged into a correlation matrix, where each element of each column and each row variable correspond to the correlation coefficients.

3.3. Machine Learning Models

This section describes the utilisation of ML tools to monitor the condition of a cobot. A correlation analysis was first used to identify which variables are significant. The most closely correlated variables of the collaborative robot were then fed to an ML tool to perform predictive analyses.

3.3.1. Regression Model Used to Predict the Quantitative Parameters

A Multiple Linear Regression(MLR) model was used to predict quantitative parameters. An MLR model predicts the linear relationship between a dependent variable and other variables. A multiple linear regression model with p predictor variables $x1$, $x2$, ... , xp and response Y, can be formulated as

$$Y = \beta_0 + \beta_1 x_1 + \beta_2 x_2 + ... \beta_p x_p + \epsilon, \tag{5}$$

where $\beta_0, \beta_1, ..., \beta_p$ are known as model coefficients or parameters and ϵ is a noise term which is a random error. Training data can be used to estimate $\hat{\beta}_0, \hat{\beta}_1, ..., \hat{\beta}_p$, and the coefficients being known, predictions can be made using the following equation:

$$\hat{y} = \hat{\beta}_0 + \hat{\beta}_1 x_1 + \hat{\beta}_2 x_2 + ... + \hat{\beta}_p x_p, \tag{6}$$

where $\hat{y}$ represents a prediction of Y on the basis of $X = x$. In the previous equation, the hat ˆ symbol refers to the estimated coefficients or predicted response. Values of

β must be estimated. The least square approach is used in this model to minimise the following equation:

$$RSS = \sum_{i=1}^{n}(Y_i - \hat{y}_i)^2 = \sum_{i=1}^{n}(Y_i - \hat{\beta}_0 - \hat{\beta}_1 x_{i1} - \hat{\beta}_2 x_{i2} - ... - \hat{\beta}_p x_{ip})^2. \tag{7}$$

The following metrics have been selected from Table 2 to evaluate regression model:

Table 2. Evaluation metrics of the linear regression model .

Metrics	Description	Formulation
Residual Standard Error (RSE)	The Residual Standard Error is a measure of the quality of a linear regression fit.	$RSE = \sqrt{\frac{1}{n-2}RSS}$
R^2	R squared is the square of the Pearson correlation coefficient between the labels and the predicted values. This metric ranges from zero to one. A higher value indicates a higher quality model.	$R^2 = 1 - \frac{(y_i - \hat{y}_i)^2}{y_i - \hat{y})^2} = \frac{SS_{reg}}{SS_{tot}}$
Adjusted R^2	This measures the proportion of variation explained by only those independent variables that really help to explain the dependent variable. In the equation, where R^2-sample R-square; p-Number of predictors; N-total sample size	$\bar{R}^2 = 1 - \frac{(1-R^2)(N-1))}{N-p-1}$
F-score	This makes it possible to compute the variance of the dependant variable, the simpler model is not able to explain as much as the complex model. In the equation k_1 and k_2 are parameters of two models	$F\text{-statistics} = \frac{(\frac{RSS_1 - RSS_2}{k_2 - k_1})}{(\frac{RSS_2}{n - k_2})}$
p-value	This is a statistical test that determines the probability of extreme results of the statistical hypothesis test, and which takes the Null Hypothesis to be correct.	

3.3.2. Automatic Classification Model to Predict Qualitative Parameters

The Automatic machine learning(AutoML) system was adopted to find the best ML model for our framework. H2O AutoML is an open source, user-friendly machine learning software that was designed not only for advanced machine learning users, but also for non experts. Recent studies show that H2O AutoML [43] performs better than other competitor tools. The authors of [44,45] assessed the robustness and efficiency of AutoML, with respect to other automatic models such as TPOT [46] and AutoKeras [47]. These authors used dirty, clean and noisy data sets to evaluate the robustness of the tool. Other studies, [48,49] have shown the effectiveness of the AutoML system, with respect to other tools, like auto-sklearn [50] and Auto-WEKA [51] using open source datasets. AutoML relies on the efficient training of H2O machine learning algorithms to produce a large number of models in a short time. H2O AutoML supports the supervised training of regression, binary classification and multi-class classification models on tabular datasets. H2O AutoML is available in Python, R, Java and Scala as well as through a web GUI.

The base models of H2O are Generalized Linear Models (GLM), Distributed Random Forests (DRF), XGBoost, Gradient Boosting Machines (GBM), and Deep Learning (NN). The used hyperparameters are chosen from a predefined search space using a grid search. H2O chooses one of the three different options. It may use just one of the base models or their hyperparameter-optimised versions. It can also choose a Best Of Family Stacked Ensemble model, which includes one model from each category and the last available option is the All Models Stacked Ensemble pipeline. After training one of the above mentioned models, AutoML uses a test dataset to evaluate the accuracy and quality of the new model, and provides a number of evaluation metrics that indicate how good the model performs on the test dataset. The following metrics were selected to evaluate the regression model:

The evaluation metrics used for the classification model are shown in the Table 3.

Table 3. Evaluation metrics of the classification model.

Metrics	Description of the Metrics
AUC PR	The area under the precision-recall (PR) curve. This value ranges from zero to one, and a higher value indicates a higher-quality model.
AUC ROC	The area under the receiver operating characteristic (ROC) curve. This ranges from zero to one, and a higher value indicates a higher-quality model.
Accuracy	The fraction of classification predictions produced by the model that were correct.
Log loss	The cross-entropy between the model predictions and the target values. This ranges from zero to infinity, and a lower value indicates a higher-quality model.
RMSE	The root-mean-square error metric is a frequently used measure of the differences between the values predicted by a model, or an estimator, and the observed values. This metric ranges from zero to infinity; a lower value indicates a higher quality model.
MSE	This is an estimator that measures the average of the squares of the errors, that is, the average squared difference between the estimated values and the actual values
Feature importance	AutoML generates tables that indicate how much each feature impacts a specific model. The values are provided as a percentage of each feature: the higher the percentage is, the more that feature impacts model training.

3.4. Description of the Case Study: Monitoring Cobot Arm Joints

This section describes a case study where a cobot (UR3) performs pick and place tasks with a human operator, considering different loads—maximum, medium and minimum. The components of the experiment consist of a human operator, a UR3 robot and a shared workspace as shown in Figure 3. Both the robot and the human worker can access all the components necessary for the assembly in the workspace, such as the base, flanges, bolts and nuts. The HRC assembly application and integrated assembly method is described in [25]. During the case study, Physical and hardware data, such as temperature, load, speed, power, and programmed stops, protective stops and so forth, were acquired from the developed data acquisition framework during the case study. The robot was used without any workpiece to indicate the minimum load. A medium load corresponds to a 1.5 kg workpiece and a maximum load to a 3.0 kg workpiece. The monitoring system monitored all the physical parameters and predicted the parameters of influence of the cobots using different ML models.

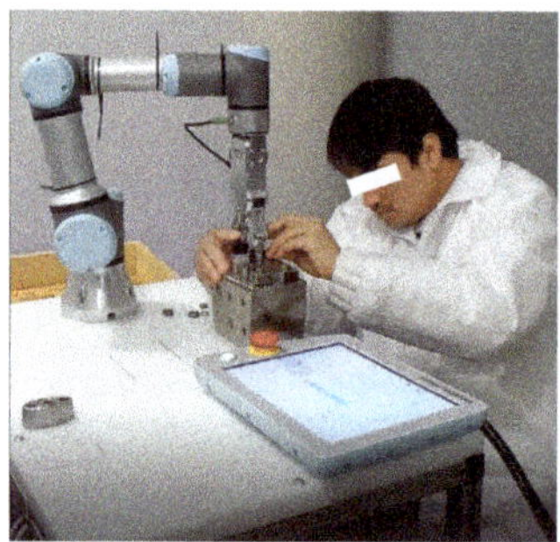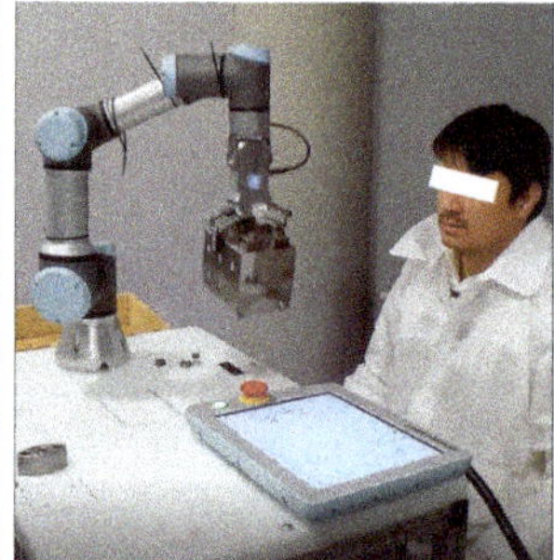

Figure 3. HRC application processes: the figure on the left demonstrates the HRC assembly process; The figure on the right demonstrates the cobot transporting assembled components to the target position.

3.5. Correlation Analyses

The acquired variables from the case study were correlated to identify any important variables of the cobot during the task execution with a human. The correlation coefficient is an important measure of the association of continuous data.

Figures 4–9 show the correlation matrix of six joints to identify the most correlated variables of the specific joints. The correlation matrix clearly shows that the temperature and load variables are very closely correlated for each joint. Current in joint 1, joint 2 and joint 3 is the next most closely correlated variable with load and temperature. Voltage and speed variables are weakly correlated with other variables. Voltage is slightly correlated only with a robot speed between Joint 0–5 with a maximum -0.78 value in Joint 1. The joint speed variables are not correlated with other variables and they are almost 0 in every joint, only with voltage 0.01 in joint 1.

The most closely correlated variables in the correlation matrix were chosen for the regression and classification models to make predictions.

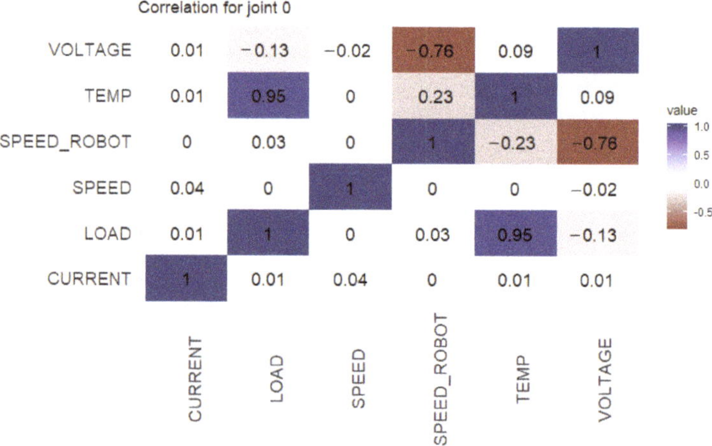

Figure 4. Correlation matrix of joint 0.

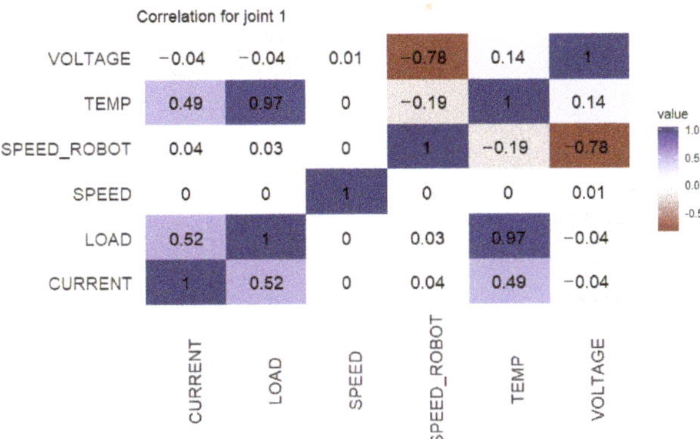

Figure 5. Correlation of joint 1.

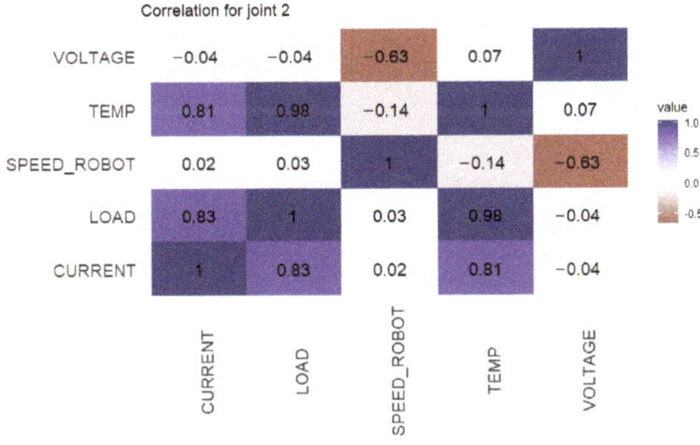

Figure 6. Correlation of joint 2.

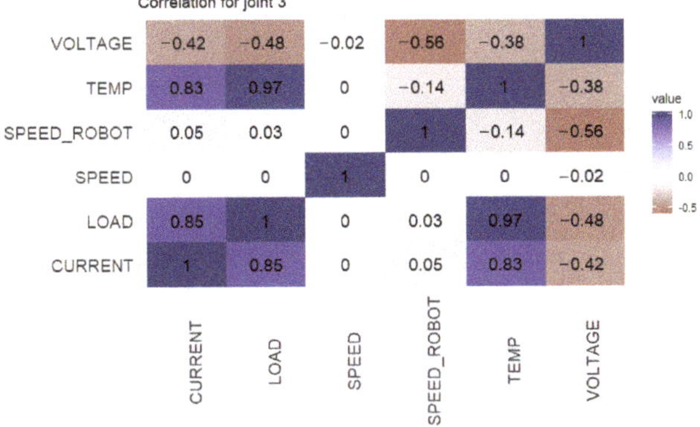

Figure 7. Correlation of joint 3.

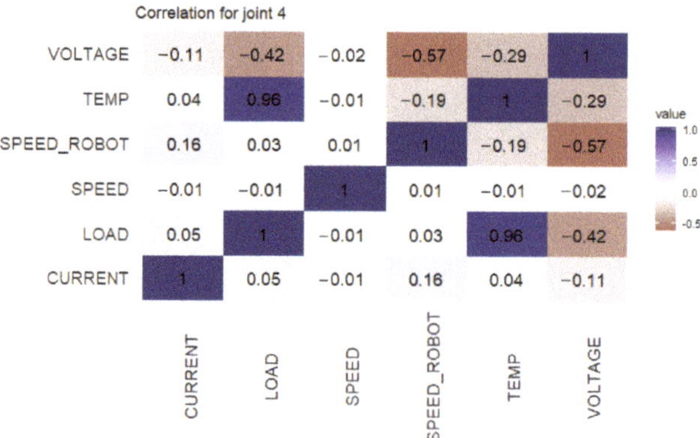

Figure 8. Correlation of joint 4.

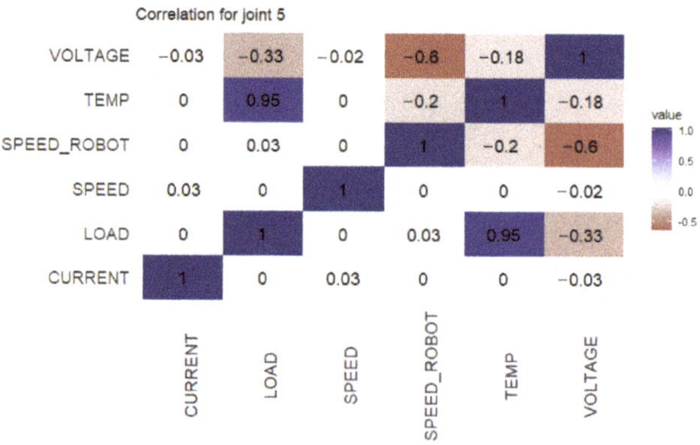

Figure 9. Correlation of joint 5.

Figure 10 shows a box plot of the temperature of each joint when working with different loads. Figure 11 shows a box-plot of the power (product of current and voltage) vs different loads. The box plots indicate that when there is a maximum load, or the end-effector of the robot is working under a full load condition, there is a risk of the temperature in some joints, especially Joints 2, 3, 4 and 5, of rising above 50 degrees. According to the datasheet of the robot, the maximum temperature should not exceed 50 degrees, and this value limits the working range of the robot.

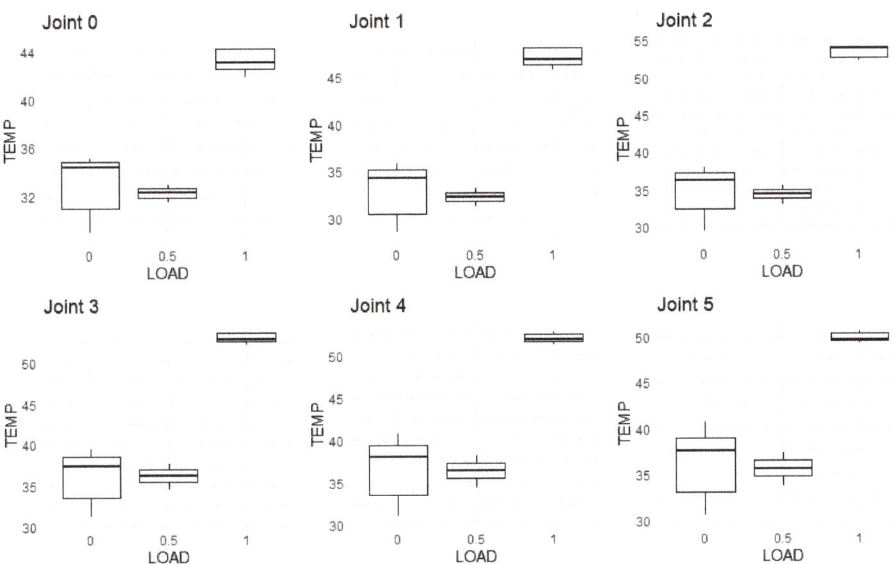

Figure 10. Box-plot of the Load with the Temperature.

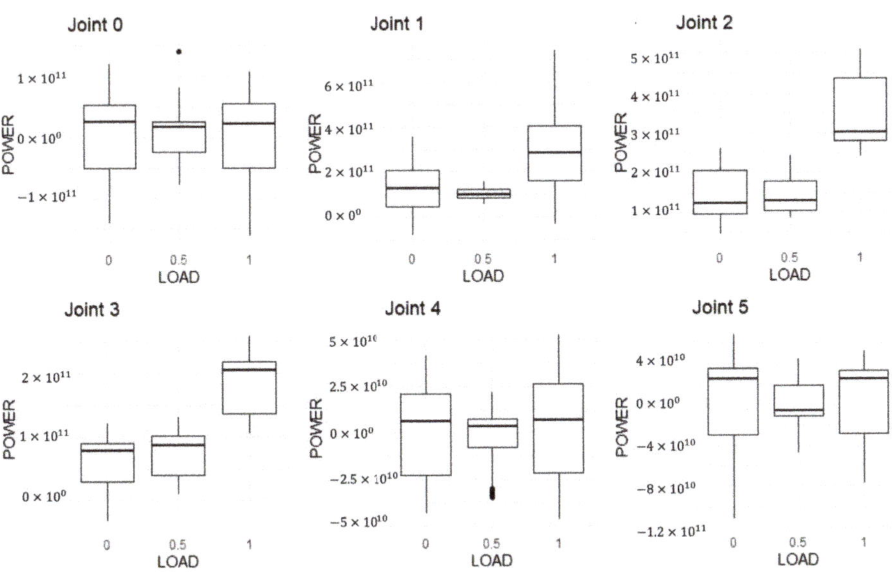

Figure 11. Box-plot of the Load and the Power.

Figure 12 displays a box-plot of the speed vs different loads, and the plot shows that varying the loads has no effect on the speed of the robot.

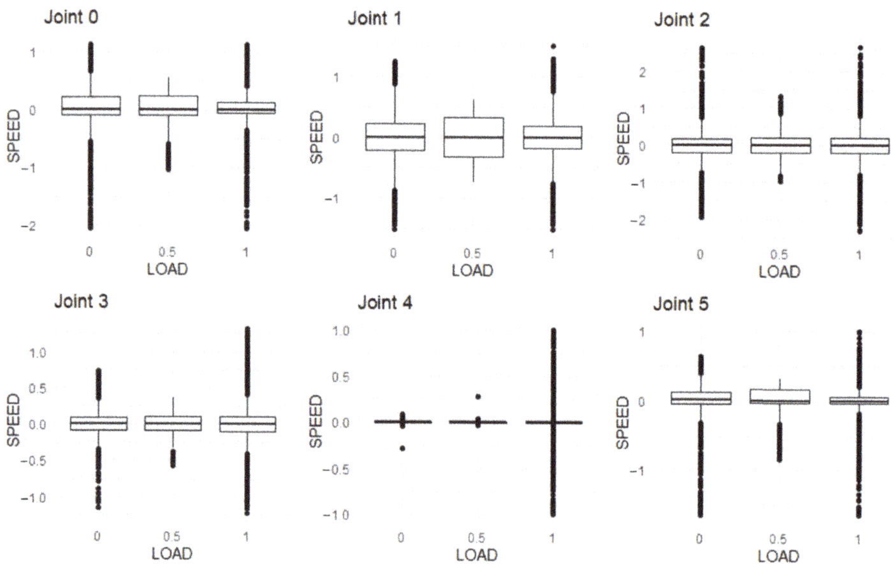

Figure 12. Box-plot of the Load with the Speed.

Correlated variables such as loads, temperature and power are passed to the ML tools to monitor the conditions of the cobot. The monitoring system is a hybrid system, as both quantitative data regression models (data driven) and qualitative parameters classification models are used. All the results appear on the analytical dashboard of the robot.

4. Results And Discussions

The results of the linear regression model, when used to predict the temperature of the robot and when AutoML is used to forecast safety stops during collaborative work are presented in this section. The correlation matrix and studies have shown that the temperature of the each joint and protective stops during collaborative tasks are important factors of the robot manipulator that can influence the reliability of collaborative work. Figure 13 shows the results of a linear regression model used to predict the temperature of a robot while performing human robot collaboration tasks with different loads. The blue data on the graph are the original data and the red data are the predicted variables. Moreover, the graph shows the critical temperature condition when the robot works with maximum loads and normal condition when the robot performs tasks with lighter loads.

The Table 4 shows the performance results of the linear regression model. The main evaluation parameters of the model were the Adjusted R squared, multiple R-squared, F-statistics and p-values. The linear model with all the data resulted in a higher error than for the other experimental setups. The performance of the LM with all the data including the max. medium and min data, resulted in an Adjusted R-squared equal to 0.9346 and a multiple R-squared of 0.9529, which are sufficient to be integrated into the monitoring system as temperature prediction models.

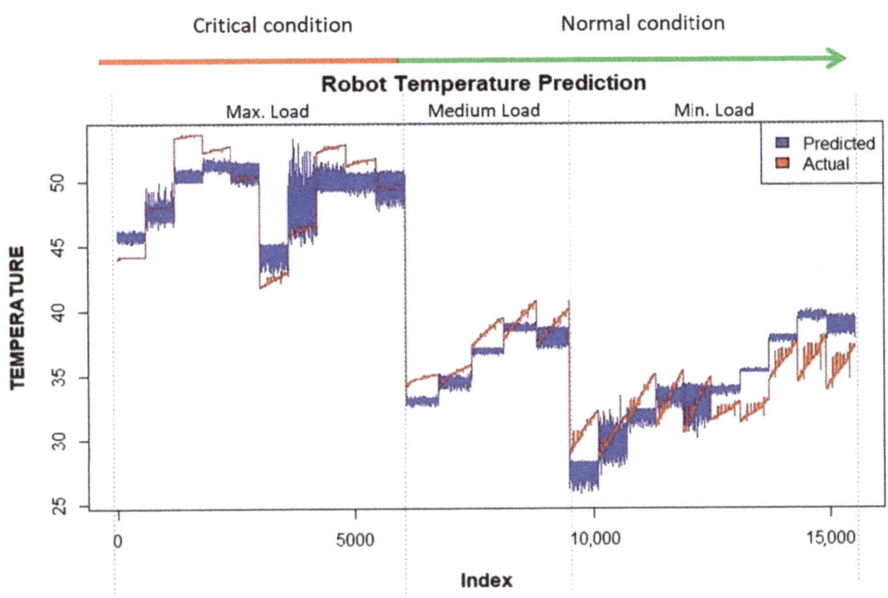

Figure 13. Temperature prediction results under different conditions.

Table 4. Performances of the general linear model when predicting the cobot temperature.

Experimental Setups	Res. Std. Err.	Multiple R-Squared	Adjusted R-Squared	F-Statistic	*p*-Value
UR3 with all the data	1.865	0.9529	0.9346	52.02	$<1.163 \times 10^{-10}$
UR3 with the Max.Load	0.1422	0.9152	0.9151	3.347×10^4	$<2.2 \times 10^{-16}$
UR3 with the Medium Load	0.364	0.1571	0.1569	578.3	$<2.2 \times 10^{-16}$
UR3 with the Min. Load	0.3832	0.3952	0.395	2027	$<2.2 \times 10^{-16}$

Table 5 demonstrates the coefficients of the Linear regression model for UR3 with all the data and important variables. It is clear, from the table, that the most influencing variable with the greatest influence on the temperature predictive model is load.

Table 5. Performances of the linear regression model when predicting the robot temperature.

Input Variables	Estimate	Std. Error	t Value	Pr($>$ | t |)
Intercept	$1.195 \times 10^{+3}$	$1.375 \times 10^{+3}$	0.869	<0.396248
LOAD	$1.208 \times 10^{+1}$	$1.041 \times 10^{+0}$	11.611	8.56×10^{-10}
JOINT_0	$1.534 \times 10^{+0}$	2.694×10^{-1}	5.693	2.13×10^{-5}
SPEED_ROBOT	-6.107×10^{-2}	1.411×10^{-2}	-4.327	0.000406
TIME	-7.458×10^{-7}	8.761×10^{-7}	-0.851	0.405837
CURRENT	$5.800 \times 10^{+1}$	$2.902 \times 10^{+1}$	1.999	0.060971
VOLTAGE	1.048×10^{-10}	2.652×10^{-10}	0.395	0.697487
POWER	-4.051×10^{-10}	2.063×10^{-10}	-1.964	0.065163

H2O AutoML is used to predict safety stops during collaborative tasks. Our dataset is trained in an H2O cluster using R, version 3.6.3. The AutoML function in H2O automates the process of building a large number of models and finds the most suitable model for a given dataset. AutoML includes a "leaderboard" of models that are trained in the process. It includes a 5-fold validated model performance, and no hyperparameters were used for our dataset. Figure 14 shows the performance of AutoML and the importance of the ten models.

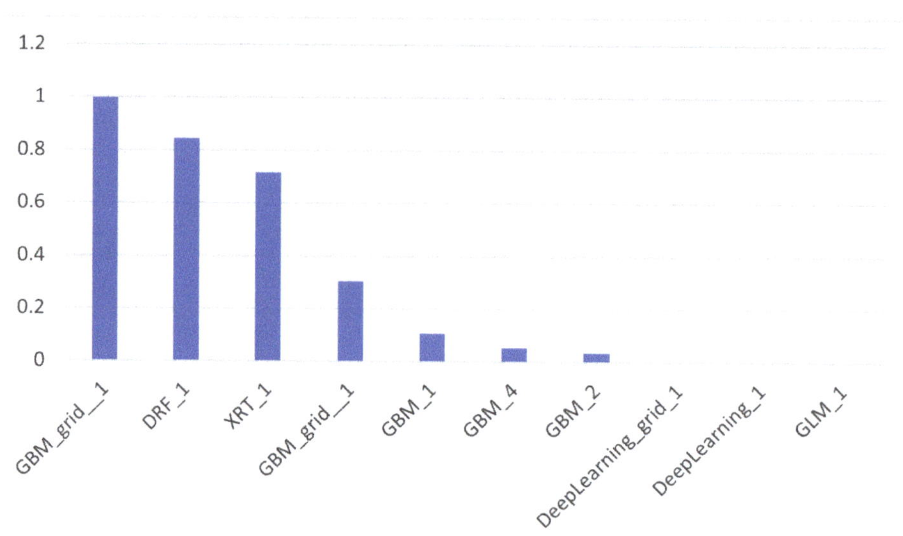

Figure 14. Performance of the AutoML algorithms.

Users can receive scores of the dataset using the leaderboard frame. Our dataset was trained in a binary classification model, and AUC metrics was used as the main model parameter.

The predictive model results of AutoML are shown in Table 6.

Table 6. Results of different models used to predict safety stops using the automatic machine learning.

Model_id	AUC	logloss	aucpr	rmse	mse
GBM_1_AutoML_20201030_160448	0.982	0.052	0.921	0.109	0.012
GBM_grid__1_AutoML_20201030_160448_model_3	0.979	0.055	0.913	0.112	0.012
XRT_1_AutoML_20201030_160448	0.979	0.055	0.927	0.106	0.011
DRF_1_AutoML_20201030_160448	0.979	0.059	0.927	0.105	0.011
DeepLearning_grid__2_AutoML_20201030_160448_model_1	0.913	0.107	0.766	0.150	0.022
DeepLearning_grid__1_AutoML_20201030_160448_model_1	0.907	0.141	0.749	0.152	0.023
DeepLearning_grid__3_AutoML_20201030_160448_model_1	0.904	0.120	0.754	0.151	0.023
DeepLearning_1_AutoML_20201030_160448	0.883	0.116	0.744	0.153	0.023
GLM_1_AutoML_20201030_160448	0.835	0.114	0.687	0.153	0.023

According to the table, the best performing algorithm is Gradient Boosting Machine (GBM). The importance of the variables for the GBM model is shown in Figure 15. According to the plot, the variables with the most influence on the GBM model are the SPEED of the robot, and CURRENT.

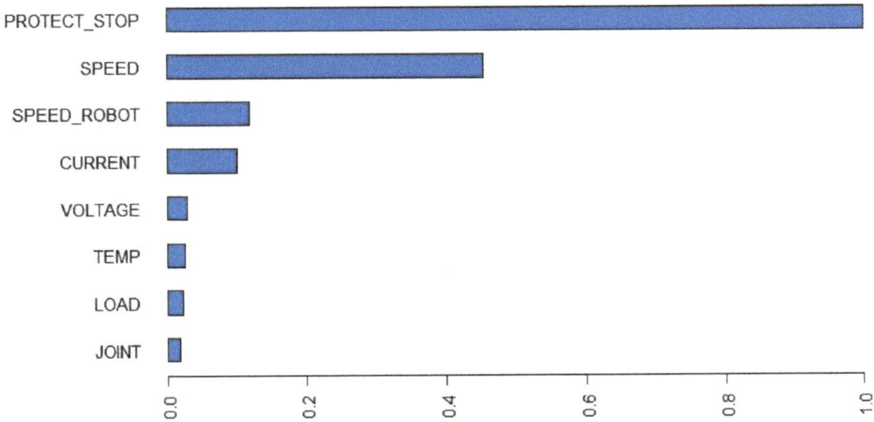

Figure 15. Importance of the variables on the different models.

5. Conclusions

A platform used to monitor the health status of collaborative robots during collaborative tasks is presented in this paper. The case study was performed on benchmark tasks for collaborative assembly processes. An automatic machine learning(ML) tool was used to perform on-line monitoring and predict outages of the industrial cobots during a human-robot collaboration process. Such an on-line monitoring system allows more reliable human robot collaboration applications to be created, unplanned downtime during task execution to be eliminated, and the trust of humans during interaction with a robot and the lifetime of the robot to be maximised. The proposed framework demonstrates data management techniques on an industrial robot that is considered as a physical cyber-system. Using an assembly case study, the parameters of a robot were collected and fed to an automatic ML model in order to identify the most significant reliability factors and to predict the necessity of safety stops of the robot. According to the results, a linear regression model was selected for certain quantitative variables such as temperature. The classification model was used to predict the qualitative variables. The linear regression model was found to be sufficiently good to be integrated in the monitoring system to predict temperature. H2O, with the AutoML function, was used to predict safety stops during collaborative tasks and the results show that GBM appears to be the best model for the considered dataset. Further improvement will involve the integration of other relevant sensors in the monitoring platform to further increase the usability of the system under variable working conditions.

Author Contributions: Conceptualization, K.A.; methodology, K.A. and D.A.; software, K.A.; validation, D.A.; formal analysis, D.A.; investigation, K.A. and D.A.; resources, K.A. and D.A.; data curation, K.A.; writing—original draft preparation, K.A. and D.A.; writing—review and editing, K.A. and D.A.; visualization, K.A.; supervision, D.A.; project administration, D.A.; funding acquisition, D.A. All authors have read and agreed to the published version of the manuscript.

Funding: This paper did not receive any external funding.

Institutional Review Board Statement: Not applicable.

Informed Consent Statement: Not applicable.

Data Availability Statement: Not applicable.

Acknowledgments: The authors would like to thank Emiliano Traini for his support with the RStudio software.

Conflicts of Interest: The authors declare no conflict of interest.

References

1. Aliev, K.; Antonelli, D. Analysis of Cooperative Industrial Task Execution by Mobile and Manipulator Robots. In *Advances in Manufacturing II*; Trojanowska, J., Ciszak, O., Machado, J.M., Pavlenko, I., Eds.; Springer International Publishing: Cham, Switzerland, 2019; pp. 248–260.
2. Grigore, L.S.; Priescu, I.; Joita, D.; Oncioiu, I. The Integration of Collaborative Robot Systems and Their Environmental Impacts. *Processes* **2020**, *8*, 494. [CrossRef]
3. Bortolini, M.; Ferrari, E.; Gamberi, M.; Pilati, F.; Faccio, M. Assembly system design in the Industry 4.0 era: A general framework. *IFAC-PapersOnLine* **2017**, *50*, 5700–5705. [CrossRef]
4. Michalos, G.; Spiliotopoulos, J.; Makris, S.; Chryssolouris, G. A method for planning human robot shared tasks. *CIRP J. Manuf. Sci. Technol.* **2018**, *22*, 76–90. [CrossRef]
5. Yang, C.; Sun, Y.; Ladubec, C.; Liu, Y. Developing Machine Learning-Based Models for Railway Inspection. *Appl. Sci.* **2020**, *11*, 13. [CrossRef]
6. Cardoso, D.; Ferreira, L. Application of Predictive Maintenance Concepts Using Artificial Intelligence Tools. *Appl. Sci.* **2020**, *11*, 18. [CrossRef]
7. Villani, V.; Pini, F.; Leali, F.; Secchi, C. Survey on human–robot collaboration in industrial settings: Safety, intuitive interfaces and applications. *Mechatronics* **2018**, *55*, 248–266. [CrossRef]
8. Magrini, E.; Ferraguti, F.; Ronga, A.J.; Pini, F.; De Luca, A.; Leali, F. Human-robot coexistence and interaction in open industrial cells. *Robot. Comput.-Integr. Manuf.* **2020**, *61*, 101846. [CrossRef]
9. ISO. *Robots and Robotic Devices–Safety Requirements for Industrial Robots—Part 1: Robots*; ISO: Geneva, Switzerland, 2011.
10. ISO. ISO 10218-2: 2011. *Robots and Robotic Devices—Safety Requirements for Industrial Robots–Part 2: Robot Systems and Integration (ISO 10218-2: 2011) Machinery*; ISO: Geneva, Switzerland, 2011.
11. ISO. *TS 15066: 2016: Robots and Robotic Devices–Collaborative Robots*; International Organization for Standardization: Geneva, Switzerland, 2016.
12. Scopus Database Scopus. Available online: www.scopus.com (accessed on 20 January 2021).
13. Matheson, E.; Minto, R.; Zampieri, E.G.; Faccio, M.; Rosati, G. Human–Robot Collaboration in Manufacturing Applications: A Review. *Robotics* **2019**, *8*, 100. [CrossRef]
14. Grahn, S.; Langbeck, B.; Johansen, K.; Backman, B. Potential Advantages Using Large Anthropomorphic Robots in Human-robot Collaborative, Hand Guided Assembly. *Procedia CIRP* **2016**, *44*, 281–286.
15. Michalos, G.; Makris, S.; Papakostas, N.; Mourtzis, D.; Chryssolouris, G. Automotive assembly technologies review: Challenges and outlook for a flexible and adaptive approach. *CIRP J. Manuf. Sci. Technol.* **2010**, *2*, 81–91. [CrossRef]
16. Nguyen, M.T.; La, H.M.; Teague, K.A. Collaborative and compressed mobile sensing for data collection in distributed robotic networks. *IEEE Trans. Control Netw. Syst.* **2017**, *5*, 1729–1740. [CrossRef]
17. Kirschner, D.; Schlotzhauer, A.; Brandstötter, M.; Hofbaur, M. Validation of relevant parameters of sensitive manipulators for human-robot collaboration. In *International Conference on Robotics in Alpe-Adria Danube Region*; Springer: Berlin, Germany, 2017; pp. 242–252.
18. Aliev, K.; Antonelli, D.; Awouda, A.; Chiabert, P. Key Performance Indicators Integrating Collaborative and Mobile Robots in the Factory Networks. In *Collaborative Networks and Digital Transformation*; Camarinha-Matos, L.M., Afsarmanesh, H., Antonelli, D., Eds.; Springer International Publishing: Cham, Switzerland, 2019; pp. 635–642.
19. Dagalakis, N.G.; Yoo, J.M.; Oeste, T. Human-robot collaboration dynamic impact testing and calibration instrument for disposable robot safety artifacts. *Ind. Robot. Int. J.* **2016**, *43*, 328–337. [CrossRef] [PubMed]
20. Polygerinos, P.; Correll, N.; Morin, S.A.; Mosadegh, B.; Onal, C.D.; Petersen, K.; Cianchetti, M.; Tolley, M.T.; Shepherd, R.F. Soft robotics: Review of fluid-driven intrinsically soft devices; manufacturing, sensing, control, and applications in human-robot interaction. *Adv. Eng. Mater.* **2017**, *19*, 1700016. [CrossRef]
21. Wang, Q.; Cheng, Y.; Jiao, W.; Johnson, M.T.; Zhang, Y. Virtual reality human-robot collaborative welding: A case study of weaving gas tungsten arc welding. *J. Manuf. Process.* **2019**, *48*, 210–217. [CrossRef]
22. Antonelli, D.; Astanin, S. Qualification of a Collaborative Human-robot Welding Cell. *Procedia CIRP* **2016**, *41*, 352–357.
23. Antonelli, D.; Astanin, S. Enhancing the quality of manual spot welding through augmented reality assisted guidance. *Procedia CIRP* **2015**, *33*, 556–561. [CrossRef]
24. Chen, S.; Qiu, T.; Lin, T.; Wu, L.; Tian, J.; Lv, W.; Zhang, Y. Intelligent Technologies for Robotic Welding. In *Robotic Welding, Intelligence and Automation*; Tarn, T.J., Zhou, C., Chen, S.B., Eds.; Springer Berlin Heidelberg: Berlin/Heidelberg, Germany, 2004; pp. 123–143.
25. Aliev, K.; Antonelli, D.; Bruno, G. Task-based Programming and Sequence Planning for Human-Robot Collaborative Assembly. *IFAC-PapersOnLine* **2019**, *52*, 1638–1643.
26. Ende, T.; Haddadin, S.; Parusel, S.; Wüsthoff, T.; Hassenzahl, M.; Albu-Schäffer, A. A human-centered approach to robot gesture based communication within collaborative working processes. In Proceedings of the 2011 IEEE/RSJ International Conference on Intelligent Robots and Systems, San Francisco, CA, USA, 25–30 September 2011; pp. 3367–3374. [CrossRef]

27. Rosi, G.; Vignali, G.; Bottani, E. A Conceptual Framework for the Selection of an "Industry 4.0" Application to Enhance the Operators' Safety: The Case of an Aseptic Bottling Line. In Proceedings of the 2018 IEEE International Conference on Engineering, Technology and Innovation (ICE/ITMC), Stuttgart, Germany, 17–20 June 2018; pp. 1–9. [CrossRef]
28. Pini, F.; Leali, F. Human-robot collaborative reconfigurable platform for surface finishing processes. *Procedia Manuf.* **2019**, *38*, 76–83.
29. Hu, J.; Kabir, A.M.; Hartford, S.M.; Gupta, S.K.; Pagilla, P.R. Robotic deburring and chamfering of complex geometries in high-mix/low-volume production applications. In Proceedings of the2020 IEEE 16th International Conference on Automation Science and Engineering (CASE), Hong Kong, China, 20–21 August 2020; pp. 1155–1160. [CrossRef]
30. Byner, C.; Matthias, B.; Ding, H. Dynamic speed and separation monitoring for collaborative robot applications–Concepts and performance. *Robot. Comput.-Integr. Manuf.* **2019**, *58*, 239–252. [CrossRef]
31. Awouda, A.; Aliev, K.; Chiabert, P.; Antonelli, D. Practical Implementation of Industry 4.0 Based on Open Access Tools and Technologies. In *Product Lifecycle Management in the Digital Twin Era*; Fortin, C., Rivest, L., Bernard, A., Bouras, A., Eds.; Springer International Publishing: Cham, Switzerland, 2019; pp. 94–103.
32. Khalid, A.; Kirisci, P.; Ghrairi, Z.; Thoben, K.D.; Pannek, J. A methodology to develop collaborative robotic cyber physical systems for production environments. *Logist. Res.* **2016**, *9*, 23. [CrossRef]
33. RTDE Protocol Universal Robots. Available online: https://www.universal-robots.com/articles/ur/real-time-data-exchange-rtde-guide/ (accessed on 20 January 2021).
34. MODBUS SERVER Universal Robots. Available online: https://www.universal-robots.com/articles/ur/modbus-server/ (accessed on 20 January 2021).
35. García, S.; Luengo, J.; Herrera, F. *Data Preprocessing in Data Mining*; Springer: Berlin, Germany, 2015.
36. Twala, B.; Cartwright, M. Ensemble missing data techniques for software effort prediction. *Intell. Data Anal.* **2010**, *14*, 299–331. [CrossRef]
37. Myrtveit, I.; Stensrud, E.; Olsson, U.H. Analyzing data sets with missing data: An empirical evaluation of imputation methods and likelihood-based methods. *IEEE Trans. Softw. Eng.* **2001**, *27*, 999–1013. [CrossRef]
38. Al Shalabi, L.; Shaaban, Z.; Kasasbeh, B. Data mining: A preprocessing engine. *J. Comput. Sci.* **2006**, *2*, 735–739. [CrossRef]
39. Kuhn, M.; Johnson, K. *Applied Predictive Modeling*; Springer: Berlin, Germany, 2013; Volume 26.
40. Daoud, J.I. Multicollinearity and regression analysis. In *Journal of Physics: Conference Series*; IOP Publishing: Bristol, UK, 2017; Volume 949, p. 012009.
41. Liu, H.; Motoda, H. Feature transformation and subset selection. *IEEE Intell. Syst. Their Appl.* **1998**, *13*, 26–28.
42. Gareth, J.; Daniela, W.; Trevor, H.; Robert, T. *An Introduction to Statistical Learning: With Applications in R*; Spinger: Berlin, Germany, 2013.
43. H2O 3 H2O.ai. Available online: https://docs.h2o.ai/h2o/latest-stable/h2o-docs/welcome.html# (accessed on 8 February 2021).
44. Halvari, T.; Nurminen, J.K.; Mikkonen, T. Testing the Robustness of AutoML Systems. *arXiv* **2020**, arXiv:2005.02649.
45. Feurer, M.; Klein, A.; Eggensperger, K.; Springenberg, J.; Blum, M.; Hutter, F. Efficient and robust automated machine learning. In *Advances in Neural Information Processing Systems*; MIT Press: Cambridge, MA, USA, 2015; pp. 2962–2970.
46. Olson, R.S.; Moore, J.H. TPOT: A tree-based pipeline optimization tool for automating machine learning. In Proceedings of the Workshop on Automatic Machine Learning, PMLR, New York, NY, USA, 24 June 2016; pp. 66–74.
47. Jin, H.; Song, Q.; Hu, X. Auto-keras: An efficient neural architecture search system. In Proceedings of the 25th ACM SIGKDD International Conference on Knowledge Discovery & Data Mining, Anchorage, AK, USA, 4–8 August 2019; pp. 1946–1956.
48. LeDell, E.; Poirier, S. H2o automl: Scalable automatic machine learning. In Proceedings of the AutoML Workshop at ICML, Nashville, TN, USA, 8–12 July 2020; Volume 2020.
49. Milutinovic, M.; Schoenfeld, B.; Martinez-Garcia, D.; Ray, S.; Shah, S.; Yan, D. On evaluation of automl systems. In Proceedings of the ICML Workshop on Automatic Machine Learning, Tahoe City, CA, USA, 17–18 July 2020.
50. Feurer, M.; Klein, A.; Eggensperger, K.; Springenberg, J.T.; Blum, M.; Hutter, F. Auto-sklearn: Efficient and robust automated machine learning. In *Automated Machine Learning*; Springer: Cham, Switzerland, 2019; pp. 113–134.
51. Kotthoff, L.; Thornton, C.; Hoos, H.H.; Hutter, F.; Leyton-Brown, K. Auto-WEKA 2.0: Automatic model selection and hyperparameter optimization in WEKA. *J. Mach. Learn. Res.* **2017**, *18*, 826–830.

Article

Achievement of Accurate Robotic Arm-based Bike Frame Quality Check Using 3D Geometry Mathematical Model

Hsiung-Cheng Lin *, Bo-Ren Yu, Jen-Yu Wang, Jun-Ze Lai and Jia-Yang Wu

Department of Electronic Engineering, National Chin-Yi University of Technology, Taichung 41170, Taiwan; sdennil9999@gmail.com (B.-R.Y.); st831209@gmail.com (J.-Y.W.); hank081083@gmail.com (J.-Z.L.); sweet911105@gmail.com (J.-Y.W.)
* Correspondence: hclin@ncut.edu.tw

Received: 29 October 2019; Accepted: 6 December 2019; Published: 8 December 2019

Abstract: Currently, the bike frame quality check (QC) mostly relies on human operation in industry. However, some drawbacks such as it being time-consuming, having low accuracy and involving non-computerized processes are still unavoidable. Apart from these problems, measured data are difficult to systematically analyze for tracking sources of product defects in the production process. For this reason, this paper aims to develop a 3D geometry mathematical model suitable for bicycle frames QC using robotic arm-based measurement. Unlike the traditional way to find coefficients of a space sphere, the proposed model requires only three check point coordinates to achieve the sphere axis coordinate and its radius. In the practical work, the contact sensor combined with the robotic arm is used to realize the compliance items measurement in shaft length, internal diameter, verticality, parallelism, etc. The proposed model is validated based on both mathematic verification and actual bike frame check.

Keywords: quality check; bike frame; mathematical model; graphical user interface

---

1. Introduction

In recent years, bike riding has become a popular leisure sport around the world. For this reason, analysts forecast that the global high-end bicycle market will grow with a compound annual growth rate (CAGR) of 4.82% during the period 2017–2021 according to the report from Research and Markets Ltd. It is known that high-end bikes demand a high quality of bike frame [1–7]. Unfortunately, frame QC still relies on the Vernier caliper, the plug gauge, the cylindrical gauge, and the thread gauge, etc. It normally takes a long time to complete the process. Consequently, the automation measurement process for the QC of the bike frame is essential in industry [8–12].

The coordinate-measuring-machine (CMM) is now being widely applied as part of workpiece inspection in the production line [13–17]. It can be used to measure the geometry of physical objects by sensing the discrete points on the object surface with a probe, including mechanical, optical, laser, and white light. Basically, it has two major advantages: (1) high precision up to 0.001 mm and (2) high reliability in both hardware and software. However, the CMM inspection planning session has been a challenging issue because of its time-consuming nature using traditional methods, e.g., expert experiences and technical documents data mining. Additionally, it may suffer from some following disadvantages: (1) Its operation speed is limited. (2) It is sensitive to the environment temperature and humidity. (3) It is not applicable to irregular shape object measurements. (4) It has a high cost. Obviously, CMM is not suitable for the bike frame measurement due to the restriction of the operation range. Alternatively, robotic arms are typically used for multiple industrial applications such as material handling, welding, thermal spraying, assembly, palletizing, drilling, and painting, etc. [18–21].

Appl. Sci. **2019**, *9*, 5355; doi:10.3390/app9245355   www.mdpi.com/journal/applsci

For instance, a platform based on a robotic arm using three degrees of freedom (DoF) principle was proposed to estimate the calibration parameters of microelectromechanical systems (MEMS) [22]. It can be placed indifferent positions for collecting a dataset of points evenly distributed. This case implies that the measurement technique using robotic arms may provide a good solution for the bike frame QC process.

2. System Description

2.1. System Structure

The proposed system structure of bike frame quality check is shown in Figure 1a, consisting of subsystems such as robotic arm, graphical user interface (GUI), programmable logic controller (PLC), contact sensor, mathematical model, database, and workbench. Each subsystem is responsible to carry out a specific task, described as follows. (1) Robotic arm can carry the contact sensor and move it to the check points. Therefore, the coordinates of check points can be found based on the robotic arm coordinate system. (2) The contact sensor can feedback a digital signal to the robotic arm immediately once it touches the surface of the check points. (3) PLC is to control the rotating disk to rotate the bike frame 90° for the robotic arm to reach every check point. (4) Graphical user interface (GUI) provides a friendly user interface for users to input data and display a real-time measurement outcome. (5) Mathematical model presents a geometry algorithm that can effectively integrate the sphere formula with the inner product of normal vector to find four parameters in the sphere formula using only three measured points. Accordingly, the center coordinate of check point and its diameter can be calculated accurately and simply. (6) A database using MySQL is used to store the measured data and export measurement data report. (7) The workbench shown in Figure 1b is designed to sustain all hardware devices. It contains: (1) a fixing frame, (2) A rotating disk, (3) a work platform, and (4) a sensor pedestal. Moreover, the XAML and C# package are used to build up the system software such as the robotic arm simulation object, the window object, the control object, and the 3D-geometry mathematical model. Through Transmission Control Protocol (TCP) and Internet Protocol (IP) (TCP/IP), the contact sensor and robotic arm can communicate with each other between different objects.

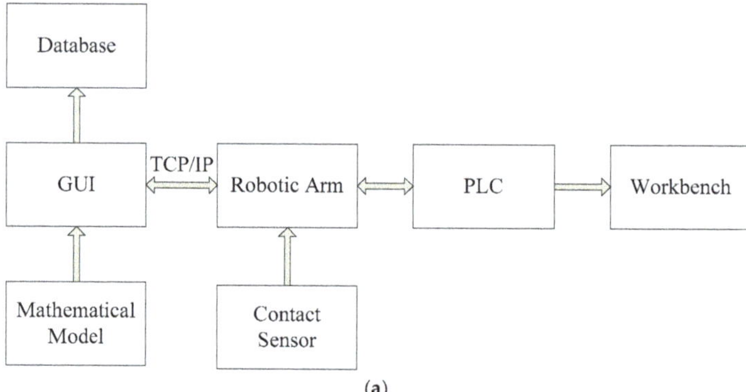

(a)

Figure 1. *Cont.*

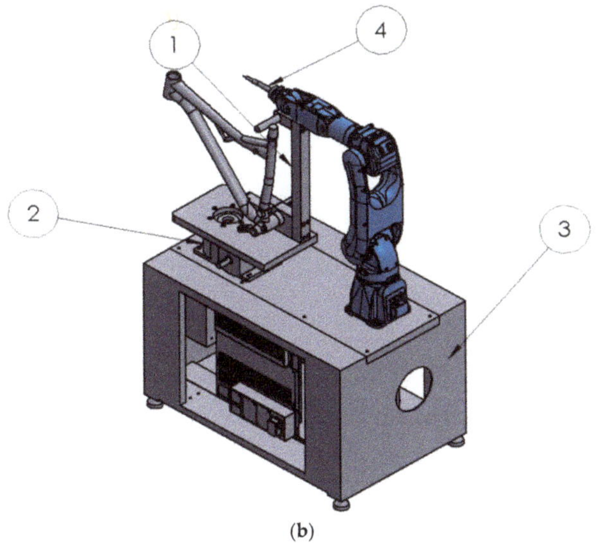

Figure 1. System structure: (**a**) system block; (**b**) profile of workbench.

In this study, we mainly focused on the development of mathematical model required for the bike frame quality check using a robotic arm. Based on the proposed mathematical model, the robotic arm is combined with the contact sensor to implement the bike frame quality check in shaft length, internal diameter, verticality, and parallelism, etc. The major devices used in the proposed system are listed as follows:

(1)   Robotic arm: YASKAWA-GP7
(2)   Contact sensor: Compact module changing touch-trigger probe (Renishaw TP20)
(3)   PLC: DELTA DVP –PM1000M
(4)   Database: MySQL

2.2. Introduction of Bike Frame

Generally, the bike frame consists of: (1) a B.B rotating shaft, (2) a S/T rotating shaft, (3) a S/T groove, (4) a T/T rotating shaft, (5) a shock absorber, (6) and H/T, where they are required for quality check, as shown in Figure 2.

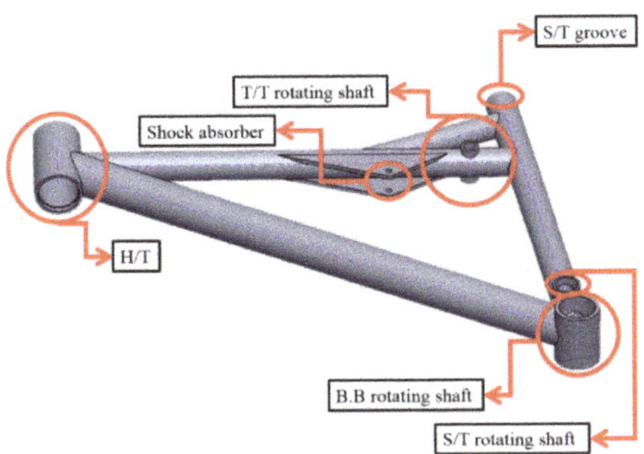

Figure 2. Profile of a bike frame.

3. Mathematical Model

The check items of bike frame for quality evaluation mainly include the shaft length, internal diameter, verticality, and parallelism located in different shafts. The proposed mathematical model provides the solutions for checked point coordinate calculation. It is described as follows.

3.1. Generation of Bike Frame Center Plane

Initially, the center plane of a bike frame should be generated from the B.B rotating shaft, as shown in Figure 3, which is used as the base of the coordinate system. The coordinate of the center point $aa3(aa3_x, aa3_y, aa3_z)$, as shown in Figure 3, can be determined from $aa1(aa1_x, aa1_y, aa1_z)$ and $aa2(aa2_x, aa2_y, aa2_z)$ as:

$$aa3_x = \frac{aa1_x + aa2_x}{2} \qquad aa3_y = \frac{aa1_y + aa2_y}{2} \qquad aa3_z = \frac{aa1_z + aa2_z}{2}$$

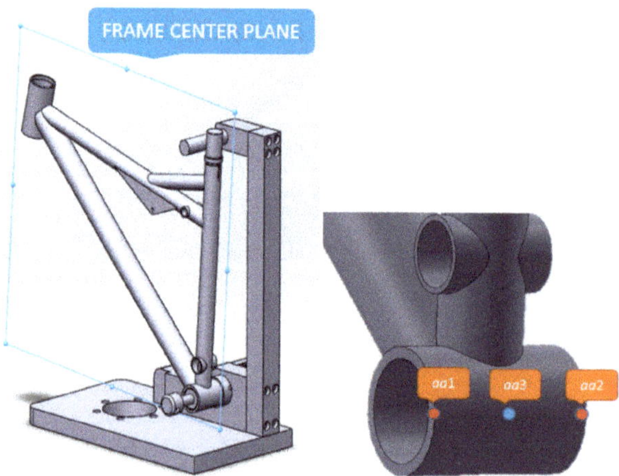

Figure 3. The center plane (**left**) and point (**right**) of bike frame.

3.2. Center Plane Offset

The center plane offset is used to check if there is a shift at the center plane. For this purpose, the y-axis $aa3_y$ taken from the center point $aa3(aa3_x, aa3_y, aa3_z)$ is regarded as the center standard plane. In Figure 4, two check points, i.e., $cc1$ and $cc22$, coordinates at the S/T rotating shaft are expressed as:

$$cc1(cc1_x, cc1_y, cc1_z) \qquad\qquad cc22(cc22_x, cc22_y, cc22_z)$$

cc is defined as the center point between $cc1$ and $cc22$ as:

$$cc\left(\frac{cc1_x + cc22_x}{2}, \frac{cc1_y + cc22_y}{2}, \frac{cc1_z + cc22_z}{2}\right)$$

$w1$ shown in Equation (1) is defined as the center plane offset at the S/T rotating shaft, and it is the distance between the y axis coordinate of the cc point and the center plane.

$$w1 = \left|\frac{cc1_y + cc22_y}{2} - aa3_y\right| \tag{1}$$

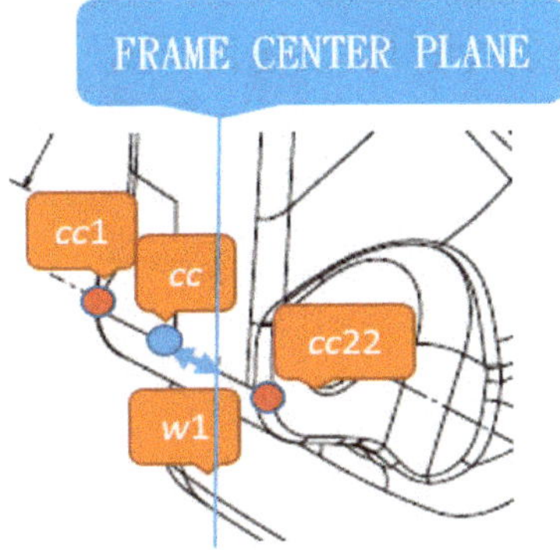

Figure 4. Center plane offset.

3.2.1. H/T Rotating Shaft Length

The H/T rotating shaft length can be determined by taking eight check points, as shown in Figure 5.

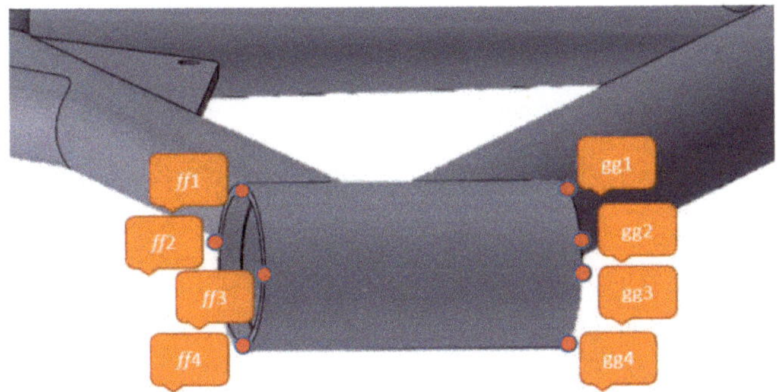

Figure 5. Check points of H/T rotating shaft length.

The coordinates of the eight check points are shown as follows:

$$ff1(ff1_x, ff1_y, ff1_z) \quad ff2(ff2_x, ff2_y, ff2_z) \quad ff3(ff3_x, ff3_y, ff3_z) \quad ff4(ff4_x, ff4_y, ff4_z)$$
$$gg1(gg1_x, gg1_y, gg1_z) \quad gg2(gg2_x, gg2_y, gg2_z) \quad gg3(gg3_x, gg3_y, gg3_z) \quad gg4(gg4_x, gg4_y, gg4_z)$$

The length $\overline{ff1gg1}$ between two check points ($ff1$ and $gg1$) is:

$$\overline{ff1gg1} = \sqrt{(gg1_x - ff1_x)^2 + \left(gg1_y - ff1_y\right)^2 + (gg1_z - ff1_z)^2} \tag{2}$$

Similarly, the lengths $\overline{ff2gg2}$, $\overline{ff3gg3}$, and $\overline{ff4gg4}$ can be formulated according to Equation (2), where the number one changes to numbers two to four, respectively.

Average length (L) between two check points ($ff2$ and $gg2$) is:

$$L = \frac{\overline{ff1gg1} + \overline{ff2gg2} + \overline{ff3gg3} + \overline{ff4gg4}}{4} \tag{3}$$

3.2.2. T/T Rotating Shaft Internal Diameter

The three check points at the T/T rotating shaft can be used to calculate the internal diameter, as shown in Figure 6.

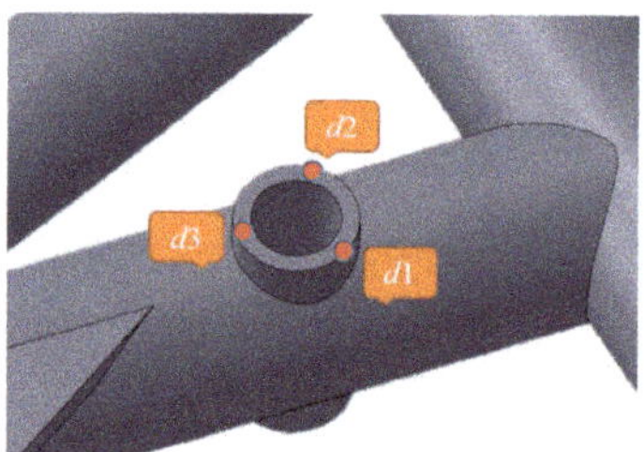

Figure 6. The three check points at the T/T rotating shaft.

The coordinates of the three check points are expressed as:

$$d1(d1_x, d1_y, d1_z) \qquad d2(d2_x, d2_y, d2_z) \qquad d3(d3_x, d3_y, d3_z)$$

The vectors $\overrightarrow{d_3d_1}$ and $\overrightarrow{d_3d_2}$ are:

$$\overrightarrow{d_3d_1} = \left((d1_x - d3_x), (d1_y - d3_y), (d1_z - d3_z)\right) \tag{4}$$

$$\overrightarrow{d_3d_2} = \left((d2_x - d3_x), (d2_y - d3_y), (d2_z - d3_z)\right) \tag{5}$$

The use cross product for the vectors $\overrightarrow{d_3d_1}$ and $\overrightarrow{d_3d_2}$, and their normal vector $\vec{n}$ can be obtained as:

$$\vec{n} = \overrightarrow{d_3d_2} \times \overrightarrow{d_3d_1} = \left(n_x, n_y, n_z\right) \tag{6}$$

where

$$n_x = \begin{vmatrix} (d2_y - d3_y) & (d2_z - d3_z) \\ (d1_z - d3_z) & (d1_y - d3_y) \end{vmatrix}$$

$$n_y = -\begin{vmatrix} (d2_x - d3_x) & (d2_z - d3_z) \\ (d1_x - d3_x) & (d1_z - d3_z) \end{vmatrix}$$

$$n_z = \begin{vmatrix} (d2_x - d3_x) & (d2_y - d3_y) \\ (d1_x - d3_x) & (d1_y - d3_y) \end{vmatrix}$$

The spherical general shown in Equation (7) is used to find the axis point coordinate and axial bore radius in Figure 6:

$$x^2 + y^2 + z^2 + dx + ey + fz + g = 0 \tag{7}$$

where the spherical axis point coordinate ($d4$) in Figure 6 is $\left(\frac{-d}{2}, \frac{-e}{2}, \frac{-f}{2}\right)$, and d, e, f, and g are real numbers.

The vector $\overrightarrow{d_4d_3}$ can be obtained as:

$$\overrightarrow{d_4d_3} = \left((d3_x) - \left(\frac{-d}{2}\right), (d3_y) - \left(\frac{-e}{2}\right), (d3_z) - \left(\frac{-f}{2}\right)\right) \tag{8}$$

The vectors $\overrightarrow{d_4d_3}$ and $\vec{n}$ are perpendicular to each other so that their inner product is zero.

$$\overrightarrow{d_4d_3} \cdot \vec{n} = 0 \tag{9}$$

Following this, we substitute the coordinates of three check points $d1(d1_x, d1_y, d1_z)$, $d2(d2_x, d2_y, d2_z)$, $d3(d3_x, d3_y, d3_z)$ into Equation (7) to form Equations (10)–(12). Additionally, Equation (13) is obtained based on Equation (9).

$$(d1_x)^2 + \left(d1_y\right)^2 + (d1_z)^2 + d(d1_x) + e\left(d1_y\right) + f(d1_z) + g = 0 \tag{10}$$

$$(d2_x)^2 + \left(d2_y\right)^2 + (d2_z)^2 + d(d2_x) + e\left(d2_y\right) + f(d2_z) + g = 0 \tag{11}$$

$$(d3_x)^2 + \left(d3_y\right)^2 + (d3_z)^2 + d(d3_x) + e\left(d3_y\right) + f(d3_z) + g = 0 \tag{12}$$

$$\begin{vmatrix} \left(d2_y - d3_y\right) & \left(d2_z - d3_z\right) \\ \left(d1_z - d3_z\right) & \left(d1_y - d3_y\right) \end{vmatrix} \left((d3_x) - \left(\tfrac{-d}{2}\right)\right) - \begin{vmatrix} (d2_x - d3_x) & (d2_z - d3_z) \\ (d1_x - d3_x) & (d1_z - d3_z) \end{vmatrix} \left(\left(d3_y\right) - \left(\tfrac{-e}{2}\right)\right)$$
$$+ \begin{vmatrix} (d2_x - d3_x) & \left(d2_y - d3_y\right) \\ (d1_x - d3_x) & \left(d1_y - d3_y\right) \end{vmatrix} \left((d3_z) - \left(\tfrac{-f}{2}\right)\right) = 0 \tag{13}$$

The parameters values (d, e, f, g) can be thus be found by solving the simultaneous equations from Equations (10)–(13).

Consequently, $d_4\left(\dfrac{-d}{2}, \dfrac{-e}{2}, \dfrac{-f}{2}\right)$ can be obtained, and the axial bore radius of T/T rotating shaft can be calculated as:

$$\overline{d_4 d_1} = \overline{d_4 d_2} = \overline{d_4 d_3} = \frac{1}{2}\sqrt{d^2 + e^2 + f^2 - 4g} \tag{14}$$

3.2.3. Parallelism

In Figure 7, the parallelism angle between B.B. and T/T rotating shafts can be calculated as follows:

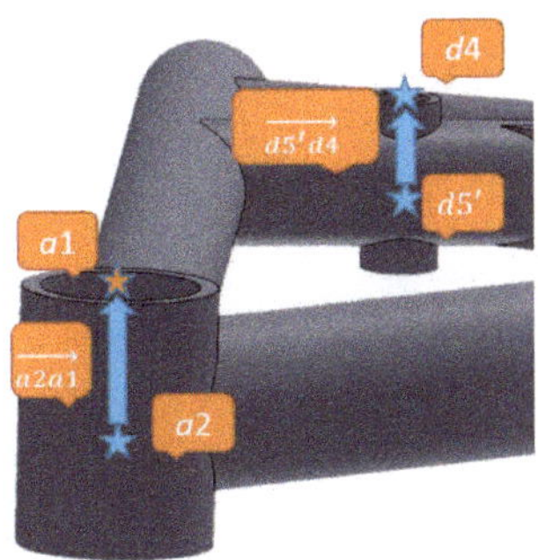

Figure 7. The parallelism between B.B. and T/T rotating shafts.

$$\overrightarrow{a2a1} = \left((a1_x - a2_x), \left(a1_y - (-333.5)\right), (a1_z - a2_z)\right) \tag{15}$$

$$\overrightarrow{d5'd4} = \left(0, \left(d4_y - \frac{\left(d1_y + d11_y\right)}{2}\right), 0\right) \tag{16}$$

Using the inner product formula, the parallelism angel θ between $\overrightarrow{a2a1}$ and $\overrightarrow{d5'd4}$ can be calculated as:

$$\cos\theta = \frac{\overrightarrow{d5'd4} \cdot \overrightarrow{a2a1}}{\left|\overrightarrow{d5'd4}\right|\left|\overrightarrow{a2a1}\right|} \tag{17}$$

3.2.4. Verticality

In Figure 8, the verticality angel between B.B. and H/T rotating shafts can be calculated as follows.

$$\overrightarrow{a2g4} = \left((g4_x - a2_x), \left(g4_y - (-333.5)\right), (g4_z - a2_z)\right) \tag{18}$$

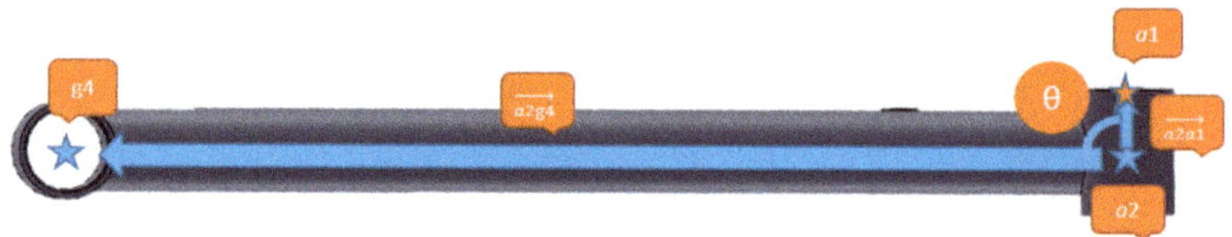

Figure 8. The verticality angel between the B.B. rotating shaft and the H/T rotating shaft.

Use the inner product formula, the verticality angel θ between $\overrightarrow{a2a1}$ and $\overrightarrow{a2g4}$ can be calculated as:

$$\cos\theta = \frac{\overrightarrow{a2a1} \cdot \overrightarrow{a2g4}}{\left|\overrightarrow{a2a1}\right|\left|\overrightarrow{a2g4}\right|} \tag{19}$$

4. Model Verification Using the Real Data

The proposed mathematical model is verified using the real data taken from the SOLIDWORKS drawing of the bike frame.

4.1. The Center Plane

From Figure 3, the two check points are:

$$aa1(139.59, -297, 109.14) \qquad\qquad aa2(139.59, -370, 108.85)$$

The center point of B.B rotating shaft is $aa3(aa3_x, aa3_y, aa3_z)$, where:

$$aa3_x = \frac{139.59 + 139.59}{2} = 139.59$$
$$aa3_y = \frac{-297 + (-370)}{2} = -333.5$$
$$aa3_z = \frac{109.14 + 108.85}{2} = 108.995$$
$$\therefore \quad aa3(139.59, -333.5, 108.995)$$

Accordingly, the center plane is located at $aa3_y = -333.5\text{mm}$.

4.2. Bike Center Plane Offset

From Figure 4, it is known that:

$$cc1(107.23, -373.75, 138.38) \qquad\qquad cc2(107.23, -327.25, 138.38)$$

The cc is located at the center point between $cc1$ and $cc2$.
Therefore,

$$cc(107.23, -350.5, 138.38)$$
$$w1 = \left|-350.5 - (-333.5)\right| = 17\text{mm}$$

As above, it is confirmed that the theoretical value matches the computational result.

4.3. The H/T Rotating ShaftLength

In Figure 5, the coordinates of eight check points are shown as follows.

$ff1(625.05, 123.58, 600.41)$ $ff2(598.67, 150.63, 594.12)$ $ff3(650.93, 151.37, 606.57)$
$ff4(624.70, 177.33, 600.32)$ $gg1(651.01, 123.77, 493.51)$ $gg2(623.60, 150.68, 486.98)$
$gg3(676.13, 149.86, 499.50)$ $gg4(649.70, 177.24, 493.20)$

Accordingly,

$$\overline{ff1gg1} = \sqrt{(651.01 - 625.05)^2 + (123.77 - 123.58)^2 + (493.51 - 600.41)^2} \cong 110\text{mm}$$

$$\overline{ff2gg2} = \sqrt{(623.60 - 598.67)^2 + (150.68 - 150.63)^2 + (486.98 - 594.12)^2} \cong 110\text{mm}$$

$$\overline{ff3gg3} = \sqrt{(676.13 - 650.93)^2 + (149.86 - 151.37)^2 + (499.50 - 606.57)^2} \cong 110\text{mm}$$

$$\overline{ff4gg4} = \sqrt{(649.70 - 624.70)^2 + (177.24 - 177.33)^2 + (493.20 - 600.32)^2} \cong 110\text{mm}$$

The average length (L) is obtained as:

$$L = \frac{\overline{ff1gg1} + ff2gg2 + ff3gg3 + ff4gg4}{4} \cong 110\text{mm}$$

As above, the calculated value is confirmed equal to the theoretical value.

4.4. T/T Rotating Shaft Internal Diameter

In Figure 6, the coordinates of three check points ($d1$, $d2$, $d3$) are shown as follows:

$d1 = (55.39, 24, 353.63)$ $d2 = (55.85, 24, 366.87)$ $d3 = (66.64, 24, 359.98)$

Therefore,

$$\overrightarrow{d3d1} = (-11.25, 0, -6.35)$$

$$\overrightarrow{d3d2} = (-10.79, 0, 6.89) \tag{20}$$

$$\vec{n} = \overrightarrow{d3d2} \times \overrightarrow{d3d1} = (n_x, n_y, n_z)$$

where;

$$n_x = \begin{vmatrix} 0 & 6.89 \\ 0 & -6.35 \end{vmatrix} = 0$$

$$n_y = \begin{vmatrix} -10.79 & 6.89 \\ -11.25 & -6.35 \end{vmatrix} = -146.029$$

$$n_z = \begin{vmatrix} -10.79 & 0 \\ -11.25 & 0 \end{vmatrix} = 0$$

$$\vec{n} = (0, -146.029, 0)$$

According to Equation (7), the axis point coordinate is $d4\left(\dfrac{-d}{2}, \dfrac{-e}{2}, \dfrac{-f}{2}\right)$.

Therefore,

$$\overrightarrow{d4d3} = \left(66.64 - \left(\frac{-d}{2}\right), 24 - \left(\frac{-e}{2}\right), 359.98 - \left(\frac{-f}{2}\right)\right) \tag{21}$$

$\overrightarrow{d4d3}$ and $\vec{n}$ are perpendicular to each other so that:

$$\overrightarrow{d4d3} \cdot \vec{n} = 0$$

$\Rightarrow$

$$0 \cdot \left(66.64 - \left(\frac{-d}{2}\right)\right) - 146.029 \cdot \left(24 - \left(\frac{-e}{2}\right)\right) + 0 \cdot \left(359.98 - \left(\frac{-f}{2}\right)\right) = 0 \tag{22}$$

$$e = -48 \tag{23}$$

Substitute the coordinates of $d1$, $d2$ and $d3$ into Equation (10), as follows.

$$(55.39)^2 + (24)^2 + (353.63)^2 + d(55.39) + e(24) + f(353.63) + g = 0 \tag{24}$$

$$(55.85)^2 + (24)^2 + (366.87)^2 + d(55.85) + e(24) + f(366.87) + g = 0 \tag{25}$$

$$(66.64)^2 + (24)^2 + (359.98)^2 + d(66.64) + e(24) + f(359.98) + g = 0 \tag{26}$$

$\Rightarrow$

$$d = -118.27901 \qquad\qquad f = -720.25544 \qquad\qquad g = 133709.17771$$

As above, it can be obtained:

$$d4\left(\frac{-d}{2}, \frac{-e}{2}, \frac{-f}{2}\right) = d4(59.1395, 24, 360.12772)$$

The radius between the axis and check point is:

$$\overline{d4d1} = \overline{d4d2} = \overline{d4d3} = \frac{1}{2}\sqrt{d^2 + e^2 + f^2 - 4g} \cong \frac{15}{2}\,(\text{mm}) \tag{27}$$

Accordingly, the internal diameter of T/T rotating shaft is:

$$= \sqrt{d^2 + e^2 + f^2 - 4g} \cong 15\text{mm}$$

As above, the calculated value is confirmed equal to the theoretical value.

4.5. Parallelism Between T/T and B.B Rotating Shafts Axes

In Figure 8, the coordinates of two check points ($d1$, $d11$) are:

$$d1(55.39, 24, 353.63) \qquad\qquad d11(55.39, -24, 353.63)$$

The coordinate of middle point $d5$ located between $d1$ and $d2$ is:

$$d5(55.39, 0, 353.63)$$

The X-axis and Y-axis coordinates of axis point $d4$ at the T/T rotating shaft are transferred to d5 to form d5′. Therefore, the vector formed by $d4$ and d5′ is $\overrightarrow{d5'd4}$.

$$d5' = (59.1395, 0, 360.12772)$$
$$\overrightarrow{d5'd4} = (0, 24, 0) \tag{28}$$

In Figure 7,

$$a1(94.98, 48, 723.32) \qquad\qquad a2(94.98, -333.5, 723.32)$$

$\overrightarrow{a2a1}$ at the B.B rotating shaft is:

$$\overrightarrow{a2a1} = (0, 48 - (-333.5), 0) = (0, 381.5, 0) \tag{29}$$

The parallelism angel (θ) between $\overrightarrow{a2a1}$ and $\overrightarrow{d5'd4}$ can be calculated as:

$$\cos\theta = \frac{\overrightarrow{d5'd4} \cdot \overrightarrow{a2a1}}{\left|\overrightarrow{d5'd4}\right|\left|\overrightarrow{a2a1}\right|} = \frac{(0, 24, 0) \cdot (0, 381.5, 0)}{\left|(0, 24, 0)\right| \cdot \left|(0, 381.5, 0)\right|} = \frac{9156}{9156} = 1 \tag{30}$$

$$\therefore \theta = 0^{\circ}$$

As above, the T/T and B.B rotating shafts axes are confirmed parallel.

4.6. Verticality Between H/T and B.B Rotating Shafts Axes

The vector $\overrightarrow{a2g4}$ from the B.B rotating shaft to H/T axis point (g4) is:

$$\overrightarrow{a2g4} = (559.7486, -5, -154.8109)$$

where $a2(94.98, -333.5, 723.32)$ and $g4(654.7286, -338.5, 568.5091)$.

The verticality angel (θ) between $\overrightarrow{a2a1}$ and $\overrightarrow{a2g4}$ can be calculated as:

$$\cos\theta = \frac{\overrightarrow{a2a1} \cdot \overrightarrow{a2g4}}{\left|\overrightarrow{a2a1}\right|\left|\overrightarrow{a2g4}\right|} = \frac{(0, 381.5, 0) \cdot (559.7486, -5, -154.8109)}{\left|(0, 381.5, 0)\right| \cdot \left|(559.7486, -5, -154.8109)\right|}$$

$$= \frac{-1907.5}{381.5 \times 580.7839} = -0.0086 \cong 0 \tag{31}$$

$$\therefore \theta = 90^{\circ}$$

As above, the H/T and B.B rotating shafts axes are confirmed vertical.

5. Practical Verification

The process of real bike frame quality check is carried out based on the proposed 3D geometry mathematical model. The real system profile is shown in Figure 9.

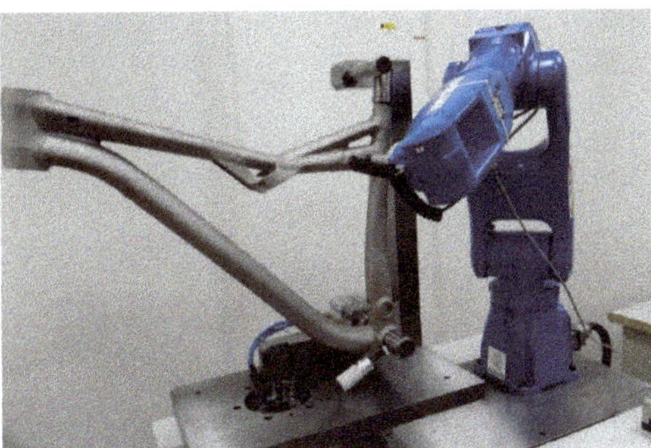

Figure 9. Profile of real measurement system.

5.1. Results with GUI

The quality check results are displayed online using GUI, as shown from Figures 10–15.

5.1.1. Bike Frame Plane

The GUI of the bike frame plane is shown in Figure 10. The performance result is –333.5 mm and that matches the theoretical value.

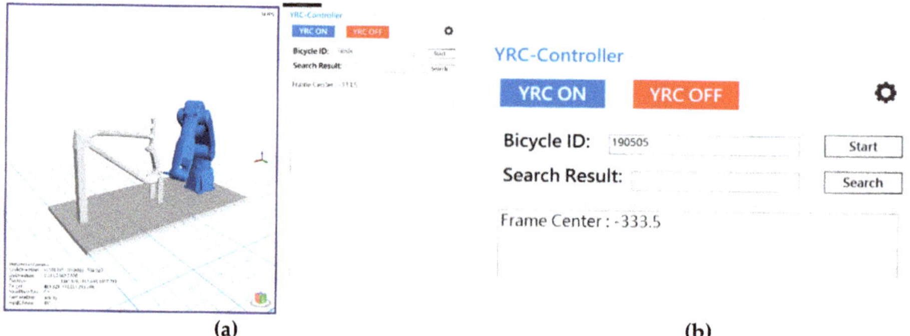

(a)   **(b)**

Figure 10. Graphical user interface (GUI) of the bike frame plane: (**a**) synchronous action; (**b**) GUI.

5.1.2. Bike Center Plane Offset

The GUI of bike center plane offset is shown in Figure 11. The performance result is 17 mm and that matches the theoretical value.

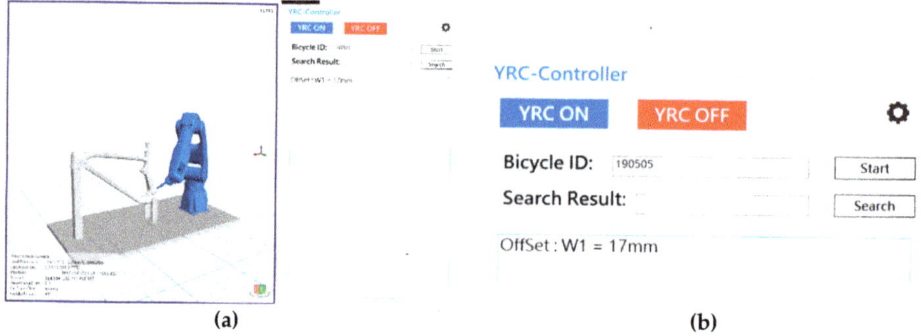

(a)   **(b)**

Figure 11. GUI of the bike center plane offset: (**a**) synchronous action; (**b**) GUI result.

5.1.3. H/T Rotating Shaft Length

The GUI of H/T rotating shaft length is shown in Figure 12. The performance result is 110 mm and that matches the theoretical value.

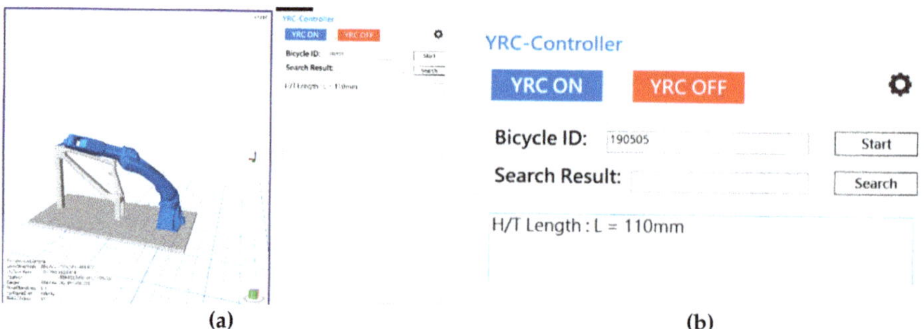

(a)   **(b)**

Figure 12. GUI of the H/T rotating shaft length: (**a**) synchronous action; (**b**) GUI result.

5.1.4. T/T Rotating Shaft Internal Diameter

The GUI of T/T rotating shaft internal diameter is shown in Figure 13. The performance result is 15 mm and that matches the theoretical value.

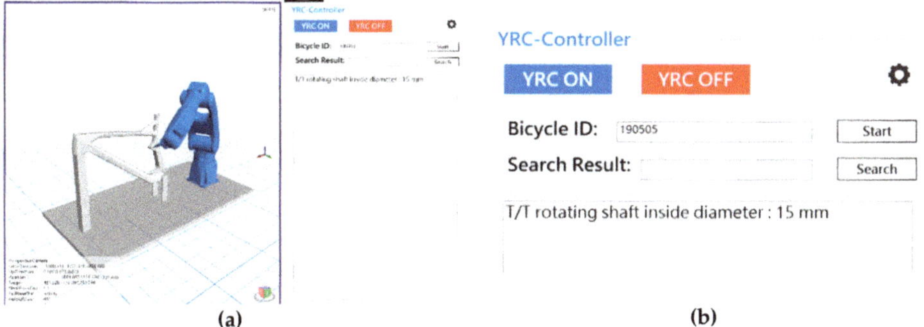

(a)   **(b)**

Figure 13. GUI of the T/T rotating shaft internal diameter: (**a**) synchronous action; (**b**) GUI result.

5.1.5. Parallelism

The parallelism between T/T and B.B rotating shafts axes using GUI is shown in Figure 14. The performance result is $\theta = 0^\circ$ that matches the theoretical value.

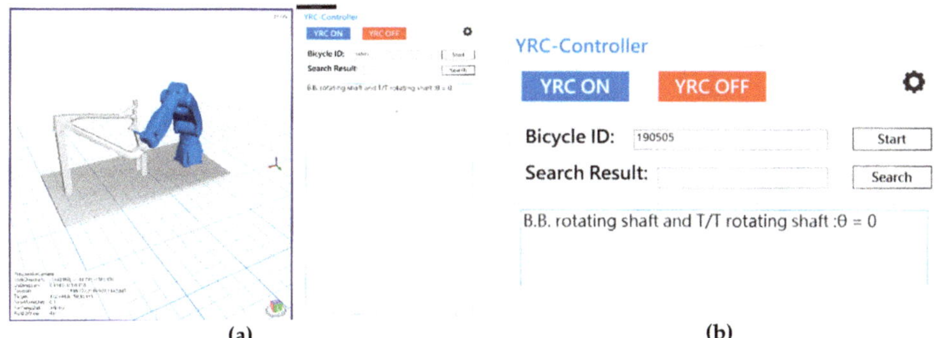

(a)   **(b)**

Figure 14. GUI of parallelism: (**a**) synchronous action; (**b**) GUI result.

5.1.6. Verticality

The verticality between H/T and B.B rotating shafts axes using GUI is shown in Figure 15. The performance result is $\theta = 90^\circ$ that matches the theoretical value.

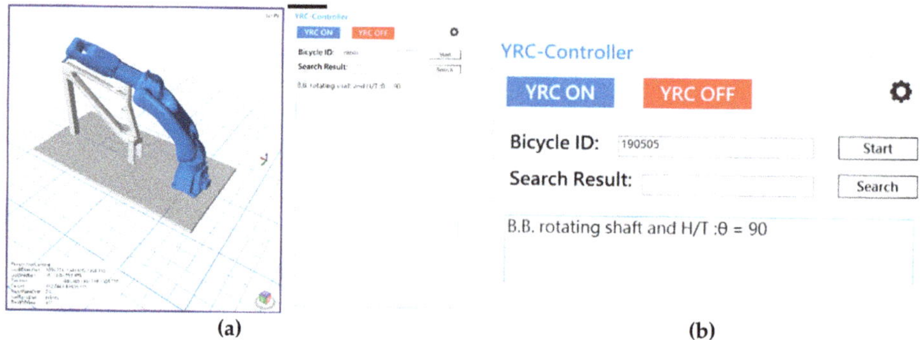

(a)   (b)

Figure 15. GUI of verticality: (**a**) synchronous action; (**b**) GUI result.

5.2. Practical Results

The measurement results from 10-times average values using the real bike frame are concluded in Table 1. Based on the same bike frame, the error between the proposed model and the Vernier caliper is below 0.05 mm, and the repeatability is at the range of 0.1 mm. This verifies that the proposed model presents both robust and stable performance. Nevertheless, the measured data reveals that the tested frame has some defects occurred in the center plane offset, parallelism and verticality.

Table 1. Measured values using the proposed model.

Check Item \ Check Point	T/T Rotating Shaft	H/T Rotating Shaft
Center plane offset	−2.71 mm	Not applicable
Parallelism	8.2°	Not applicable
Verticality	Not applicable	114.65°
Internal diameter	14.86 mm	Not applicable
Length	Not applicable	109.89 mm

To clarify the uncertainty of the measurement, the estimated standard deviation for a series of n measurements is expressed mathematically as:

$$s = \sqrt{\frac{\sum\limits_{i=1}^{n}(x_i - \bar{x})^2}{n-1}} \tag{32}$$

where x_i is the result of the ith measurement and $\bar{x}$ is the arithmetic mean of the n measurement results.

When a set of several repeated readings has been taken, the mean, $\bar{x}$, and estimated standard deviation, s, can be calculated. The measurement uncertainty, u, of the mean is therefore defined as:

$$u = \frac{s}{\sqrt{n}} \tag{33}$$

where n is the number of measurements in the set.

The estimated standard deviation and measurement uncertainty based on 10 measurements for S/T rotating shaft, T/T rotating shaft, and H/T rotating shaft is shown in Tables 2–4, respectively. From the statistics, it is obvious that both standard deviation (s) and measurement uncertainty (u) for all shaft measurements present a very low value no more than 0.018. Thus, accuracy and robustness of the proposed model is thus confirmed.

Table 2. Standard deviation (s) and measurement uncertainty (u) at S/T rotating shaft.

Check Item / Estimated Topic	Standard Deviation (s)	Measurement Uncertainty (u)
Center Plane Offset	0.007	0.002
Internal Diameter	0.009	0.003
Length	0.006	0.002
Parallelism	0.010	0.003

Table 3. Standard deviation (s) and measurement uncertainty (u) at T/T rotating shaft.

Check Item / Estimated Topic	Standard Deviation (s)	Measurement Uncertainty (u)
Center Plane Offset	0.005	0.002
Internal Diameter	0.018	0.006
Length	0.007	0.002
Parallelism	0.011	0.003

Table 4. Standard deviation (s) and measurement uncertainty (u) at H/T rotating shaft.

Check Item / Estimated Topic	Standard Deviation (s)	Measurement Uncertainty (u)
Internal Diameter	0.007	0.002
Length	0.005	0.002
Verticality	0	0

6. Conclusions

Traditional methods for the QC of bike frame products usually use general jigs or Vernier calipers. However, this kind of measurement process may take tens of minutes to complete. Another disadvantage is that it is difficult to analyze the measured data due to lack of computerization. For these reasons, the proposed 3D geometry mathematical model has successfully developed an accurate bike frame measurement based on a robotic arm with a contact sensor. In this study, the proposed model requires only three simultaneous equations to find the axis coordinate and its radius instead of four equations in a space sphere. It verifies that the measured data obtained from the model performance is consistent with the SOLIDWORKS drawing, including H/T rotating shaft length, T/T rotating shaft internal diameter, parallelism, and verticality, etc. Accordingly, it is applicable for industrial QC applications in a variety of bike frames. Other than these advantages, the stylus probe used in this proposed model presents both simple and accurate performance. However, successful measurement depends on the activity range of robotic arm that the certain features of bike frames should be reached by the stylus probe. In the future work, the optical sensors used in CMM may provide an alternative solution, although more complex signal processing algorithm should be addressed.

Author Contributions: All authors conceived the study. H.-C.L. led the project and wrote this article. B.-R.Y. built the mathematical model. J.-Y.W. (Jen-Yu Wang) designed the system software. J.-Z.L. wrote the robotic-arm control program. J.-Y.W. (Jia-Yang Wu) designed the system mechanism.

Funding: This work was supported by the Ministry of Science and Technology, Taiwan (grant number MOST 108-2637-E-167-001).

Conflicts of Interest: The authors declare no conflict of interest.

References

1. Cicero, S.; Lacalle, R.; Cicero, R.; Fernández, D.; Méndez, D. Analysis of the cracking causes in an aluminium alloy bike frame. *Eng. Fail. Anal.* **2011**, *18*, 36–46. [CrossRef]
2. Collotta, M.; Solazzi, L.; Pandini, S.; Tomasoni, G. New design concept of a downhill mountain bike frame made of a natural composite material. In *The Institution of Mechanical Engineers, Part P: Journal of Sports Engineering and Technology, Proceedings of the Institution of Mechanical Engineers*; The Institution of Mechanical Engineers: London, UK, 2018; pp. 50–56.
3. Priadythama, I.; Suhardi, B.; Adiasa, I. Further study on a short wheel base recumbent bike frame using simulated finite element analysis. In Proceedings of the 2016 2nd International Conference of Industrial, Mechanical, Electrical, and Chemical Engineering, Yogyakarta, Indonesia, 6–7 October 2016. [CrossRef]
4. Chang, C.T.; Huang, Y.C.; Chen, S.Y. Design optimization and automation of metal and composite bike frame. In Proceedings of the International SAMPE Technical Conference, Baltimore, MD, USA, 18–21 May 2015; pp. 1–18.
5. Collins, P.K.; Leen, R.; Gibson, I. Industry case study: Rapid prototype of mountain bike frame section. *Virtual. Phys. Prototyp.* **2016**, *11*, 295–303. [CrossRef]
6. Vdovin, D.; Chichekin, I.; Ryakhovsky, O. Quad bike frame dynamic load evaluation using full vehicle simulation model. *IOP Conf. Ser. Mater. Sci. Eng.* **2019**, *589*, 1–9. [CrossRef]
7. Hull, A.; O'Holleran, C. Bicycle infrastructure: can good design encourage cycling? *Urban, Plan. Transp. Res.* **2014**, *2*, 369–406. [CrossRef]
8. Shin, S. Trend of Process Automation and Factory Automation. In Proceedings of the 2006 SICE-ICASE International Joint Conference, Busan, Korea, 18–21 October 2006.
9. Merdan, M.; Lepuschitz, W.; Axinia, E. Advanced process automation using automation agents. In Proceedings of the 5th International Conference on Automation, Robotics and Applications, Wellington, New Zealand, 6–8 December 2011.
10. Zhou, C.; Huang, S.; Xiong, N.; Yang, S.H.; Li, H.; Qin, Y.; Li, X. Design and Analysis of Multimodel-Based Anomaly Intrusion Detection Systems in Industrial Process Automation. *IEEE Trans. Systs. Man. Cybern. Systs.* **2015**, *45*, 1345–1360. [CrossRef]
11. Jiao, Z.J.; He, C.Y.; Wang, J.; Zhao, Z. Development and application of automation control system to plate production line. In Proceedings of the 2010 11th International Conference on Control Automation Robotics & Vision, Singapore, 7–10 December 2010.
12. Lin, H.C.; Li, L.L.; Lee, V.C.S. Multiple Autonomous Robots Coordination and Navigation. *J. Robot.* **2019**, *2019*, 1–2. [CrossRef]
13. Ito, S.; Kikuchi, H.; Chen, Y.; Shimizu, Y.; Gao, W.; Takahashi, K.; Kanayama, T.; Arakawa, K.; Hayashi, A. A Micro-Coordinate Measurement Machine (CMM) for Large-Scale Dimensional Measurement of Micro-Slits. *Appl. Sci.* **2016**, *6*, 156. [CrossRef]
14. Pawar, M.G.; Nandeshwar, B.S.; Borikar, V.N.; Jaiswal, S.B. Design of Portable Coordinate Measuring Machine. *Int. J. Emerg. Eng. Res. Technol.* **2015**, *3*, 1–9.
15. Puertas, C.; Luis Pérez, J.; Salcedo, D.; León, J.; Luri, R.; Fuertes, J.P. Precision Study of a Coordinate Measuring Machine Using Several Contact Probes. *Procedia. Eng.* **2013**, *63*, 547–555. [CrossRef]
16. Anagnostakis, D.; Ritchie, J.; Lim, T.; Sung, R.; Dewar, R. Automated Coordinate Measuring Machine Inspection Planning Knowledge Capture and Formalization. *J. Comput. Inf. Sci. Eng.* **2018**, *18*, 1–12. [CrossRef]
17. Mian, S.H.; Al-Ahmari, A. New developments in coordinate measuring machines for manufacturing industries. *Int. J. Metrol. Qual. Eng.* **2014**, *5*, 1–10.
18. Majeed, A.; Lee, S. A Fast Global Flight Path Planning Algorithm Based on Space Circumscription and Sparse Visibility Graph for Unmanned Aerial Vehicle. *Electron* **2018**, *7*, 375. [CrossRef]
19. Plaku, E.; Plaku, E.; Simari, P. Direct Path Superfacets: An Intermediate Representation for Motion Planning. *IEEE Robot. Autom. Lett.* **2016**, *2*, 350–357. [CrossRef]
20. Jahnavi, K.; Sivraj, P. Teaching and learning robotic arm model. In Proceedings of the 2017 International Conference on Intelligent Computing, Instrumentation and Control Technologies (ICICICT), Kannur, India, 6–7 July 2017.

21. Elhoseny, M.; Shehab, A.; Yuan, X. Optimizing robot path in dynamic environments using Genetic Algorithm and Bezier Curve. *J. Intell. Fuzzy. Syst.* **2017**, *33*, 2305–2316. [CrossRef]
22. Juan, B.V.; David, M.V.; Luis, C.L.; Luis, M.G. A low-cost platform based on a robotic arm for parameters estimation of Inertial Measurement Units. *Measurement* **2017**, *110*, 257–262.

applied sciences

Article

Influence of Contamination of Gear Oils in Relation to Time of Operation on Their Lubricity

Leszek Gil [1], Krzysztof Przystupa [2], Daniel Pieniak [1], Edward Kozłowski [3], Katarzyna Antosz [4,*], Konrad Gauda [1] and Paweł Izdebski [1]

[1] Transport and Computer Science Faculty, University of Economics and Innovation in Lublin, Projektowa 4, 20-209 Lublin, Poland; leszek.gil@wsei.lublin.pl (L.G.); daniel.pieniak@wsei.lublin.pl (D.P.); konrad.gauda@wsei.lublin.pl (K.G.); p-izdebski@wp.pl (P.I.)

[2] Mechanical Engineering Faculty, Lublin University of Technology, Nadbystrzycka 36, 20-618 Lublin, Poland; k.przystupa@pollub.pl

[3] Faculty of Management, Lublin University of Technology, Nadbystrzycka 36, 20-618 Lublin, Poland; e.kozlovski@pollub.pl

[4] Faculty of Mechanical Engineering and Aeronautics, Rzeszow University of Technology, Aleja Powstańców Warszawy 8, 35-959 Rzeszow, Poland

[*] Correspondence: katarzyna.antosz@prz.edu.pl

Citation: Gil, L.; Przystupa, K.; Pieniak, D.; Kozłowski, E.; Antosz, K.; Gauda, K.; Izdebski, P. Influence of Contamination of Gear Oils in Relation to Time of Operation on Their Lubricity. *Appl. Sci.* **2021**, *11*, 11835. https://doi.org/10.3390/app112411835

Academic Editors: Stephanos Theodossiades and Luis Lugo

Received: 8 November 2021
Accepted: 9 December 2021
Published: 13 December 2021

Abstract: The quality and reliability of consumables, including gear oils, results in the failure-free operation of the transmission components in heavy trucks. It is known that oil viscosity is essential for all lubricated tribopairs for wear and friction reduction in all vehicles with a gearbox. Viscosity may be influenced by the contamination that wear products can impart on the oil. Oil contamination can also affect lubrication efficiency in the boundary friction conditions in gearboxes where slips occur (including bevel and hypoid gearboxes). The present research focused on this issue. An obvious hypothesis was adopted, where it was theorized that exploiting the contaminants that are present in gear oil may affect how the lubricating properties of gear oils deteriorate. Laboratory tests were performed on contaminants that are commonly found in gear oil using the Parker Laser CM20. The study was designed to identify a number of different solid particles that are present in oil. At the second stage, friction tests were conducted for a friction couple "ball-on-disc" in an oil bath at 90 °C on a CSM microtribometer. The quantitative contamination of the gear oils that contained solid particles and the curves representing the friction coefficients of fresh oils with a history of exploitation were compared. The test results were statistically analysed. Exploitation was shown to have a significant impact on the contamination of gear oils. It was revealed that the contamination and the mileage had no effect on the tested oils.

Keywords: lubricity; gear oil; wear; operational reliability

1. Introduction

Ensuring that machines and devices are able to conduct high-quality work and maintain operational reliability is not only a very important issue in chemical applications [1–3], e.g., transport [4,5], electronic systems [6], or scientific works, e.g., for chemical purposes [7], it is also of great importance for applications that are related to the production of high-quality food products [8]. Lubrication is essential for all sliding pairs in all tribosystems [9–11].

The physical essence of lubrication processes is the conversion of adverse external friction to friction that takes place inside of the tribofilm [12–14]. In order for good lubrication to be maintained, the grease must have high adhesion to the frictional surfaces in question, and the grease layer that is between these components must be maintained at a certain thickness. This should be maintained regardless of friction speed, pressure, and temperature [15], and this is usually difficult to achieve. The formation of a layer of grease on the friction surface is associated with the phenomena of the physical absorption of polar particles and the chemical absorption of boundary films as well as a hydrodynamic effect [16]. However, load transfer occurs through a layer of grease that is generated by the hydrodynamic effect in the machine kinematic nodes that are under increasing motion [12]. High-pressure lubrication takes place in concentrated contacts, including in gear contacts [17]. High pressure in the contact area increases the viscosity of the lubricant as well as the elastic deformation of the surfaces that are in contact with each other indirectly through the lubricant film. This is the case for elastohydrodynamic lubrication. The term elasthydrodynamic film is used to refer to the intermediate lubricant film [15]. In gear transmissions, lubrication conditions are not favorable for the formation of a lubricating film. It seems that gullet pressure is one of the determinants. However, as stated in [18], high pressure in the contact area increases the viscosity of the lubricant and the elastic deformation of the contact surfaces in an indirect way through the lubricant film. Moreover, pressure leads to teeth bending, and when teeth pairing occurs, the tips of the teeth belong to the powered wheel rub the grease from the surface of the powering wheel. Modifying the teeth only partially eliminates this phenomenon [15]. Moreover, high circumferential speeds result in oil being removed from the contact area. It is known that the formation of a lubricating film in oil is intended to prevent metal surfaces from coming into direct contact with each other, but it also refers to a situation in which wear and high friction occur between sliding surfaces. In light of the above situation, this becomes difficult in the context of gear boxes. In addition, oil properties change during exploitation, and this represents another disadvantage. It can be assumed that the contamination of the used oil may impair the ability to form a permanent tribofilm that reduces the friction and wear of sliding pairs [19]. The lubricant that is used in closed-circuit mechanical systems is subject to aging [18]. It undergoes oxidation because it comes into contact with air. The particles, which are a product of tribological wear, also get into the oil. The contamination of gear oils with wear products may result in power losses, among other consequences, as well as decreases in the flow resistance of the lubricant [19]. Therefore, the quality and condition of the lubricant affects the friction resistance in gear boxes and may affect the efficiency of tribomechanical systems [20,21]. The applications that oil is used for may lead to changes in its viscosity [19]. On the other hand, as shown in paper [21], oil viscosity results in the power losses in meshing when operating under a working load. It is for this reason that periodic oil changes are performed [15].

Synthetic oils are usually used in the gear boxes of modern vehicles. This is because of the many advantages of these oils. One of them is the reduction of the friction coefficient during meshing by up to 25% compared to when mineral oils are used as a lubricant [21]. The use of such oils is beneficial, although the problem of the operational quality of these oils is also important, with the preservation of their lubricating properties being one of the main issues, especially since these properties can be affected by the operating time as well as by the level of contamination resulting from exploitation. On the basis of the above, it can be concluded that the overworking and contamination of gear oil may affect its lubrication efficiency. This observation is especially important when considering that the gear box elements need to be protected against excessive wear and the boundary lubrication condition, which can be seen in gears where slips occur (including in bevel and hypoid gear boxes). The present research focused on this issue. The hypothesis that operational contaminants may affect the deterioration of the lubricating properties of the synthetic gear oils was adopted. The main goal of this paper was to anlayze how operational contaminants affect the deterioration of the lubricating properties of synthetic gear oils. The article consists of the introduction followed by a chapter describing the experiment—the Materials and Methods section. Finally, the results that were achieved through the experiment are presented and compared.

2. Materials and Methods

The following gear oils were used for the laboratory tests: class SAE 75W-140 Scania Oil STO 2:0 A (oil from the axle of a truck), class 80W-14 Scania Oil AXLE STO 1:0 (oil from the axle of a truck), class 75W-90 and Scania Oil 2:0 G of (oil from the gearbox of a truck).

Lubrication tests were conducted on the CSM microtribometer (CSM Switzerland). The tests were performed using a ball-on-disk module, such as the one shown in Figure 1. A steel friction node was installed. Both elements of the friction pair were made of 100Cr6 steel. Friction tests were conducted in an oil bath at 90 °C, which supposedly corresponded to the operating temperature of the oil in real working conditions. The load in the friction test was constant and equaled 5 N. The sliding distance was 630 m, and the linear speed was 60 mm/s. During the friction tests, the friction coefficient was recorded at the frequency of 10 Hz as a function of the friction path.

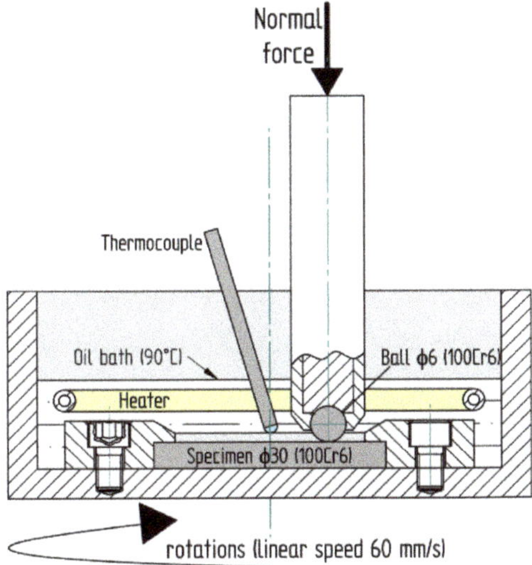

Figure 1. Picture of the tribometer test set up for the lubricating oil tests.

The laboratory tests determining the oil contaminants were performed using the Parker Laser CM20 device, which is designed to identify the number of solid particles in oil and to classify them with the use of the optical scanning method. The measurement procedure was in accordance with PN-ISO 4406: March 2005 [22].

The methods that were used to observe and count the number of contaminants are shown in Figure 2.

The measurement accuracy of this type of transmitter is an important issue. It should be noted that a quality of measurements better than 5% was obtained when the ISO MTD (ISO Medium Test Dust) and ISO ACFTD (ISO Air Cleaner Fine Test Dust) procedures were used. As a result, it is possible to obtain results that are in accordance with the ISO standard and that are in the range of 7–22 µm; NAS and SAE measurements were also obtained in the ranges of 0–12 µm and 0–12 µm, respectively. In the conducted research, it was assumed that the contaminants would be classified into six sections: 4; 6; 14; 21; 38; and 70 µm. Figure 3 shows the measuring device and an exemplary printout of the measurement results.

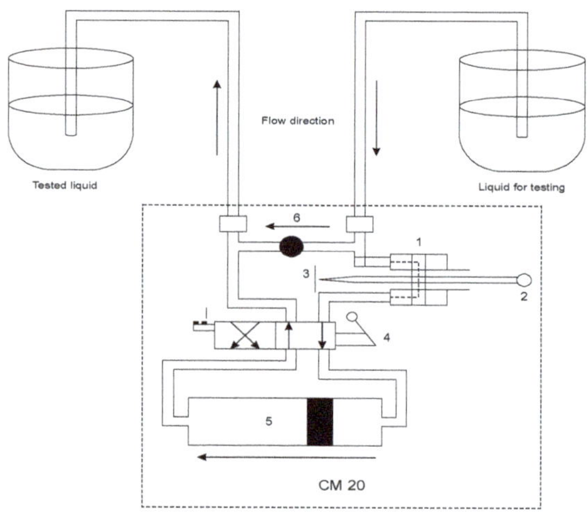

Figure 2. Diagram of the measuring system, 1—measuring chamber, 2—laser light source, 3—optical scanner, 4—switching valve (hydraulic), 5—dosing pump, 6—flow sensor.

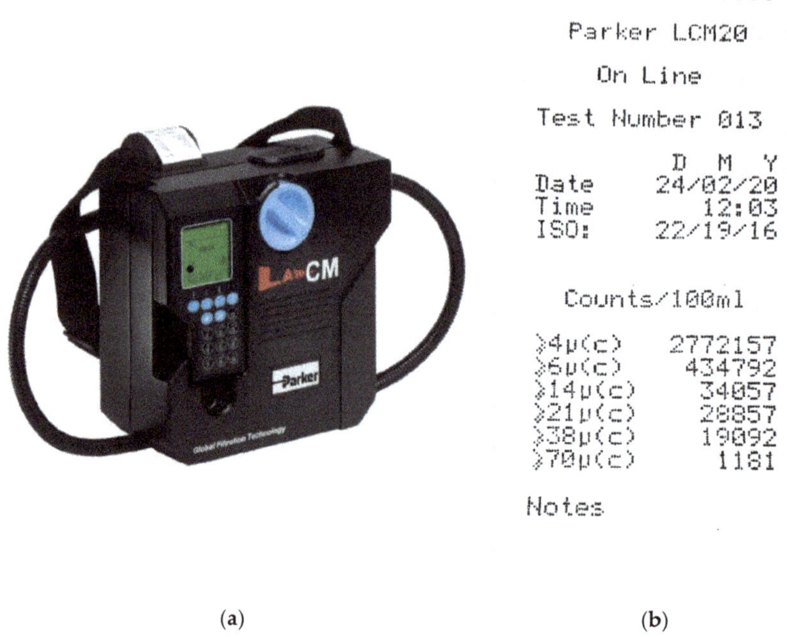

(**a**)   (**b**)

Figure 3. Parker Laser CM20 automatic particle number meter (**a**) device and (**b**) sample printout of measurements.

3. Results and Discussion

The test results were statistically processed. Table 1 presents descriptive statistics of the test results for the friction coefficient and contains the minimum (min), maximum (max), and average (mean) values and the standard deviation (std. dev.).

Table 1. Descriptive statistics of the friction coefficient obtained in the tests determining the lubricity of gear oils.

Oil	Mileage (km)	Gearbox	Min	Max	Mean	Std. Dev.
75W-140	fresh oil	axle	0.0600	0.1374	0.0979	0.0142
75W-140	350,000	axle	0.0298	0.1193	0.1020	0.0086
80W-140	fresh oil	axle	0.0177	0.0984	0.0671	0.0136
80W-140	220,000	axle	0.0032	0.0582	0.0523	0.0091
75W-90	fresh oil	gearbox	0.0753	0.1566	0.1046	0.0075
75W-90	210,000	gearbox	0.0345	0.0908	0.0655	0.0102

The statistical values of the friction coefficient of the fresh oils and the oils with a service history differ. In the case of the used oils, the work of which was expressed as the mileage of a vehicle between 220,000 km and 210,000 km, the friction coefficient demonstrated lower average values than the friction coefficient that was observed for the fresh oils. Oil changes were planned for after these mileages were achieved. T Figures 4–6 present graphs of the friction coefficient based on different paths. The graph presenting linear wear is shown in Figure 7. The graph shows the variability of linear wear depending on the number of friction cycles. It should be noted that the presented curves depend on the sliding wear of the friction pair, but this is not the only thing that should be taken into account. The thermal expansion of the ball and disc that are heated by the conditioned oil was also influenced. Both factors of the experiment have a synergistic influence on the shape of the curves. The variability tests of the friction coefficient can function as a measure for the lubricity of gear and diesel oils [23,24].

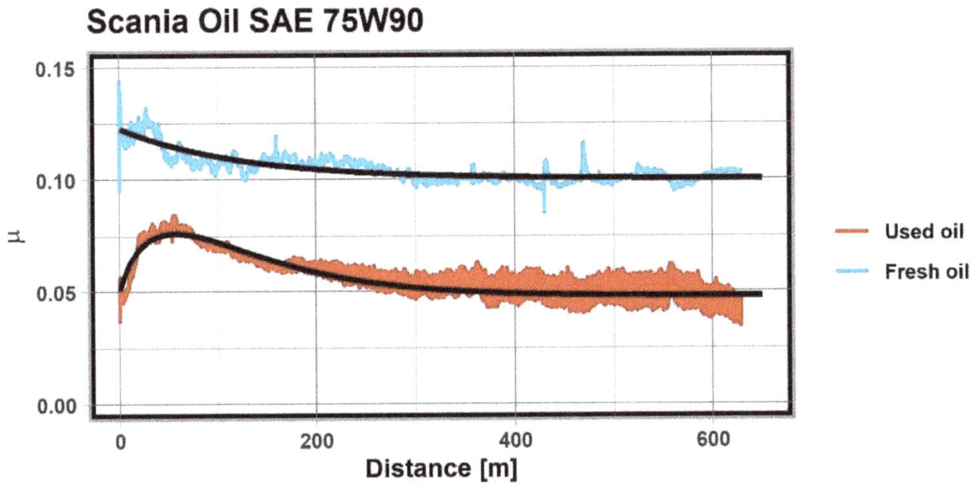

Figure 4. The curve of the friction coefficient depending on the friction path for oil SAE 75W-90 and fitting Function (1).

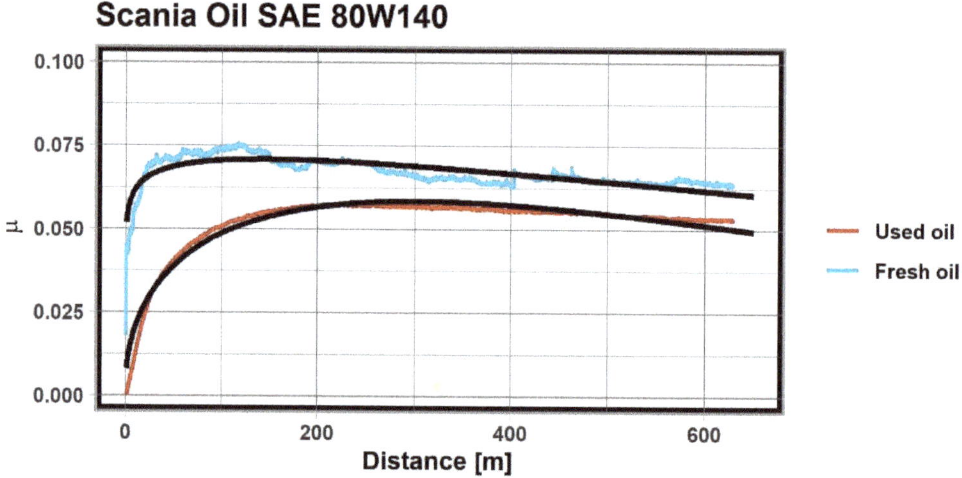

Figure 5. The curve of the friction coefficient depending on the friction path for oil SAE 80W-140 and fitting Function (1).

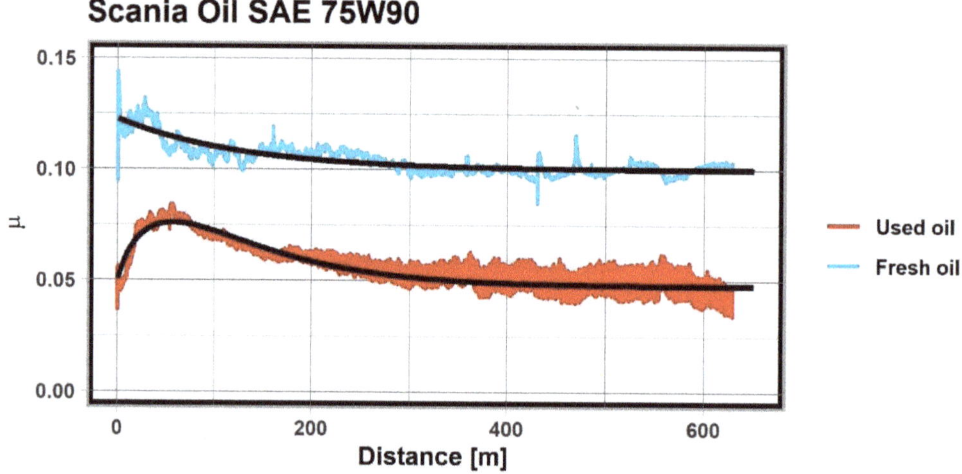

Figure 6. The curve of the friction coefficient depending on the friction path for oil SAE 75W-90 and fitting Function (1).

The nonlinear dependence between the friction coefficient and distance (friction path) is defined as follows:

$$y = \alpha_0 + \alpha_1 x^{\alpha_2} e^{-\alpha_3 x} + \varepsilon, \tag{1}$$

where y denotes the friction coefficient, x—distance, and ε—disturbances with a normal distribution N(0, σ2) and is connected with measurement. The shapes of the curves for the friction coefficients indicate a gradual friction process. All of the tested oils are characterized by friction coefficient having an increasing curve when friction begins, and then a slight decrease at the first stage of friction. At the next stage, a steady-state friction regime with slight deviations was observed. However, for some oils, the course of the friction coefficient decreased slightly—75W-90 (fresh oil), 75W-140 (after 210,000 km)—at the second stage of friction. The most stable friction curve, which demonstrated the smallest amount of fluctuation, was characterized by oil 80W-140 (after 220,000 km). In this case, the variation seen in the friction function was similar than that of the theoretical model, which, as we will see later, was explained by the dependence between the number of particles and the diameters of those particles.

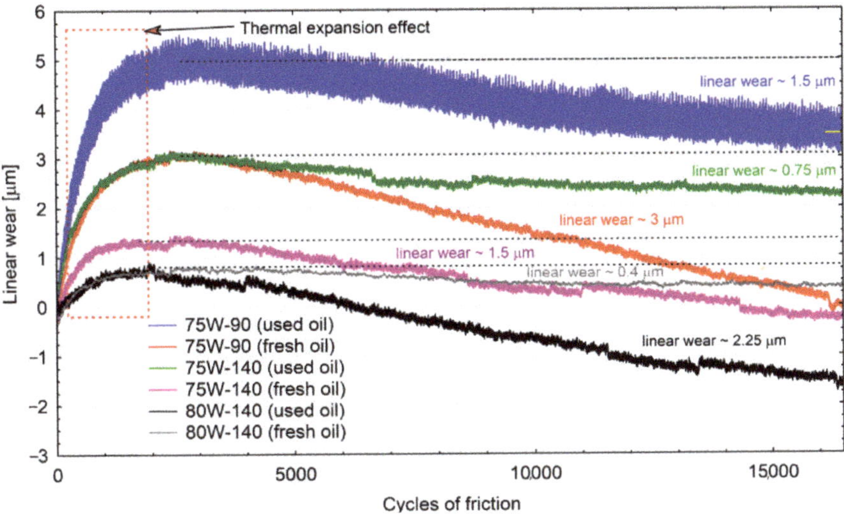

Figure 7. Linear wear of friction pairs in oil bath.

For each oil, the parameters $\alpha_0, \ldots, \alpha_3$ were estimated by applying the least squares method. Parameter α_2 corresponds to the shape of the curve, but value α_3 corresponds to the scale of descent. The values of these parameters are presented in Table 2. The fitting of model (1) to the data is marked with a black curve in Figures 4–6. Additionally, the basic indices of fitting function (1) were determined using the sum of squares (*SSE*)

$$SSE = \sum_{i=1}^{n} \left(y_i - \hat{\alpha}_0 - \hat{\alpha}_1 x_i^{\hat{\alpha}_2} e^{-\hat{\alpha}_3 x_i} \right)^2,$$

where $\hat{\alpha}_0, \ldots, \hat{\alpha}_3$ denote the estimator of unknown parameters, and the sum of squares total (*SST*) is a sum of the squared differences between the observed dependent variable and its mean

$$SST = \sum_{i=1}^{n} \left(y_i - \bar{y} \right)^2.$$

Table 2. Parameter values of Function (1) for different types of oils and values for fitting this function to real measures.

Type	75W90	75W90	80W140	80W140	75W140	75W140
Category	Used oil	Fresh oil	Used oil	Fresh oil	Used oil	Fresh oil
α_0	0.04784	0.09997	0.00000	0.00025	0.01341	0.08129
α_1	0.00276	0.02276	0.00798	0.05168	0.07399	0.03602
α_2	0.76353	0.00000	0.42336	0.07891	0.05117	0.03766
α_3	0.01342	0.00801	0.00141	0.00055	0.00034	0.00386
SSE	0.31497	0.32592	0.06434	0.13916	0.21841	0.33773
SST	2.07269	1.22027	1.38345	0.38203	0.35047	2.93344
R^2_{pseudo}	0.84804	0.73291	0.95349	0.63573	0.37680	0.88487

SST represents the deviance of the intercept-only model, but *SSE* represents the deviance of the fitted nonlinear model. According to [25], we can calculate the pseudo R^2, which presents the goodness of fit model to the data as follows:

$$R^2_{pseudo} = 1 - \frac{SSE}{SST}.$$

The figures that are presented below present the fitting of Function (1) to the measured data for the different oils.

In the literature, the stage at which the curve of the friction coefficient is constant is called stationary or normal [26]. The stability of the friction process is important when assessing oils. However, the lubricity of the oils that were tested is of practical significance. The SAE (Society of Automotive Engineers) defines lubricity as a measure of the difference in friction when comparing the properties of different oils with the same viscosities under the same conditions [27]. In the results from the tests that were conducted in the current research, the nominal viscosities of the fresh oil and used oils were the same, as the assumed mileages (km) were not exceeded. Three different oils were tested in these tests. When comparing the gear oils with different SAE viscosity classes, the following definition of lubricity proposed in [28] may be relevant: lubricity is the liquid's ability to cause low static resistance when moving solid surfaces and high resistance when bringing them together under a normal load. According to [29], lubricity is the ability of a substance to provide better lubricating properties in conditions where the lubricant film is so thin that its action is not only determined by viscosity. This is probably related to the occurrence of mixed friction in many steel friction nodes. The approach presented in these works is utilitarian, and the value of the friction coefficient is of practical importance when assessing oils. According to the criterion for assessing the condition of oil, i.e., lubricity, the condition of these oils can be considered suitable for use with high probability. A similar relationship was demonstrated for engine oils composed of mineral oils that were used in heavy trucks [30,31]. After 350,000 km, the used oil demonstrated a friction coefficient that was slightly higher than that of the average value of the friction coefficient that was obtained for fresh oil, and the standard deviation of the used oil's friction coefficient was clearly lower. The results indicate that the used oil 75W-140, which performed greater operational work than oils 80W-140 and 75W-90, had worse lubrication properties and was closer to reaching the limit state. It is worth adding that in [32], the limit state of the object is defined as a technical condition in which further operation of the object is not recommended. It should be kept in mind that the quality of a product is determined by its degree of compliance with requirements [30], and the technically justified service life of oil should ensure the maximum use of the potential of oil quality [33]; if this is true, then used oil 75W-140 should not be considered to be suitable for use. Such a performance of this oil is also confirmed by the linear wear results that are presented in Figure 7 (the curve marked in blue). The highest difference in average values of the friction coefficient for the fresh and used oil was found for the oil with the lowest nominal viscosity of 75W-90. The difference was ~37%. The same relationship was also shown in the linear wear measurements. In addition, the average friction coefficient for fresh oil 75W-90 was the highest among all of the tested oils. The linear wear fresh oil 75W-90 was also the highest of all of the fresh oils. It is possible that the condition of the used oil 75W-90 also depended on the type of a gearbox that the oil was worked in. The two other oils worked in axles. Bevel gearboxes are used in the rear axles, and this type of gearbox is characterized as having a much greater degree of slipping compared to hypoid gear boxes, which means that the working conditions of the oil in the axles are more demanding [19]. A different extortion spectrum may be reflected in the dimension of the qualitative changes that were observed in the oils used in axles.

The analysis also concerned the impact of the degree of contamination on lubricating properties. An analysis of the relationship between the number of particles and the diameter was conducted. The purpose of this analysis was to identify and compare the trends that were obtained for the fresh and used oils. For each of the oil states (clean and used), the relationship between $log_{10}n$ (logarithm of the decimal number of particles depending on the diameter of x) was identified. For this purpose, a linear model with a particular transformation of the dependent variable was considered as follows:

$$(log_{10}n)^b = \alpha_0 + \alpha_1 x + \varepsilon,$$

(2)

where ε is a random variable with a normal distribution of $N(0, \sigma^2)$. In the paper, b = 0.2 was assumed (for this parameter, the highest determination indicators of R^2 were obtained for both the clean and used oils). From (2), we can see that the dependence between the number of particles and the diameters of these particles is nonlinear.

Linear regression plots (2) are presented in Figures 8–10. The parameters of the linear regression model are presented in Tables 3–5. The results of the particle content tests indicated that the number of particles with the largest diameters is clearly higher in the used oils. This means that these oils contain more contaminants. To estimate the unknown parameters in model (2), the least squares method was applied. The linear models of the dependences between the diameters of the particles in the oil samples and the number of these particles are well matched to the empirical data. The values of the coefficients of determination are close to or above 0.9. The best fit of the linear regression model was demonstrated for fresh oil 75W-90. The same oil also had the highest average friction coefficient in the lubricity tests.

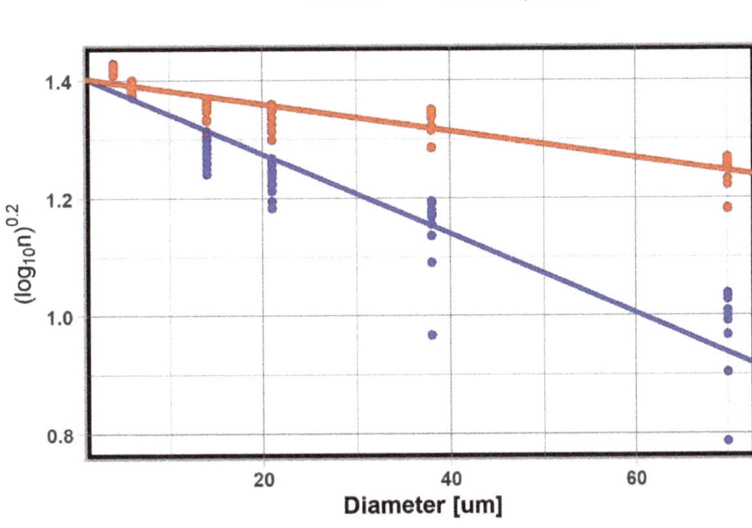

Figure 8. Regression model of particle content as a function of particle diameter of fresh and used (350,000 km) gear oil SAE 75W-140.

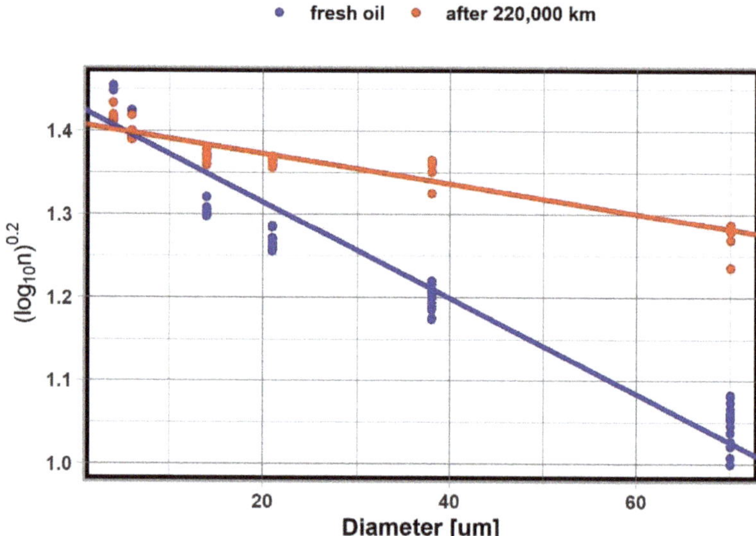

Figure 9. Regression model of particle content as a function of particle diameter of fresh and used (220,000 km) gear oil SAE 80W-140.

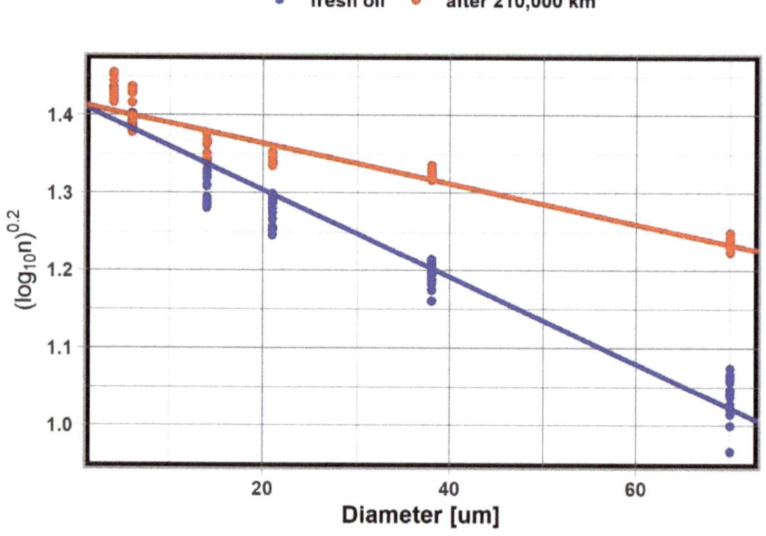

Figure 10. Regression model of particle content as a function of particle diameter of fresh and used (210,000 km) gear oil SAE 75W-90.

Table 3. Basic parameters of the linear regression model for oil 75W-140.

	Fresh Oil	Oil after 350,000 km
α_1	−0.00674 ***	−0.00227 ***
(Std. Error)	(0.00024)	(0.00011)
α_0	1.40932 ***	1.40508 ***
(Std. Error)	(0.00812)	(0.0037)
Observations	89	90
R^2	0.89971	0.83432
Adjusted R^2	0.89856	0.83244
Residual Std. Error	0.05120 (df = 87)	0.02341(df = 88)
F Statistic	780.524 *** (df = 1; 87)	443.1412 *** (df = 1; 88)

Note: *** $p < 0.01$.

Table 4. Basic parameters of the linear regression model for oil SAE 80W-140.

	Fresh Gear Oil	Oil after 220,000 km
α_1	−0.00578 ***	−0.00181 ***
(Std. Error)	(0.00017)	(0.00007)
α_0	1.43028 ***	1.40962 ***
(Std. Error)	(0.00599)	(0.00224)
Observations	90	90
R^2	0.92529	0.89762
Adjusted R^2	0.92444	0.89645
Residual Std. Error	0.03792 (df = 88)	0.01415 (df = 88)
F Statistic	1.089.854 *** (df = 1; 88)	771.511 *** (df = 1; 88)

Note: *** $p < 0.01$.

Table 5. Basic parameters of the linear regression model for oil 75W-90.

	Fresh Gear Oil	Oil after 210,000 km
α_1	−0.00561 ***	−0.00260 ***
(Std. Error)	(0.00013)	(0.00011)
α_0	1.41607 ***	1.41604 ***
(Std. Error)	(0.00449)	(0.00369)
Observations	90	90
R^2	0.95418	0.86863
Adjusted R^2	0.95366	0.867174
Residual Std. Error	0.02843 (df = 88)	0.02339 (df = 88)
F Statistic	1.932.421 *** (df = 1; 88)	581.8797 *** (df = 1; 88)

Note: *** $p < 0.01$.

The comparison of the results of the lubricity tests and the amount of particles due to the amount of contaminants does not indicate a correlation between the degree of oil contamination and lubricity. A higher share of particles with the size of several dozen micrometres was found in the used oils. It is possible that these large particles, that have also been found in other wear products, are suspended in oil. This is also confirmed by the information contained in PN-ISO 4406:2005 [22], where it is stated that particles that are larger than 4 micrometres are treated as a reference value for suspended substances. It is believed that in the case of suspensions, the presence of solid particles in a liquid additionally gives the liquid a non-Newtonian liquid character, which is associated with various types of viscosity anomalies [34]. However, the content of large particles, which was much higher in the tested oils with a history of exploitation, did not adversely affect the friction coefficient in the kinematic test pair. It is possible that the content of small particles from fractional parts with diameters from 1 micrometre to 5µm in size is important. In paper [35], it was stated that the share of oil contaminants with such dimensions can be as high as 96%. The share of these particles was the highest in the volume of all of the tested gear oils. A similar situation occurred in the fresh oils and in the oils with a history of exploitation. Additives are often used during the operation of oils; these can be microscopic particles of soft metals ranging in size from 5 to 155 µm. In study [36], a beneficial effect of such particle additives on the lubricating properties of gear oils is shown. It is possible that this factor had a decisive impact on the lack of deterioration of the lubricating properties of the tested oils. The fact emphasized in other papers, including [37,38], that large particles with diameters ranging from a dozen to several dozen micrometres in size correspond to the dynamic clearance, thus determining the thickness of a lubricating film should also be noted. They are harmful because the dynamic clearance in gears that are in the contact area of the meshing teeth should be from 0.1 to 1 micrometres [39]. Large dirt particles may affect the continuity of a lubricating film.

4. Conclusions

The current paper presents an analysis of the operational contaminants that may affect the deterioration of the lubricating properties of synthetic gear oils. The main objective of the research was not only to compare the properties of oils, but to also determine the relationship between the physical properties in order to determine the condition of the oil. Two models were considered in the present work. One was the path friction coefficient. The second was the relationship between the number of contaminants and the diameter of the contaminants. For each oil, these models were fitted to the empirical data.

Based on the conducted research, the following conclusions were formulated:

1. The used oils can be characterized by a significant number of large contaminant particles. According to normative requirements and operational experience, this indicates their inability to be used to their full potential. The results of the friction tests indicate that the exploitation use of oil 75W-140 occurred after 350,000 km of mileage. It also applies to oil 75W-90, which had a much lower mileage but worked in a gearbox. Moreover, the comparison of the results of the lubricity tests and the amount of particles due to the amount of contaminants does not indicate a correlation between the degree of oil contamination and lubricity. A higher share of particles that were several dozen micrometres in size was found in the used oils. It is possible that these large particles, which also present in other products of wear, are also suspended in oil.

2. In the light of the conducted research, it seems reasonable to hypothesize that the use of fluids in the expected operational runs does not cause a critical deterioration of the lubricating and anti-wear properties. The deterioration in the properties is non-catastrophic.

3. Unfortunately, the current research has not allowed us to check how long the liquid work time in a gearbox and axle must be in order to reach critical deterioration.

The conducted research showed that the presence of contaminants is not catastrophic and that in order to fully examine the oils and to determine the critical moment, the oils with a much greater operational mileage should be tested in order to establish the relationship between the number of particles and their tribological properties. After testing a larger number of samples with different mileages, the second model supports the development of a classifier that allows the oil mileage to be estimated in technical devices. This will be the subject of further research.

Author Contributions: Conceptualization, L.G., D.P., K.P. and E.K.; methodology, L.G., K.P., D.P., K.G., E.K. and K.A..; software, E.K. and K.A.; validation, K.P., D.P., K.A. and K.G.; formal analysis, D.P., L.G., K.P. and E.K.; investigation, D.P., E.K. and L.G.; resources, K.P., K.A., K.G. and P.I.; data curation, D.P., E.K. and P.I..; writing—original draft preparation, D.P., K.P., L.G. and K.G.; writing—review and editing, E.K., K.A. and K.P.; visualization, K.P. and K.G.; supervision, K.A., K.P. and L.G.; project administration, K.A., L.G., K.P., D.P. and K.G.; funding acquisition, K.A. and K.P. All authors have read and agreed to the published version of the manuscript.

Funding: This project was financed in the framework of the project Lublin University of Technology-Regional Excellence Initiative, funded by the Polish Ministry of Science and Higher Education (contract no. 030/RID/2018/19).

Institutional Review Board Statement: Not applicable.

Informed Consent Statement: Not applicable.

Conflicts of Interest: The authors declare no conflict of interest.

References

1. Lonkwic, P.; Przystupa, K.; Krakowski, T.; Ruta, H. Case Study of Support Frame Optimization Using a Distant Load. *Sustainability* **2020**, *12*, 974. [CrossRef]
2. Pieniak, D.; Przystupa, K.; Walczak, A.; Niewczas, A.M.; Krzyzak, A.; Bartnik, G.; Gil, L.; Lonkwic, P. Hydro-Thermal Fatigue of Polymer Matrix Composite Biomaterials. *Materials* **2019**, *12*, 3650. [CrossRef]
3. Ahmadi Moghaddam, H.; Mertiny, P. Stochastic finite element analysis framework for modelling thermal conductivity of particulate modified polymer composites. *Results Phys.* **2018**, *11*, 905–914. [CrossRef]
4. Pytka, J.; Budzyński, P.; Józwik, J.; Michałowska, J.; Tofil, A.; Łyszczyk, T.; Błażejczak, D. Application of GNSS/INS and an Optical Sensor for Determining Airplane Takeoff and Landing Performance on a Grassy Airfield. *Sensors* **2019**, *19*, 5492. [CrossRef] [PubMed]
5. Kozłowski, E.; Mazurkiewicz, D.; Żabiński, T.; Prucnal, S.; Sęp, J. Assessment model of cutting tool condition for real-time supervision system. *Eksploat. Niezawodn.* **2019**, *21*, 679–685. [CrossRef]
6. Wang, J.; Kochan, O.; Przystupa, K.; Su, J. Information-measuring System to Study the Thermocouple with Controlled Temperature Field. *Meas. Sci. Rev.* **2019**, *19*, 161–169. [CrossRef]
7. Mikhalieva, M.; Odosii, L.; Shabatura, Y.; Lunkova, H.; Hots, N.; Przystupa, K.; Atamaniuk, V. Electrical method for the cyberphysical control system of non-electrical objects. *Przegląd Elektrotechniczny* **2019**, *12*, 200–203. [CrossRef]
8. Bansal, V.; Kim, K.H. Review of PAH contamination in food products and their health hazards. *Environ. Int.* **2015**, *84*, 26–38. [CrossRef] [PubMed]
9. Krzeminski, A.; Wohlhüter, S.; Heyer, P.; Utz, J.; Hinrichs, J. Measurement of lubricating properties in a tribosystem with different surface roughness. *Int. Dairy J.* **2012**, *26*, 23–30. [CrossRef]
10. Kandeva, M.; Panov, I.V.; Dochev, B. Effects of nanomodifiers on the wear resistance of aluminum-silicon alloy alsi18 in tribosystems in case of reversive friction and lubrication. *J. Balk. Tribol. Assoc.* **2020**, *26*, 637–652.
11. Vojtov, V.A.; Kravtsov, A.G.; Tsymbal, B.M. Evaluation of Tribotechnical Characteristics for Tribosystems in the Presence of Fullerenes in the Lubricant. *J. Frict. Wear.* **2020**, *41*, 521–525. [CrossRef]
12. Skoć, A.; Spałek, J.; Markusik, S. *Podstawy Konstrukcji Maszyn Tom 2*; WNT: Warsaw, Poland, 2008.
13. Bapat, S.; Malshe, A.P. Understanding the fundamentals of the texture and tribofilm evolution. *Procedia Manuf.* **2020**, *48*, 223–229. [CrossRef]
14. Deng, F.; Zhu, Y.; Li, K.; Feng, Q.; Yang, K.; Huang, F. Tetraphenylethylene/tea saponin derivative nanocomposites for tribofilm visualization and efficient lubrication. *Comp. Part B Eng.* **2021**, *215*, 108787. [CrossRef]
15. Bair, S.; Thirteen, C. Application to Elastohydrodynamic Friction. In *High Pressure Rheology for Quantitative Elastohydrodynamics*, 2nd ed.; Bair, S., Ed.; Elsevier: Amsterdam, The Netherlands, 2019; pp. 327–353.
16. Thangarasu, V.; Anand, R. Physicochemical fuel properties and tribological behavior of aegle marmelos correa biodiesel. In *Advances in Eco-Fuels for a Sustainable Environment*; Azad, K., Ed.; Woodhead Publishing Series in Energy; Woodhead Publishing: Sawston, UK, 2019; pp. 309–336.
17. Hutchings, I.; Shipway, P. Lubricants and lubrication. In *Tribology*, 2nd ed.; Hutchings, I., Shipway, P., Eds.; Butterworth-Heinemann: Oxford, UK, 2017; pp. 79–105.
18. Zajezierska, A. Lubricating of rolling bearings. *Nafta-Gaz* **2009**, *65*, 982–992. (In Polish)
19. Oleksiak, S. Energy efficient gear oils. *Nafta-Gaz* **2016**, *72*, 1137–1143. (In Polish) [CrossRef]
20. Przystupa, K.; Ambrożkiewicz, B.; Litak, G. Diagnostics of Transient States in Hydraulic Pump System with Short Time Fourier Transform. *Adv. Sci. Technol. Res. J.* **2020**, *14*, 178–183. [CrossRef]
21. Spałek, J.; Kwaśny, M.; Bochenek, S. Analysis of power losses in the meshing of a cylindrical gear. *Szybkobieżne Pojazdy Gąsienicowe* **2010**, *1*, 63–70. (In Polish)
22. *ISO 4406:2021 Hydraulic Fluid Power—Fluids—Method for Coding the Level of Contamination by Solid Particles*; ISO: Geneva, Switzerland, 2021.
23. Gil, L.; Pieniak, D.; Walczak, M.; Ignaciuk, P.; Sawa, J. Impact of acid number of fuels on the wear process of apparatus for fuel injection in diesel engines. *Adv. Sci. Technol. Res. J.* **2014**, *8*, 54–57.
24. Ignaciuk, P.; Gil, L. Damages to injectors in diesel engines. *Adv. Sci. Technol. Res. J.* **2014**, *8*, 58–61.
25. Cameron, A.C.; Windmeijer, F.A. An R-squared measure of goodness of fit for some common nonlinear regression models. *J. Econom.* **1997**, *77*, 329–342. [CrossRef]

26. Krawiec, S. *Compositions of Plastic and Solid Lubricants in the Process of Friction of Steel Machine Nodes*; PWR: Wrocław, Poland, 2011. (In Polish)
27. Harperscheid, M. Engine Oils. In *Encyclopedia of Lubricants and Lubrication*; Mang, T., Ed.; Springer: Berlin/Heidelberg, Germany, 2014. [CrossRef]
28. Mang, T. (Ed.) Lubricity. In *Encyclopedia of Lubricants and Lubrication*; Springer: Berlin/Heidelberg, Germany, 2014. [CrossRef]
29. Bartels, T.; Bock, W. Gear Lubrication Oils. In *Lubricants and Lubrication*; Dresel, W., Mang, T., Eds.; Wiley: Hoboken, NJ, USA, 2017. [CrossRef]
30. Naikan, V.N.A.; Kapur, S. Reliability modelling and analysis of automobile engine oil. *Proc. Inst. Mech. Eng. Part D J. Automob. Eng.* **2006**, *220*, 187–194. [CrossRef]
31. Totten, G.E.; Tung, S.C. *Automotive Lubricants and Testing R-428*; SAE International and ASTM: West Conshohocken, PA, USA, 2012; p. 504.
32. PKN. *PN-EN 61014:2007 Programmers for Reliability Growth*; PKN: Warsaw, Poland, 2007.
33. Chmielewski, Z.; Stobiecki, J.; Górska, M. The concept of evaluation of reliability of modern engine oils in operation. *Autobusy–Technika, Eksploatacja, Systemy Transportowe* **2018**, *19*, 337–340. [CrossRef]
34. Mang, T. (Ed.) Oil Analyses for Used Oils. In *Encyclopedia of Lubricants and Lubrication*; Springer: Berlin/Heidelberg, Germany, 2014. [CrossRef]
35. Dziubak, T. Operating fluids contaminantions and their effect on the wear of elements of a motor vehicle's combustion engine. *Arch. Automot. Eng.* **2016**, *72*, 43–72.
36. Laber, A. Modyfikowanie warunków pracy węzła tarcia olejami z dodatkami eksploatacyjnymi na bazie środków smarnych stałych. *Tribologia* **2011**, *5*, 137–145.
37. Bloch, H. (Ed.) *Practical Lubrication for Industrial Facilities*; The Fairmont Press, Inc.: Lilburn, GA, USA, 2009.
38. Becker, E.P. Trends in tribological materials and engine technology. *Trib. Int.* **2004**, *37*, 569–575. [CrossRef]
39. Syedhidayat, S.; Wang, Q.; Mohsen, A.H.M.; Wang, J. Choice and Exchange of Lubricating Oil for Injection Molding Machine. *Recent Pat. Mech. Eng.* **2019**, *12*, 378–382. [CrossRef]

Article

Availability Estimation of Air Compression and Nitrogen Generation Systems in LNG-FPSO Depending on Design Stages

Youngkyun Seo [1], Jung-Yeul Jung [2,*], Seongjong Han [1] and Kwangu Kang [1]

[1]   Offshore Industries R&BD Center, Korea Research Institute of Ships & Ocean Engineering (KRISO), Geoje 53201, Korea; ykseo@kriso.re.kr (Y.S.); sjhan@kriso.re.kr (S.H.); kgkang@kriso.re.kr (K.K.)

[2]   Maritime Safety and Environmental Research Division, Korea Research Institute of Ships & Ocean Engineering (KRISO), Daejeon 34103, Korea

*   Correspondence: jungjy73@kriso.re.kr

Received: 21 October 2020; Accepted: 27 November 2020; Published: 3 December 2020

Abstract: This study estimated availability of an air compression system and a nitrogen generation system in liquefied natural gas—floating production storage and offloading unit (LNG-FPSO) with different design stages to investigate the gap between the availability at the early design stage and that at the late design stage. Although availability estimation in the early design stage is more important than the late design stage, it is difficult to estimate the availability accurately in the early design stage. The design stage was divided into three depending on the design progress. Monte Carlo simulation technique was employed for the availability estimation. The results of the availability estimation showed that there was 0.434% difference between the early and late design stages. This meant that the availability in the early design stage was underestimated due to limited information. A sensitivity analysis was performed to investigate critical factors affecting the results. The investigated factors were failure rate, repair time, redundant equipment, and modified preventive maintenance schedule. The most critical factor was redundant equipment. It increased 0.486% availability.

Keywords: air compression system; nitrogen generation system; utility module; availability; sensitivity analysis

---

1. Introduction

Various factors are considered in system design, such as efficiency, costs, safety, and environmental effect. Availability is also one of the important issues in the system design. The definition of the availability from BS4778-3.1 (British standards, quality vocabulary, availability, reliability, and maintainability terms.) Guide to concepts and related definitions is the ability of an item under the combined aspects of reliability, maintainability, and maintenance support to perform its required function at a specified instant or for a specified period [1]. The availability indicates that how much a system approaches ideal operation without production loss caused by equipment failures or undesired external events. Availability estimation is frequently performed in the oil and gas, chemical, and power plant industries to find the optimum design option, to predict the production level, and to evaluate maintenance and operating policies.

Many previous studies conducted the availability estimation for various systems to improve their designs. Basker and Martin [2] estimated the availability of production and electrical systems using the developed numerical method. They considered failure and repair rates following the non-exponential distribution. Keller and Stipho [3] conducted the availability estimation for two similar chlorine production plants which were located in different environmental conditions (Iraq and Switzerland). They employed the concept of "delayed time" to take into account the additional time required to reach full production rate. Bosman [4] estimated the availability of a natural gas compressor plant to investigate its unavailability. Since the plant had no backup systems, the unavailability estimation was crucial. They concluded that the availability analysis provided useful information to determine the optimum number of spares. Aven [5] indicated the methodologies for the availability estimation of oil/gas production and transport systems. He described not only an analytical approach but also a simulation method for the availability estimation. Khan and Kabir [6] conducted the availability estimation for an ammonia plant using both analytical and simulation method. They concluded that the performance of the plant could be improved by changing the overhaul strategy and plant configuration. Hajeeh and Chaudhuri [7] analyzed the availability of a reverse osmosis (RO) plant for producing potable water from seawater through desalination. They employed failure mode effect analysis (FMEA) and fault tree analysis (FTA) techniques to investigate the downtime pattern and failure. Zio et al. [8] assessed the availability of an offshore installation using Monte Carlo simulation. Marquez et al. [9] suggested a general approach for the reliability and availability assessment of complex systems by employing Monte Carlo simulation. They validated the proposed approach by performing a case study for cogeneration plants. Michelassi and Monaci [10] estimated the availability of a gas re-injection plant for the oil and gas production. They utilized reliability block diagram (RBD) techniques in conjunction with Monte Carlo simulation. They also considered the leak because the plant should be stopped when the leak was detected. Chang et al. [11] estimated the availability of conventional and novel propulsion systems with a BOG handling system of an LNG carrier. They estimated the availability depending on the required function to prevent rough evaluation: design propulsion load, emergency propulsion load, and BOG utilization. Görkemli and Ulusoy [12] suggested a new modeling approach for predicting the availability of a production system. They considered not only machine failures but also the material supply, management, and set-up in the proposed method. They also investigated the uncertainties caused by the various production environment using a fuzzy Bayesian method. Seo et al. [13] predicted the availability of CO_2 liquefaction processes for a ship-based carbon capture and storage (CCS) chain and they converted the availability to unavailability cost to calculate the life-cycle cost (LCC). Seo et al. [14] estimated the availability of LNG fuel gas supply systems to evaluate economics of them. They concluded that one of the significant factors was mechanical devices. Gowid et al. [15] reviewed the studies related with the profitability, reliability, and condition based monitoring of liquefied natural gas-floating production storage and offloading unit (LNG-PFSO). They assumed that the efficiency of LNG-PFOS depends on LNG liquefaction process type, system reliability, and maintenance approach, and reviewed the paper at theses points. They concluded that the literature was not sufficient to improve efficiency of LNG-FPSO. Hwang et al. [16] developed the condition-based maintenance system to perform proactive maintenance in advance to avoid the abnormal states. They addressed the system architecture, main components, diagnostics, and prognostic methods of the system.

The methodologies for the availability estimation has been improved to increase its accuracy and to apply to various systems. Precise availability estimation is important because it directly influences the owner's decision. The availability estimation is performed several times depending on the design stages (conceptual design, basic design, and detailed design stages). In the early design stage, the results of availability estimation are effective for design improvement, but it is hard to estimate it precisely due to the limited data. On the contrary, accurate availability estimation is possible at the end of the design stage using sufficient data, but it accompanies high costs for the system modification. Therefore, it is an important to estimate the availability in the early design stage accurately. Although many studies improved the methods to increase their accuracy, there was little effort to practically estimate the availability in the early design stage

The purpose of this study is to investigate the availability gap between in the early and late design stages by estimating it with the design stages to find practical manner of availability estimation in the early design stage. The structure of this study is as follows. The target systems are described. In Section 3, methodologies for the availability estimation are discussed. The results of the availability estimation and the sensitivity analysis are indicated in Section 4. Finally, the conclusions are presented.

2. Description of Target System

In this study, two systems in LNG-FPSO is selected as a target for the availability estimation. These are air compression and nitrogen generation systems in LNG-FPSO. LNG-FPSO is a huge facility for LNG production in offshore, and its concern has been increased because of the growing demand for LNG. LNG-FPSO is a floating unit for production, processing, storage, and offloading of LNG in remote offshore gas fields. Conventionally, the natural gas in an offshore field is transported by pipeline to onshore for processing. LNG-FPSO does not require the pipeline because it processes the natural gas itself in offshore. It is specialized for small scale gas field. Topside modules of the LNG-FPSO can be categorized into two: a processing module and a utility module. The processing module handles the primary hydrocarbon, whereas the utility module deals with utilities including energy, water, air, and diesel oil. The utility module provides utilities to the processing system for safe and stable operation. Some failure of the utility module can be critical because safety systems for preventing an accident are operated by the utility module.

The topside of LNG-FPSO can be divided into ten modules as shown in Figure 1. A produced feed gas come up through a turret, and it is transported to an inlet facility module. Slug in the feed gas is removed by a slug catcher, and liquid is separated by a separator. CO_2, Hg, and H_2O in the feed gas is removed in a pre-treatment module. The treated natural gas is liquefied by a liquefaction module, and a refrigeration module supplies the refrigerant to the liquefaction module. The heavier components than methane like ethane, butane, and propane are separated by a fractionation module. Some amounts of natural gas are transferred to a fuel gas compression system, and it is utilized for power generation. The liquefied natural gas is stored in storage tanks with LPG and condensate. A condensate stabilizer module separates the relatively light components for safe operation. Condensate is mainly composed of propane, butane, pentane, and heavier hydrocarbon. When condensate contains light components like methane and ethane, it can be vaporized and increase the pressure of a storage tank during storage. These light components should be separated before storage. A blowdown module treats combustion fluids in emergency situations. The utility module supplies various utilities to other modules for the operation.

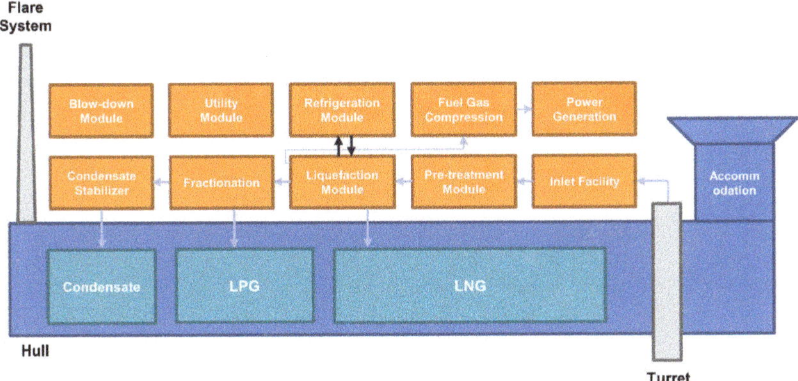

Figure 1. Topside modules of liquefied natural gas—floating production storage and offloading unit (LNG-FPSO).

In this study, air compression and nitrogen generation systems are analyzed in the utility module because those are important systems for stable and safe operation. A general utility module contains an instrument and service air system, a nitrogen generation system, a cooling water system, a seawater system, a hot oil system, a portable water system, a produced and wastewater system, and a diesel oil system. The instrument and service air system compresses the air up to approximately 10 bar for the usage of the instrument and others. The nitrogen generation system supplies nitrogen to the customers. The cooling water system is used to provide the cooling medium for all of the topside modules. The sea water system provides the seawater to various systems. The hot oil system increases the temperature of the oil within a specified range. It utilizes waste heat from flue gas using waste heat recovery units installed in a power generation system. The portable water system distributes water to topside eyewash and safety shower, and hot and cold water for personal usage. The produced and wastewater system removes the oil in the produced water from topside separators. The diesel oil system distributes the diesel oil to customers by transferring and purifying it.

Figure 2 indicates the air compression and nitrogen generation systems. The systems mainly consist of three pieces of equipment; an air compressor, an air dryer, and a nitrogen generator. Air is compressed by the air compressor, and then the small amount of water in the compressed air is dehydrated by the air dryer. The dry air is sent to a customer requiring the instrument air and to the nitrogen generator. The nitrogen generator separates the nitrogen from the dry air.

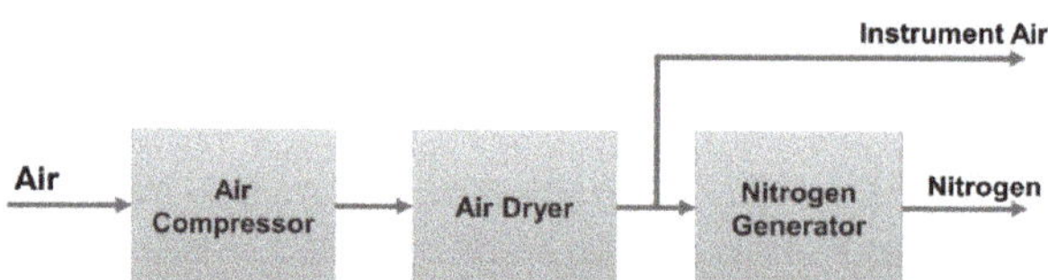

Figure 2. Block diagram of air and nitrogen systems.

Figures 3–5 show process diagrams with different design stages. Figure 3 is a process flow diagram (PFD) of the air compression and nitrogen generation systems. PFD shows the main equipment in the system. The preliminary process and instrument diagram (P&ID) is indicated in Figures 4 and 5. P&ID includes not only the main equipment but also piping, instrumentation, and control devices. In this study, the piping information is not contained because it is unnecessary for the availability estimation.

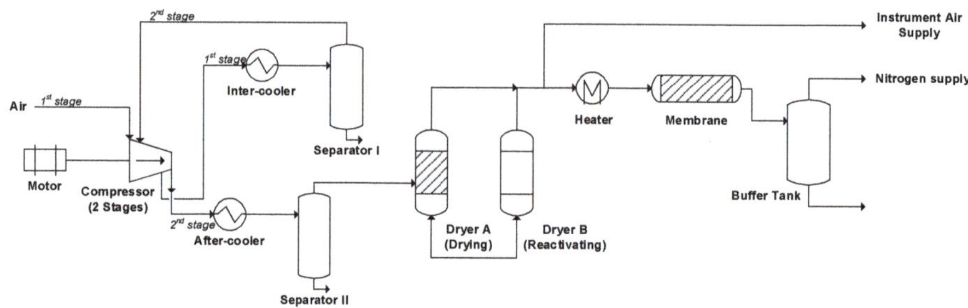

Figure 3. Process flow diagram (PFD) of air and nitrogen system.

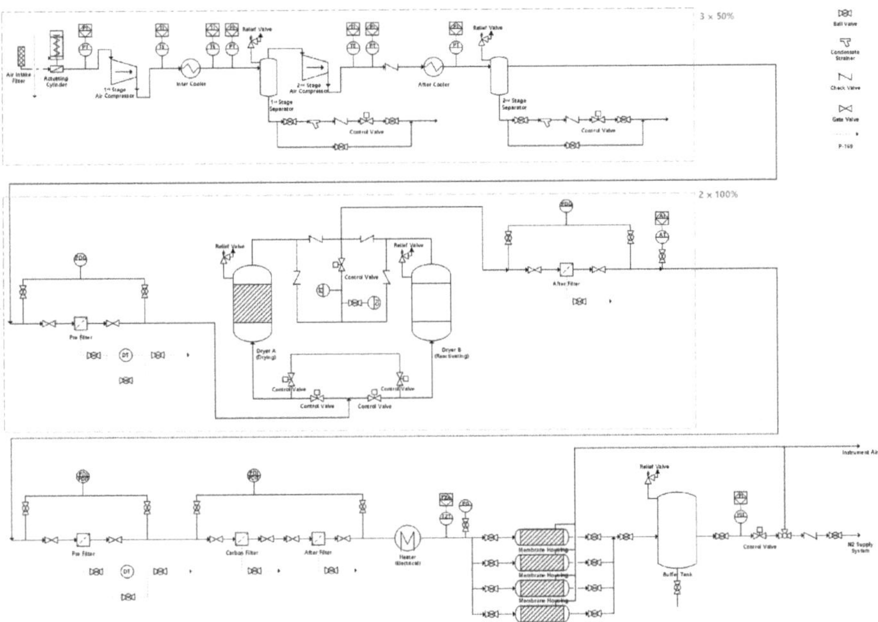

Figure 4. The preliminary process and instrument diagram (P&ID) of air and nitrogen system.

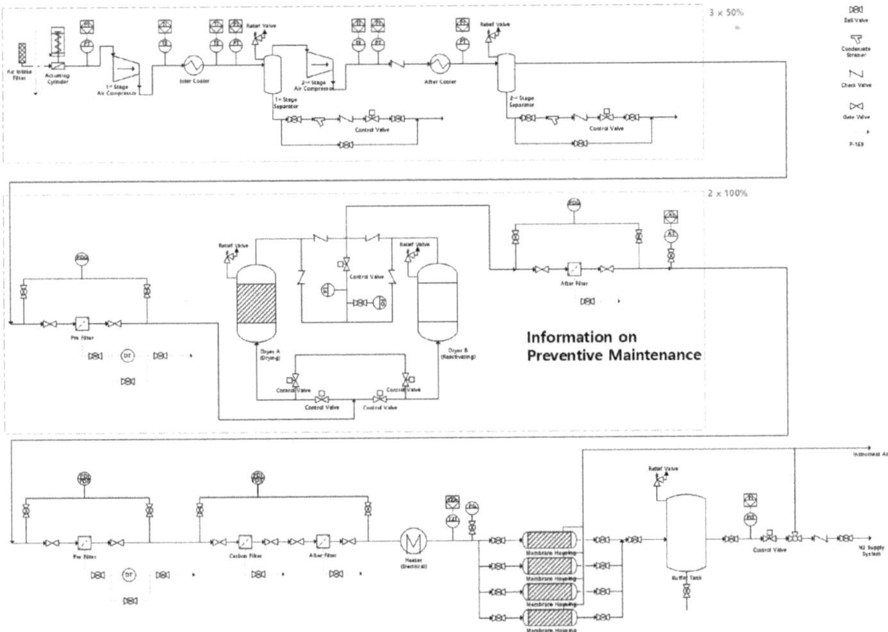

Figure 5. P&ID of air and nitrogen system with information on preventive maintenance.

The design stage considered in this study are three. The first stage is PFD and the second stage is the preliminary P&ID. The third stage is preliminary P&ID with the information on preventive maintenance.

1. Stage I—PFD
2. Stage II—Preliminary P&ID
3. Stage III—Preliminary P&ID + Information on Preventive Maintenance

3. Methodology

Several methods are available for the availability estimation: reliability block diagram (RBD), Markov model, and Monte Carlo simulation [17,18]. The former two are an analytical approach whereas the latter one is a simulation approach. The analytical approach calculates the availability using mathematical equations, while the simulation technique estimates it by generating scenarios. When the system is complex, the analytical approaches like RBD and Markov model are unrealistic. They are additionally difficult to apply to the system, which has nonconstant failure and repair rates. However, the Monte Carlo simulation approach can handle inconstant failure/repair rates and multi-state systems. One of the drawbacks of the Monte Carlo simulation is the long simulation time, but it can be overcome by the advanced simulation techniques. In this study, Monte Carlo simulation is employed for the availability estimation.

Figure 6 shows the procedure for the availability estimation using Monte Carlo Simulation. First of all, the target system is analyzed, and then the reliability block diagram is drawn for the modeling of the system. The data for reliability and maintainability is collected from the data sources. The availability of the target system is estimated using the Monte Carlo Simulation. The followings are the details of each step.

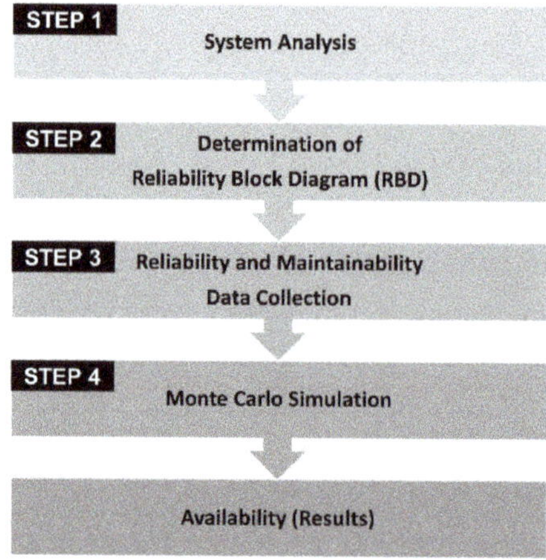

Figure 6. Procedure for availability estimation using Monte Carlo simulation.

3.1. STEP 1 System Analysis

First, the information required for the availability estimation is gathered, and the system is analyzed. The boundary and a level of the system analysis determined in this step. The given operating conditions and assumptions for the availability estimation are determined. Those include the lifespan of the system, number of simulations, distribution function of failure, distribution function of repair time, unit of failure rate, and unit of repair time. Table 1 tabulates the information.

3.2. STEP 2 Determination of Reliability Block Diagram (RBD)

RBD is a block structure to show success logic of a system. The blocks represent equipment or components of the system to fulfill a specified function. Success path can be visually verified so that it can be easily understood. In this step, the RBD of the system is determined based on Step 1's results. The followings indicate the RBD with the different design development states.

Table 1. Operation conditions and assumption.

Items	
Lifespan	20 years
Number of simulations	250
Distribution function of failure	Exponential
Distribution function of repair time	Constant
Unit of failure rate	Number of failure/10^6 h
Unit of repair time	Hours

3.2.1. RBD at Stage I (PFD Stage)

Figure 7 shows the RBD at Stage I. It is divided into three parts as shown in Figure 6: air compression, air dryer, and nitrogen generation parts. The configuration of the air compressor part is $3 \times 50\%$. It means that three compressors are installed, and the capacity of each compressor is 50%. Two compressors are in operation, and one compressor is on standby for a failure of the operating compressors. The air dryer part has $2 \times 100\%$ configuration. One air dryer is redundancy. In the nitrogen generation part, the membrane has the $4 \times 33\%$ configuration. Three membranes are operated, and the remaining membrane stands by for a failure.

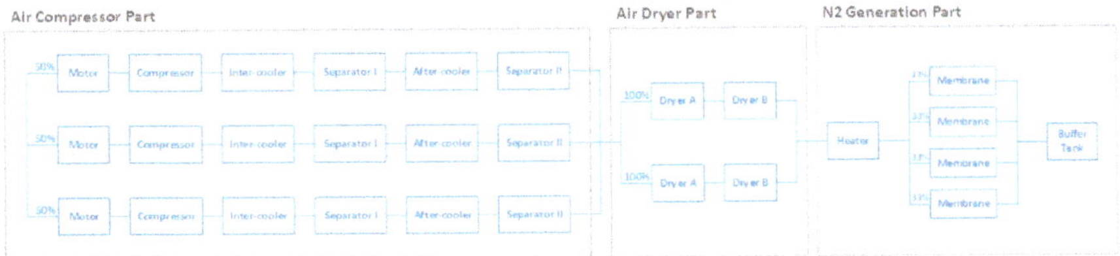

Figure 7. Reliability block diagram at Stage I (PFD stage).

3.2.2. RBD at Stage II (Preliminary P&ID)

Figures 8–10 indicate the RBD at Stage II (for the preliminary P&ID stage). Figures 8–10 show the RBD for the air compression, air dryer, and nitrogen generation parts, respectively.

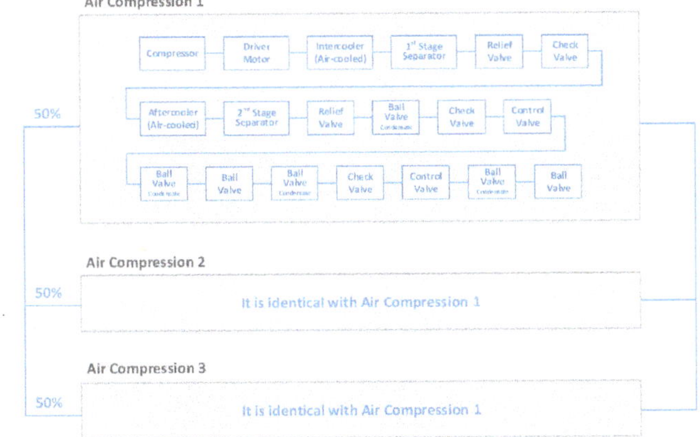

Figure 8. Reliability block diagram at Stage II (preliminary P&ID stage)—Air compression part.

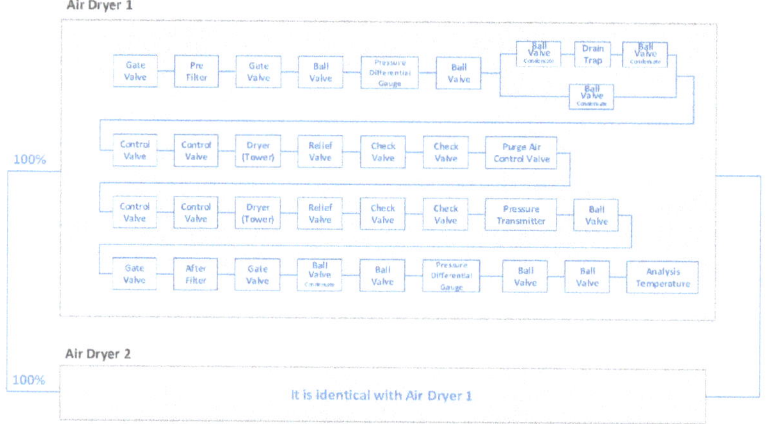

Figure 9. Reliability block diagram at Stage II (preliminary P&ID stage)—Air dryer part.

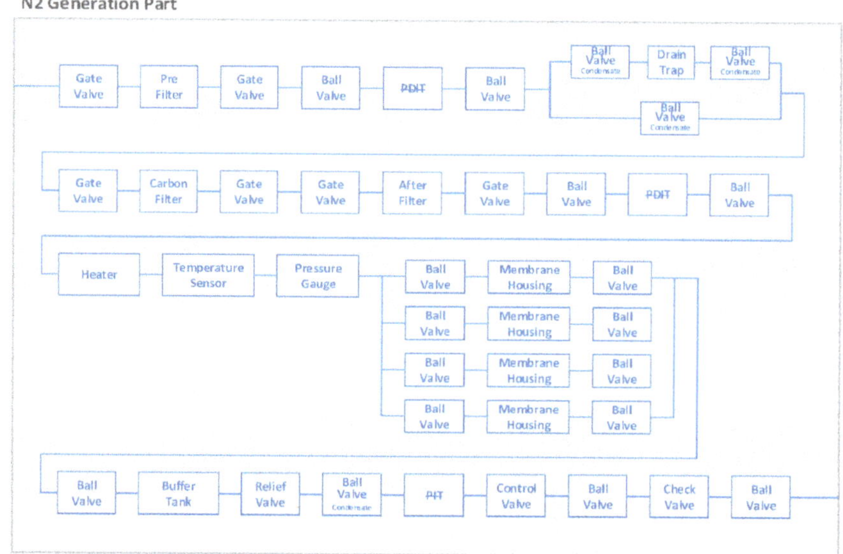

Figure 10. Reliability block diagram at Stage II (preliminary P&ID stage)—Nitrogen generation part.

3.2.3. RBD at Stage III (Preliminary P&ID + Information on Preventive Maintenance)

RBD at Stage III is almost identical with that for stage II excepting the additional information on preventive maintenance. One block for the preventive maintenance is added for stage III.

3.3. Step 3 Data Collection

The reliability and maintenance data are required for the availability estimation. Since the results of the availability estimation are significantly influenced by the reliability and maintenance data, they are important. Reliability data is linked to the failure rate. The maintenance data is associated with the corrective maintenance time (repair time) and the preventive maintenance time. When the failure occurs, the corrective maintenance is conducted to a system. Preventive maintenance is performed on the basis of maintenance policies and strategies. The data can be categorized into three depending on the kinds of sources: Open data (from open books and reports), vendor data, and in-house data. This study uses the OREDA (Offshore and onshore reliability data) and vendor data. OREDA is offshore and onshore reliability data handbook sponsored by oil and gas companies. It is considered a unique data source in the offshore industry. OREDA is employed in this study because it is the most suitable for it [19,20]. Vendor data is taken from a manufacturer of air compression and nitrogen generation systems. Table 2 indicates the reliability and maintenance data employed in this study.

Table 2. Reliability and maintainability data for air and nitrogen system.

Items	Source	Failure Rate			Active Repair Time	
		Lower	Mean	Upper	Mean	Max.
Compressor	OREDA 2009	-	140.84	779.34	15	98
Electric motor	OREDA 2015	0.87	7.52	19.63	16	25
Heat exchanger	OREDA 2015	0.28	64.94	243.9	28	96
Separator	OREDA 2015	0.34	73.49	271.39	6.4	12
Dryer	OREDA 2015	18.32	29.22	44.37	6.2	11
Heater	OREDA 2015	244.7	349.24	484.57	14	84
Membrane	OREDA 2015	18.32	29.22	44.37	6.2	11
Filter	* OREDA 2015	-	4.67	-	1.105	-
Relief valve	OREDA 2015	0.04	2.07	6.41	6.9	13
Check valve	OREDA 2015	0.01	2.47	9.24	2	2
Ball valve (Utilities)	OREDA 2015	2.73	11.65	25.65	5	6
Ball valve (Condensate processing)	OREDA 2015	12.08	72.09	226.9	27	39
Control valve	OREDA 2015	0.99	19.8	93.96	9	9
Gate valve	OREDA 2015	0.04	3.74	12.37	20	72
Control logic unit	OREDA 2015	0.08	17.4	64.59	-	-
Pressure input device	OREDA 2015	0.01	1.09	3.73	7	12
Trap	Vender	-	16.31	22.83	2	-

* This data is regenerated using the component data in Offshore and onshore reliability data (OREDA) 2015.

Table 3 indicates the information on the preventive maintenance. The preventive maintenance is conducted to prevent unexpected future failure. It is classified into four categories: age-based, clock-based, condition-based, and opportunity maintenance [18]. In the age-based maintenance, the preventive maintenance is performed at the defined age of the system (e.g., the number of take-offs/landings for an airplane). The clock-based maintenance is carried out at specified calendar time so that it is scheduled by administers. In the condition-based maintenance, the preventive maintenance is initiated by measuring condition variables. The opportunity maintenance is carried out when the system is stopped by the other failure. In this study, the clock-based maintenance is taken into account for the preventive maintenance, and the data is collected from the vendor of the air compression and nitrogen generation systems.

Table 3. Preventive maintenance information on air compression and nitrogen generation systems.

Equipment		Periodic (month)	Maintenance Time (hour)
Air compressor	Component replacement 1	6	0.5
	Component replacement 2	24	3
	Main maintenance	36	72
Air dryer	Component replacement 1	6	0.5
	Component replacement 2	24	1
	Main maintenance	36	24
Nitrogen generator	Component replacement 1	6	0.5
	Component replacement 2	24	1
	Main maintenance	36	24

3.4. Step 4 Monte Carlo Simulation

Monte Carlo simulation is employed to estimate the availability. Figure 11 shows the flowchart of the Monte Carlo simulation [21]. First of all, components, their states, and their configuration are defined. Moreover, the next transition time for each component is estimated by the random number generation. The transition time is the time when the phase of a component in the system is changed from normal to failure. In this step, the generated random number is converted into a value of time using a conversion method at a cumulative distribution function. Figure 12 shows how the generated random number is transferred to the value of time by the conversion method. The cumulative distribution function for the exponential distribution is indicated in Equation (1).

$$F(x) = 1 - e^{-\lambda x} \tag{1}$$

where λ is the failure rate, and x is a value of time.

$$R = F(x) = 1 - e^{-\lambda x} \tag{2}$$

where R is the random number between 0 and 1. R^* is a generated random number between 0 and 1.

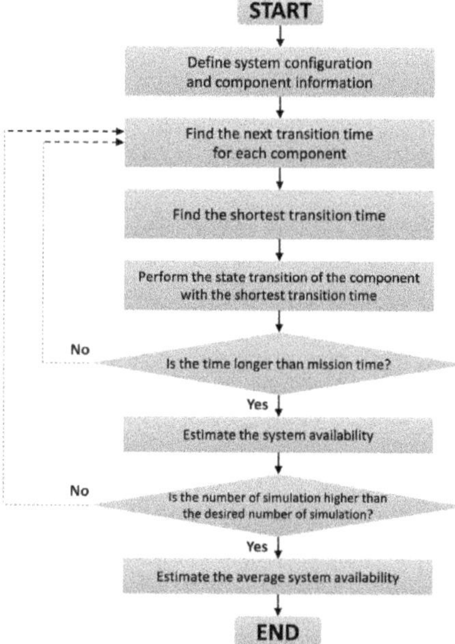

Figure 11. Procedure for availability estimation using Monte Carlo simulation [21].

The shortest transition time is found among all of the predicted times, and then the system time is changed to the shortest transition time. If the time is shorter than the mission time, the transition times for all component are estimated again. The mission time is total operation time required to the system like lifespan. When the time is longer than the mission time, the system's availability is calculated. This process is just one simulation. If the number of simulations is lower than the desired number of simulations, the next simulation is repeatedly performed. The desired number of simulations is determined as referring the convergence of results. When a result converges sufficiently, the number of simulations is selected as the desired number of simulations. The desired number of simulations is determined as setting a sufficiently high number of simulations or determining the number of simulations after the initial simulation. When the number of simulations is the same as the desired number of simulations, the average system availability is calculated finally. The average system availability is the result after the last simulation, while the system availability is the result of each simulation.

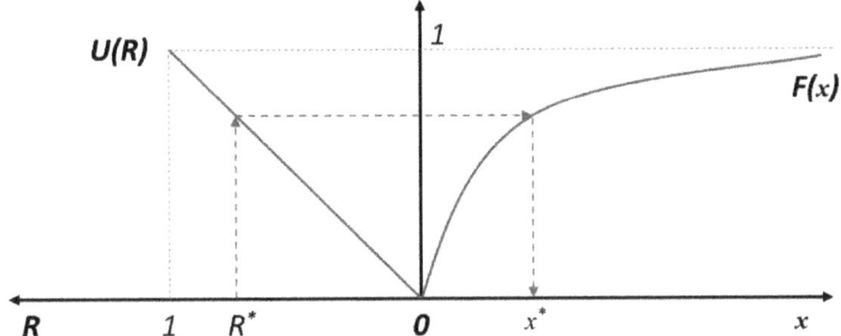

Figure 12. Conversion method to transfer random number to value of time at cumulative distribution.

The predicted time from the generated random number is shown in Equation (3).

$$x = F^{-1}(R) = -\frac{1}{\lambda}\ln(1-R) \tag{3}$$

4. Results and Discussion

4.1. Availability

Figure 13 shows the availability of the air compression and nitrogen generation systems depending on the design stages. The availability decreased with the increment of the design stages because the system in the late design stage was more complex than that in the early design stage. A complex system has more factors decreasing the availability of the system than a simple system. The availability is decreased by 0.331% when the design stage was changed from Stage I (PFD) to Stage II (P&ID). This meant that the instrument system occupies 0.331% of the system's availability. When the design stage was transferred from Stage II (P&ID) to Stage III, the availability was decreased by 0.103%. The preventive maintenance influenced about 0.103% of the availability. The availability difference between Stage I and Stage III was 0.434%. It showed that the availability in the early design stage was underestimated compared to the late design stage. The unavailability (1—availability) in the late design stage (0.972%) is approximately 1.8 times severe than that in the early design stage (0.535%). We can predict that the unavailability estimated in the late design stage is 1.8 times serious than that in the early design stage. The availability difference between early and late design stages can be dissimilar with the target system. However, this result provides meaningful information to guess the actual availability in the early design stage.

4.2. Component Criticality

Figure 14 shows the component criticality depending on the design stages. The component criticality shows the important component of the availability, and it is the ratio of the component's failure time to the system failure time. The most crucial component at Stage I was the heater, which accounted for 90.3% criticality. The heater and ball valve (condensate) were critical in Stage II and Stage III. The heater and ball valve (condensate) had 50.7% and 20.0% criticality at the design Stage III, respectively. The preventive maintenance occupied about 10% on the criticality at Stage III. The most critical component was the heater regardless of the design stages. The availability of the system can be significantly increased as installing redundant heaters. The results of the component criticality analysis guide a designer or a decision maker to select additional components to installed to increase availability.

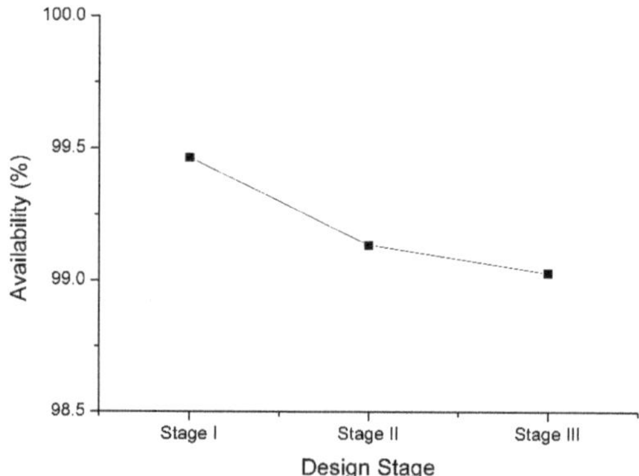

Figure 13. Availability depending on the stages.

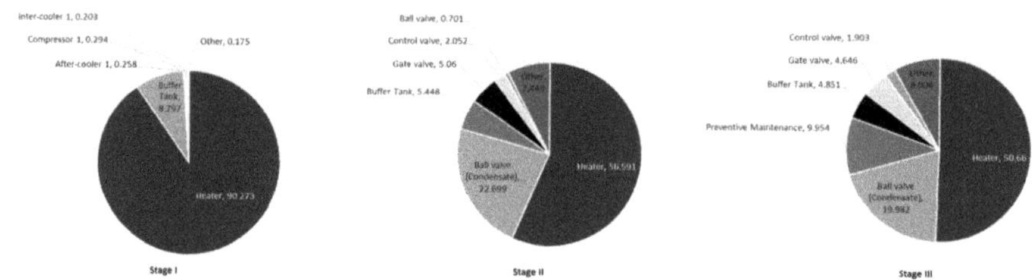

Figure 14. Component criticality depending on the stages.

4.3. Sensitivity Analysis

This study performed the sensitivity analysis to investigate the factors affecting the results. It is important to analyze the correlation between the factors and the results because the results can be changed depending on the variation of the factors. In this study, four factors are investigated for the sensitivity analysis: failure rate, repair time, redundant equipment, and modified preventive maintenance schedule. The reliability data used in this study are mainly from OREDA, and its mean value is utilized. The values can be different depending on the target conditions. OREDA predicts the failure rate with 90% confidence interval. The confidence interval describes the amount of uncertainty associated with a sample of a population. The sensitivity analysis was performed for the lower and upper limits of the failure rates. The repair times utilized in this study were also mostly from OREDA. The employed active repair time considers only the time when actual repair work is being done. It does not contain time to shut down the unit, issue the work order, wait for spare parts, start-up after repair. Some variation exists between the active repair time and the actual downtime. (The reason why OREDA only considers the active repair time is that the required time for the preparation and return to the normal operation are different depending on the location of the installation.) The additional repair time is taken into account. The availabilities with and without redundant equipment are estimated to examine its effect on the availability. Finally, the availability is calculated with different preventive maintenance schedules.

4.3.1. Lower amd Upper Failure Rates

Figure 15 indicates the availability depending on the design states with different failure rates: lower, mean, and upper failure rates. As the failure rate was increased from lower to upper, the availability was decreased. In the case of lower and mean failure rates, the availability was slightly decreased with the design stages. In contrast, the availability was significantly reduced in the case of upper failure rate. When the design stage was changed from Stage I (PFD) to Stage II (P&ID), the availability was dramatically decreased in the case of upper failure rate. This indicated that the instrument devices gave a critical impact on the availability. The availabilities are 99.506% (lower) and 97.819% (upper) at Stage III. The upper means that the result is derived using upper failure rate in Table 2, and the lower is the reverse. This meant that the most optimistic availability is 99.506% and the most pessimistic availability is 97.819%.

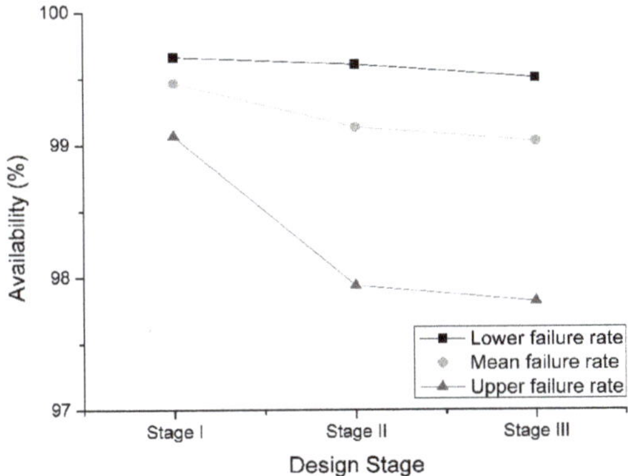

Figure 15. Availability with different failure rate.

4.3.2. Additional Repair Time

Figure 16 shows the availability depending on the design stages with the additional repair time. Three additional repair times assumed in this study are 1, 3, and 5 h to investigate the impact of the delayed repair time. The availability decreased with the increment of the repair time. When additional 1, 3, and 5 h were considered at Stage III, the availabilities were 98.969%, 98.823%, and 98.701%, respectively. This result presented that one additional hour in the repair time decreased the availability by 0.065%.

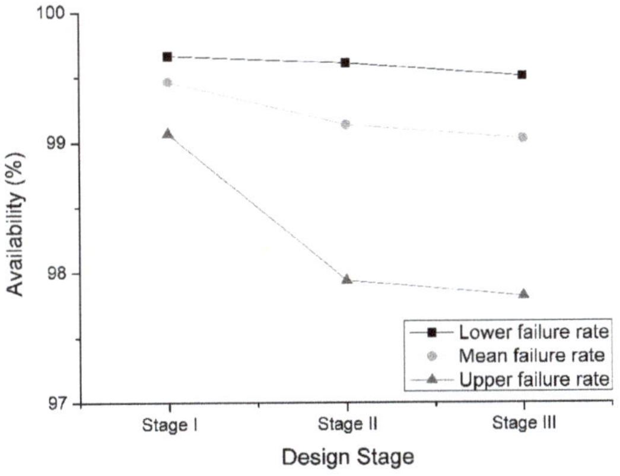

Figure 16. Availability with additional repair time.

4.3.3. Installation of Redundant Heater

Figure 17 presents the availability depending on the design stages with the installation of the redundant heater. As mentioned in Section 4.1, the most critical component in the availability was the heater regardless of the design stages. The availability was estimated depending on the installation of the redundant heater or not. The availability was considerably increased when the redundant heater is installed. The availability is 99.028% without the redundant heater at Stage III, whereas it is 99.514% with the redundant heater. That is, the redundant heater increased the availability by 0.486%. Although 0.486% availability seems to be low, it is not a negligible value in the system (LNG-FPSO).

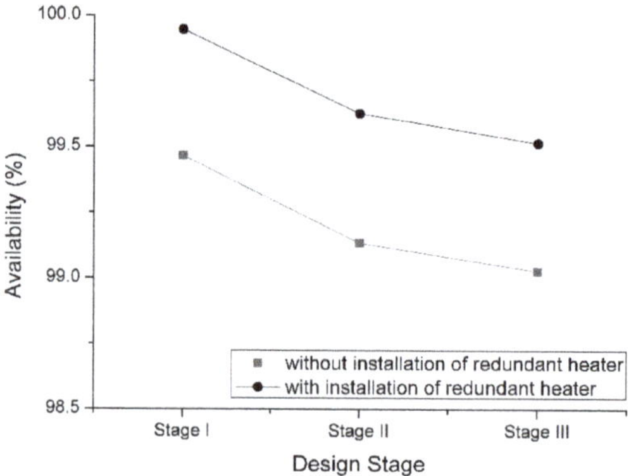

Figure 17. Availability with installation of redundant heater.

4.3.4. Modified Preventive Maintenance Schedule

Figure 18 shows the availability depending on the design stages with the modified preventive maintenance. As mentioned in Section 3, the preventive maintenance is conducted to prevent the critical failures. There are various activities for the preventive maintenance as indicated in Table 3. These activities are individually conducted depending on their inherent periodic. When the activities have different schedule, some activities can be merged to increase the availability. Although simultaneous preventive maintenance increases the availability, it requires new engineers to conduct the activities at the same time. Since all components have the same preventive maintenance schedule, different schedules were assumed in the modified schedule. The result showed that the availability was decreased by 0.076% through the modified preventive maintenance schedule. Since the preventive maintenance was not considered at Stages I and II, the values at those stages were unchanged.

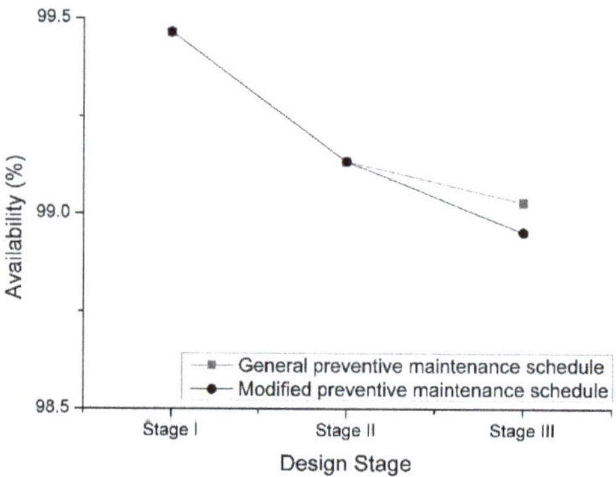

Figure 18. Availability with modified preventive maintenance schedule.

5. Conclusions

This study estimated the availability of air and nitrogen systems depending on the design stages to analyze the gap between early and late design stages. Three design stages were considered: Stages I–III. Stage I was the process flow diagram (PFD) stage and Stage II was the piping and instrument diagram (P&ID) stage. In Stage III, the preventive maintenance was additionally considered comparing to Stage II. The Monte Carlo simulation approach was employed for the availability estimation. The results presented that the availabilities were decreased with the design progress. It is obvious because the system was more complex with the design development. The availability difference between Stage I and Stage II was 0.331%, and that was 0.103% between Stage II and Stage III. These indicated that the instrument system and the preventive maintenance occupied 0.331% and 0.103%, respectively. This result also presented that the availability in the early design stage (Stage I) was underestimated compared to the late design stage (Stage III). The unavailability at the late design stage was 1.8 times higher than the early design stage. We could guess the availability at the late design stage using the result at the initial design stage. The most critical component in the air and nitrogen systems was the heater regardless of design stages. The sensitivity analysis was conducted to analyze the key factors on the results. The most crucial factor was the redundant equipment. When the redundant heater was installed, the availability was increased by 0.486% at Stage III. The factors for the modified maintenance schedule and additional repair time (1 h) were not significant in the system compared to other factors. Since this study investigated only two systems (air and nitrogen systems) among lots of systems in LNG-FPSO, future studies are required for the whole system (LNG-FPSO). Although this study did not consider the whole system (LNG-FPSO), this gives the important guide to progress the next step for the accurate availability estimation in the early design stage.

Author Contributions: Conceptualization, J.-Y.J.; methodology, Y.S.; formal analysis, S.H.; validation, K.K.; writing—original draft preparation, Y.S.; writing—review and editing, J.-Y.J. All authors have read and agreed to the published version of the manuscript.

Funding: This research was supported by a grant from Endowment Project of "Technology development of material handling and risk management for operation and maintenance service of offshore plant (PES3470)" and "Development of Evaluation Model for Hydrogen Offshore Supply Chain and Test Technologies for Hydrogen Equipment (PES3510)" funded by Korea Research Institute of Ships and Ocean Engineering.

Conflicts of Interest: The authors declare no conflict of interest.

References

1. British Standards Institution (BSI). *BS 4778-3.1:1991 Quality Vocabulary: Part 3: Availability, Reliability and Maintainability Terms, Section 3.1 Guide to Concepts and Related Definitions*; BSI: London, UK, 1991.
2. Basker, B.; Martin, P. Availability prediction by using a method of simulation. *Microelectron. Reliab.* **1977**, *16*, 135–141. [CrossRef]
3. Keller, A.; Stipho, N. Availability studies of two similar chlorine production plant. *Reliab. Eng.* **1981**, *2*, 271–281. [CrossRef]
4. Bosman, M. Availability analysis of a natural gas compressor plant. *Reliab. Eng.* **1985**, *11*, 13–26. [CrossRef]
5. Aven, T. Availability evaluation of oil/gas production and transportation systems. *Reliab. Eng.* **1987**, *18*, 35–44. [CrossRef]
6. Khan, M.R.; Kabir, A.Z. Availability simulation of an ammonia plant. *Reliab. Eng. Syst. Saf.* **1995**, *48*, 217–227. [CrossRef]
7. Hajeeh, M.; Chaudhuri, D. Reliability and availability assessment of reverse osmosis. *Desalination* **2000**, *130*, 185–192. [CrossRef]
8. Zio, E.; Baraldi, P.; Patelli, E. Assessment of the availability of an o shore installation by Monte Carlo simulation. *Int. J. Press. Vessels Pip.* **2006**, *83*, 312–320. [CrossRef]
9. Crespo, A.; Heguedas, A.S.; Iung, B. Monte Carlo-based assessment of system availability. A case study for cogeneration plants. *Reliab. Eng. Syst. Saf.* **2005**, *88*, 273–289. [CrossRef]

10. Michelassi, C.; Monaci, G. Availability assessment of a raw gas re-injection plant for the production of oil and gas. In Proceedings of the 16th IMEKO TC4 International Symposium Exploring New Frontiers of Instrumentation and Methods for Electrical and Electronic Measurements, Florence, Italy, 22–24 September 2008; pp. 777–783.
11. Chang, D.; Rhee, T.; Nam, K.; Chang, K.; Lee, D.; Jeong, S. A study on availability and safety of new propulsion systems for LNG carriers. *Reliab. Eng. Syst. Saf.* **2008**, *93*, 1877–1885. [CrossRef]
12. Görkemli, L.; Ulusoy, S.K. Fuzzy Bayesian reliability and availability analysis of production systems. *Comput. Ind. Eng.* **2010**, *59*, 690–696. [CrossRef]
13. Seo, Y.; You, H.; Lee, S.; Huh, C.; Chang, D. Evaluation of CO2 liquefaction processes for ship-based carbon capture and storage (CCS) in terms of life cycle cost (LCC) considering availability. *Int. J. Greenh. Gas Control.* **2015**, *35*, 1–12. [CrossRef]
14. Seo, S.; Chu, B.; Noh, Y.; Jang, W.; Lee, S.; Seo, Y.; Chang, D. An economic evaluation of operating expenditures for LNG fuel gas supply systems onboard ocean-going ships considering availability. *Ships Offshore Struct.* **2014**, *11*, 213–223. [CrossRef]
15. Gowid, S.; Dixon, R.; Ghani, S. Profitability, reliability and condition based monitoring of LNG floating platforms: A review. *J. Nat. Gas Sci. Eng.* **2015**, *27*, 1495–1511. [CrossRef]
16. Hwang, H.; Lee, J.-H.; Hwang, J.; Jun, H.-B. A study of the development of a condition-based maintenance system for an LNG FPSO. *Ocean Eng.* **2018**, *164*, 604–615. [CrossRef]
17. Verma, A.K.; Srividya, A.; Karanki, D.R. *Reliability and Safety Engineering*; Springer: London, UK, 2010; ISBN 978-1-84996-231-5.
18. Rausand, M.; Høyland, A. *System Reliability Theory: Models, Statistical Methods, and Applications*, 2nd ed.; John Wiley & Sons, Inc.: Hoboken, NJ, USA, 2004; ISBN 0-7803-3274-1.
19. Oreda Project. *OREDA: Participants Offshore Reliability Data Hand-Book*, 5th ed.; DNV: Oslo, Norway, 2009; ISBN 8214027055.
20. Oreda Project. *OREDA: Participants Offshore and Onshore Reliability Data Handbook*, 6th ed.; DNV GL: Oslo, Norway, 2015.
21. Seo, Y.; Han, S.; Kang, K.; Noh, H.-J.; Park, S.; Jung, J.-Y.; Chang, D.; Čepin, M.; Briš, R. Availability estimation of utility module in offshore plant depending on system configuration. In *Advances and Trends in Engineering Sciences and Technologies II*; Al Ali, M., Platko, P., Eds.; CRC Press: London, UK, 2017; pp. 2195–2202.

Publisher's Note: MDPI stays neutral with regard to jurisdictional claims in published maps and institutional affiliations.

 *applied*
sciences

Article

Investigation of Roller-Tape Contact Pair Used in Precision Mechatronic System

Antanas Fursenko [1], Artūras Kilikevičius [2], Kristina Kilikevičienė [1], Sergejus Borodinas [3], Albinas Kasparaitis [1] and Jonas Matijošius [2,*]

[1]   Department of Mechanical and material Engineering, Vilnius Gediminas Technical University, J. Basanavičiaus str. 28, LT-03,224 Vilnius, Lithuania; antanas.fursenko@vgtu.lt (A.F.); kristina.kilikeviciene@vgtu.lt (K.K.); albinas.kasparaitis@vgtu.lt (A.K.)
[2]   Institute of Mechanical Science, Vilnius Gediminas Technical University, J. Basanavičiaus str. 28, LT-03,224 Vilnius, Lithuania; arturas.kilikevicius@vgtu.lt
[3]   Department of Applied Mechanics, Vilnius Gediminas Technical University, Saulėtekio av. 11, 10,223 Vilnius, Lithuania; sergejus.borodinas@vgtu.lt
*   Correspondence: jonas.matijosius@vgtu.lt; Tel.: +370-684-04-169

Received: 12 May 2020; Accepted: 9 June 2020; Published: 11 June 2020

Abstract: Smoothness of tape movement and stability of the tape area where elements are generated are very important in precision mechatronic devices where precise elements are generated on a steel tape, controlling them in real time. During movement, deformations and vibrations form in the steel tape area where elements are generated as a result of imperfections of movement equipment, contact between the roller surface and the tape, and errors arising in the movement process. This article is based on the need for a detailed theoretical and experimental research of the effects occurring during the movement of the precision steel tape used in measuring systems with precision elements generated on the tape, including an investigation of the roller-tape contact. The article also aims to develop a model of the system for measuring the displacement of the tape in a raster formation device, to investigate and assess possible effects of external and internal factors on steel tape parameters. The article presents experimental research conducted for determining dynamic variables forming during the movement of a steel tape, assessing the factors that may cause raster generation errors in dynamic mode.

Keywords: precision steel tape; tape transportation; roller-tape interactions; roller-tape contact pair

---

1. Introduction

Steel tapes with certain symbols (rasters, special markings, etc.) generated on their surface are often used for metrological and technological purposes, for example, to measure displacement. These symbols are generated by contact movement of a steel tape and a light source beam acting on either the tape surface or its special coatings. Usually, the tape moving between feed and reception points in conveyor devices has a fixed axial tension and a constant movement velocity, stabilizing the tape at the guide rails from the bottom and the sides [1–4]. The smoothness of the movement of a moving precision tape and the stability of the tape area where symbols are generated are very important factors that affect the quality parameters of a generated tape. One of the most important characteristics of the raster generation device is the smoothness of the stretching of the tape, which is important for the overall operation of the system. The smoothness of the stretching of the tape affects errors in the position of the raster element being generated and the control of activation of the laser beam. Determining not only the amplitude, but also the frequency of the belt stretching oscillations is important [5].

Adverse deviations in the position or the shape of symbols being generated that way depend on uncontrolled deviations of parameters of the said movement of the tape and the light beam from

285

the set parameter values [6]. Tape vibrations forming as a result of the system tape movement mechanism and other internal and external sources of excitation of vibrations are one of the most important components of such deviations [7–9] They affect the occurring uncontrollable changes in tape's position, the displacement velocity and tape deformations that have an adverse impact on the quality of the structures being generated, also affecting the accuracy and reliability of active process control done in real time [10]. This impact on the system of defined parameters depends on vibration parameters—frequencies, amplitudes and other statistical characteristics. Knowing these parameters is important for developing the structure and optimizing the processes of formation of the structure and the symbols [2,11–13].

Lateral movement of the tape, which can be caused by a roller tilt, tape defects, roller unevenness and other factors is one of the most important factors [14,15]. Lateral movement of the tape can be suppressed to a certain extent by using flanged rollers, but they can deform the edge of the tape and cause high frequency low amplitude tape vibrations. On the other hand, the use of flangeless rollers can eliminate the above problems but can enhance low frequency lateral tape movement [16–19]. Modeling the tape movement process can be one of the key factors that help to understand the effects of deficiencies of the lateral tape dynamics [20].

The presented research analyses the tape movement system consisting of electromechanical tape pulling and its constant stretching mechanisms as well as a tape deflection mechanism, which operates in a sliding friction [21]. This system was mounted on a massive granite base placed on a foundation using passive vibration insulation supports. A research, a data processing method and the results of experimental research of a mock-up system were developed. The article examines the raster generation method and the generation device. This method may be used to produce a precision metrological scale on a stainless steel tape. The generation process takes place in the dynamic mode because both the steel tape and the laser raster generation head are constantly moving during the process.

The main aim of the research is to develop a system for measuring the displacement of the tape in the raster generation device, to examine the model of that system and to evaluate the possible impact of external and internal factors on raster generation in the dynamic mode.

Since a large number of impulses is generated with a tape moving through the angle transducer due to a very high resolution, information may be lost due to a limited speed of electric elements, which would be an essential cause of errors in the position of raster elements [22–25].

The article examines experimental research results of one of the main components (a tape displacement measuring unit) of the new precision raster generation system and possibilities for analytical modeling.

2. Research Object

The study analyses the raster generation and control system. The system was designed to have a constant linear relationship between the rotation angle of the drum and the linear displacement of the bar. This was achieved by minimizing the kinematic and geometric errors of the measuring system and the tape, which directly affects errors of the raster generation system. A focused laser beam raster is formed based on the number of pulses in the angle measuring system.

The raster generation bench consists of a massive fine-structured granite base and the tape movement devices mounted thereon presented in Figure 1, the laser raster generation components and the raster scale error management system components. Figure 1 illustrates the raster generation bench: 1—tape unwinding mechanism, 2—granite base, 3—tape displacement using angle measuring system, 4—deflection node, 5—error monitoring unit, 6—pulling—stretching system, 7—control panel, 8—tape winding mechanism, 9—mechanism for covering tape in a protective band, 10—indoor partition, 11—microscopes with CCD camera, 12—scanner, 13—optical laser beam deflection system, 14—laser, 15—constant belt tension mechanism, 16—device computing control set and laser and scanner freezer, 17—tape stretch control and management system, 18—vibration isolation supports. The length of the granite base is 2 m in the Y direction.

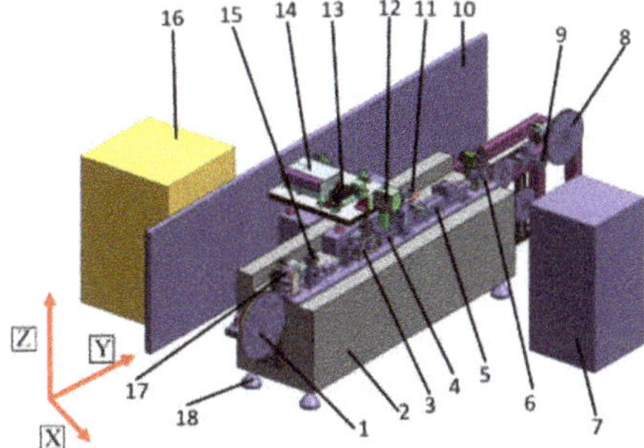

Figure 1. Raster generation bench (block diagram of a metal scale production device).

The main factors that adversely affect the accuracy of the dynamic calibration process are vibrations from external and internal sources. Vibrations from external sources—technological devices, vehicles, fans, etc. are usually transmitted through the floor. Internal sources of vibration include motors, reducers, rotary motion transmitter and laser system scanners.

One of the most important components of the device is the tape displacement measuring system (Position 3 in Figure 1). Figure 2 presents its 3D images in cross-sections: 1—angle transducer base, 2—angle transducer limbo, 3—optical angle transducer heads, 4—precision tape displacement measurement shaft, 5—system base, 6—hood, 7,8—rotary system bearings, 9—tape clamping shaft, 10—tape clamping mechanism. The circumference of the steel band tape measured in the precision tape displacement measurement shaft is 200 mm. Accordingly, the diameter of this shaft at the point of contact with the tape is 63.662 mm.

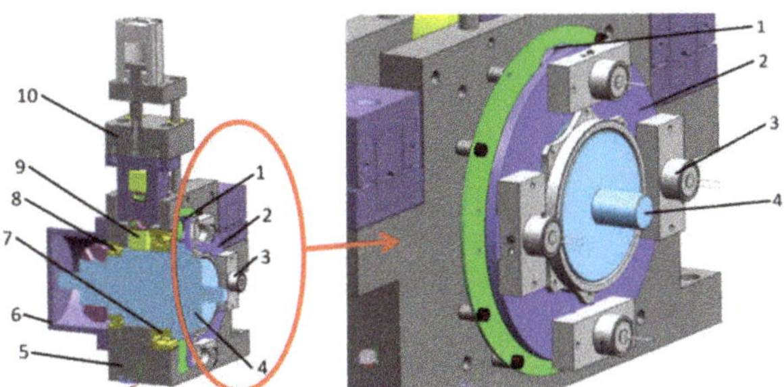

Figure 2. Three dimensional images of the tape displacement measuring system in cross-sections.

Tape displacement measuring errors have a high frequency harmonic component. One of the reasons causing such errors is the so-called internal step error of the angle measurement transducer. The latter, in turn, depends on deviations in the primary electrical signals from the regular sine and cosine form, which are substantially affected by deviations in the shape of the limbo workpiece where a raster track is generated. In order to minimize this error, workpieces were specially polished at high precision till the maximum possible accuracy under Lithuanian manufacturing conditions was achieved. Maximum precision was also pursued by moving the limbo raster drawing from the original onto a product. A limbo raster drawing error was calibrated.

Figure 3 presents a graphical illustration of a parametric function of its two realizations of the tape displacement using the angle measuring system error and approximation of its means. 6 measurements were made, and 6 realizations were obtained; Figure 3 presents error graphs of 2 realizations and an approximation of the means of the 6 realizations.

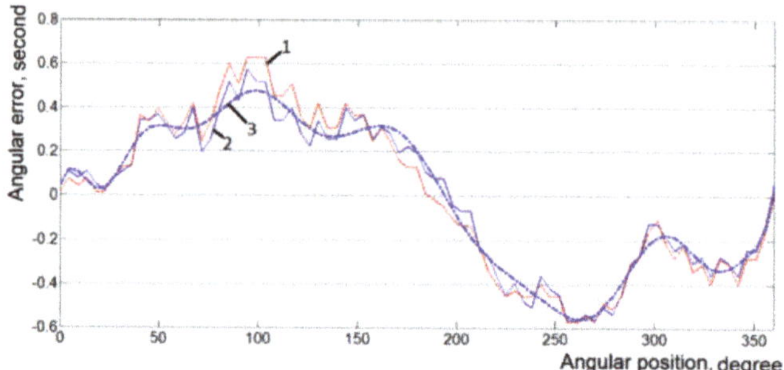

Figure 3. Parametric function of two realizations 1, 2 of the tape displacements using angle measuring system and approximation of 6 realizations means, 3 graphs.

The value of the standard deviation with respect to the means of dissemination of calibration results of the 6 realizations is $S_D = 0.087$ (see Formula 2). In pursuit of the maximum accuracy, a shaft was mounted in special, extra-high-precision bearings, using the method of four-head primary electrical signal generation. Calibration was performed autonomously by attaching a polygon to the tape displacement using an angle measuring system (polygon calibrated with PTB and its known errors), measuring it using an autocollimator, and entering an error correction.

3. Experimental Research of the Tape-Displacement Measuring System

Tape displacement measuring systems is a major part of the raster generation and control system, which directly affect the accuracy of the scales being produced. A bench was developed for the study of dynamic processes of the tape displacement measuring system.

Maintaining constant velocity is very important while working. For the purpose of this experiment, fluctuating velocity of the tape point was measured as absolute vibro-velocity.

Three parameters were measured when determining dynamic properties of the research object (Figure 4b): 1—displacement of the steel tape in the vertical direction Z (Figure 4c); 2—velocity of the steel tape in the longitudinal direction Y (Figure 4a); 3—displacement of the tape–angle transducer contact point was derived from angle sensor results. Capacitive displacement sensors CS02 were used to measure vertical displacements, measuring displacements at two points (A and B) of the tape (Figure 4c). Point A was at the raster generation point (there was an additional support under the tape) and point B—on the tape between supports. Figure 4a presents the image of the measurement of absolute steel tape vibrations (variation of point velocity) in longitudinal direction Y.

Brüel & Kjær vibration meters were used to measure vibration parameters and Micro-Epsilon meters were used to measure displacement. Displacement meters (Figure 5a,b): a capacitive displacement sensors CS02; b controller DT6220 and demodulator DL6220.Vibration meters (Figure 5c,d): c portable metering result processing, storage and control equipment 3660-D; d triaxial accelerometers 4506. Figure 6 illustrates the block diagram of the research equipment and its relation to the research object (tape displacement measurement system).

The obtained measurement results were processed using the software package Origin, also calculating statistical parameters:

arithmetic average:

$$\bar{x} = \frac{1}{n} \sum_{i=1}^{n} x_i \tag{1}$$

standard deviation:

$$S_D = \sqrt{\frac{1}{n-1} \sum_{i=1}^{n} (x_i - \bar{x})^2} \tag{2}$$

range

$$x_{range} = x_{max} - x_{min} \tag{3}$$

where n is the number of measurement results, x_i–measurement result i.

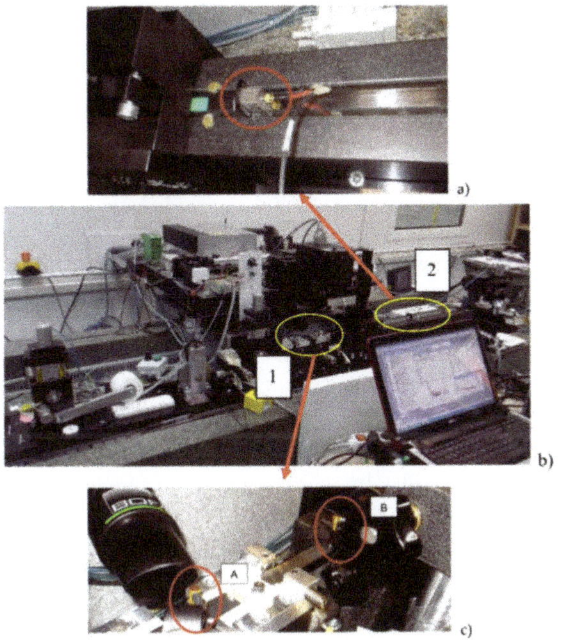

Figure 4. Measurement of vertical tape point displacements (**a**),longitudinal tape (**b**) velocity and displacement of the steel tape in the vertical direction Z (**c**).

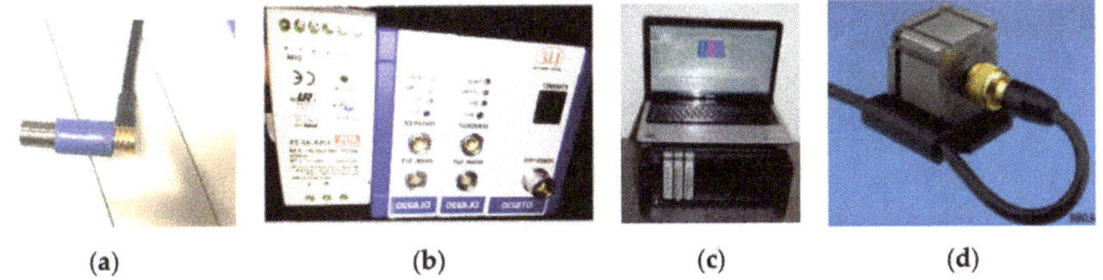

(**a**)   (**b**)   (**c**)   (**d**)

Figure 5. Instruments for measuring and analyzing vibration and displacement parameters: capacitive displacement sensor CS02 (**a**) with controller DT6220, demodulator DL6220 ad power source (**b**); triaxial accelerometer 4506 (**c**); portable metering result processing, storage and control equipment 3660-D with DELL computer (**d**).

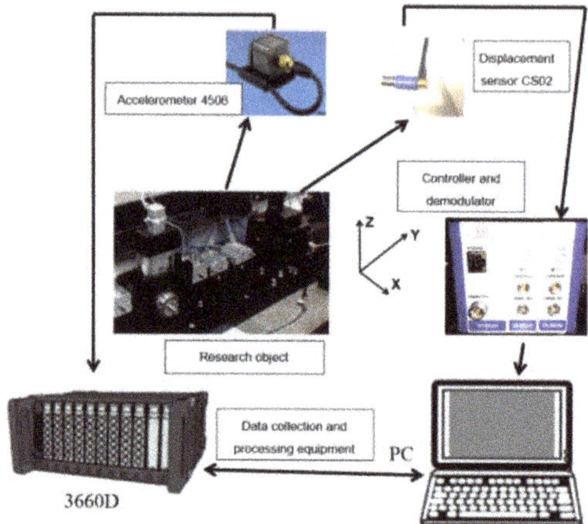

Figure 6. Block diagram of the test bench of vertical tape displacement (direction Z) and velocity (direction Y).

The received measurement results are presented in Figures 7–9. Figure 7 illustrates results of vertical displacements of the two points (A and B) (also see Figure 4c) of the tape. Figure 7a presents displacement results of points A and B (also see Figure 4c) in operation (a generated raster distance of 500 mm), and Figure 7b,c illustrates the curve and mean of three realizations, additionally presenting histograms of the mean. Table 1 presents statistical parameters of three realizations of the displacement (A1, A2 and A3 and B1, B2 and B3) of points A and B (also see Figure 4c).

The analysis of the results presented in Figure 7 and Table 1 allowed determining that the value of the displacement of tape point A in vertical direction varies by about 37.5 μm, and the value of the displacement of tape point B varies by about 53.5 μm. The value of displacement of point B is greater because the lower tape part of point B is not supported. The frequency of tape oscillations is very low (0.04 Hz), and oscillations of such frequency are likely to have been caused by deviations in the shape of the tape.

Table 1 presents the analysis of the results where the standard deviation of the measurements ranges between 0.09% and 0.87%.

Errors in the displacement measuring system directly affect the accuracy of raster formation on steel tape. Vertical displacements (see Figure 7) up to 50 μm of significant steel tape points (points A and B (also see Figure 4c)) were observed when stretching the tape. The analysis of vertical displacements of point B revealed that they have a periodic shape and vertical displacements of point A have three peaks at 48, 161 and 180 s. Points of support of the tape affect such a change of vertical displacements of the significant points A and B.

Figure 8 presents the results of the measurement of the displacement of the tape–angle transducer contact point, which is calculated using the results of the angle sensor. Figure 8a illustrates the time graph and the histogram, Figure 8b presents the spectral density graph and Figure 8c—the spectral graph part expanded to 50 Hz.

Table 1. Statistical parameters of three realizations of displacement (A1, A2 and A3 from Figure 7b and B1, B2 and B3 from Figure 7c) of points A and B (also see Figure 4c).

Measurement Point	Mean	Standard Deviation	Statistical Parameters, µm		
			Minimum	Maximum	Range (Maximum–Minimum)
Point A (Figure 7a)					
Displacement A1	4.127	8.753	−4.322	33.283	37.605
Displacement A2	3.899	8.821	−4.505	32.904	37.409
Displacement A3	3.920	8.830	−4.519	33.046	37.565
Point B (Figure 7a)					
Displacement B1	17.996	13.109	−7.910	45.683	53.593
Displacement B2	20.887	13.097	−4.701	48.791	53.492
Displacement B3	19.865	13.029	−5.784	47.738	53.522

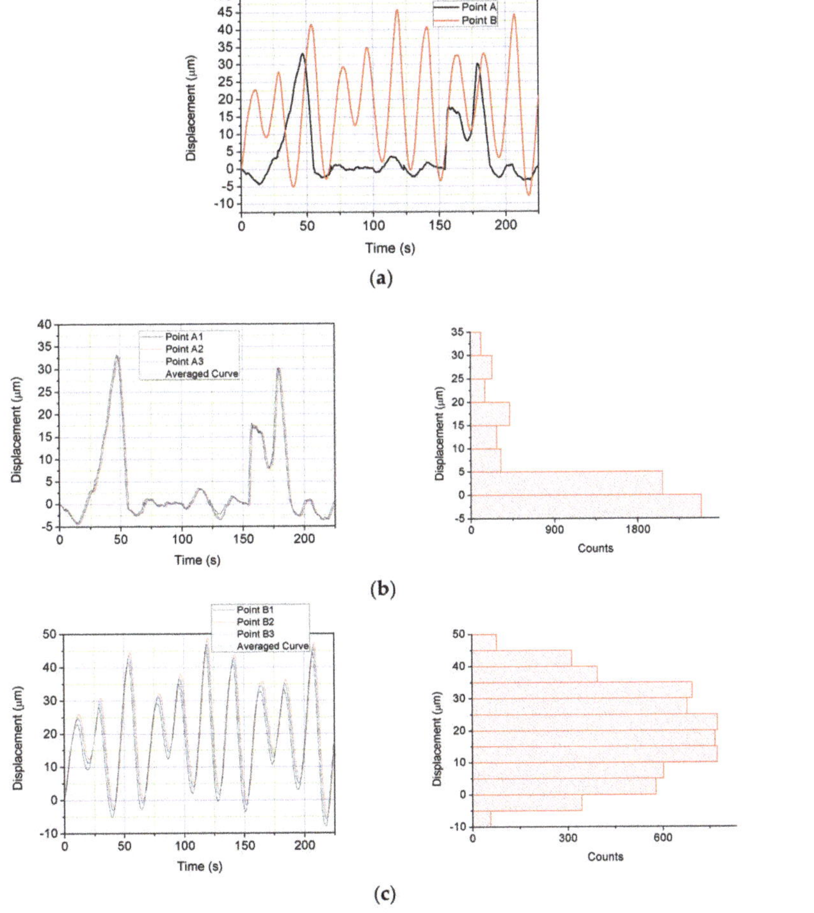

Figure 7. Results of vertical displacements of two steel tape points A and B (also see Figure 4c): displacement results of points A and B (**a**); the curve and mean of three realizations, additionally presenting histograms of the mean: points A1-A3 (**b**) and points B1-B3 (**c**).

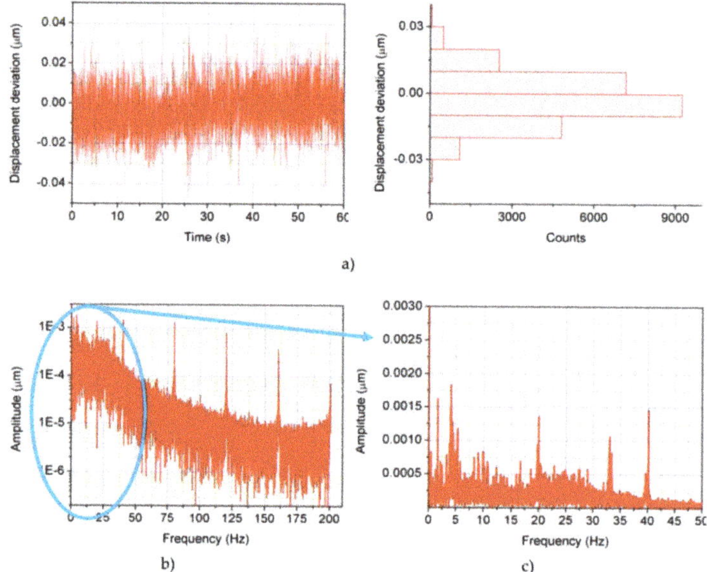

Figure 8. Results of measurements of the displacement (from precision angle transducer) of the tape–angle transducer contact point: (**a**) time graph and histogram, (**b**) spectral density graph, (**c**) spectral graph part expanded to 50 Hz.

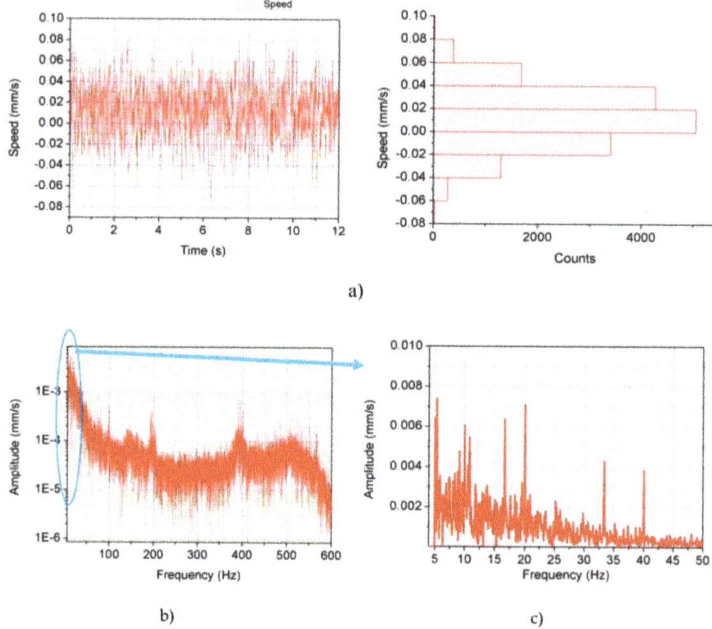

Figure 9. Results of the measurement of the steel tape velocity (also see Figure 4a): (**a**) time graph and histogram, (**b**) spectral density graph, (**c**) spectral graph part expanded to 50 Hz.

Figure 9 illustrates the results of the measurement of the steel tape velocity (i.e., speed mismatch at 2 mm/s) in the longitudinal direction Y. Figure 9a presents the time graph and the histogram,

Figure 9b—the spectral density graph and Figure 9c—the spectral graph part expanded to 50 Hz. Table 2 presents statistical parameters of the results of the measurement of the displacement of the tape–angle transducer contact point and velocity in the longitudinal direction Y.

Table 2. Statistical parameters of the results of the measurement of the displacement of the steel tape–angle transducer contact point and velocity in the longitudinal direction Y.

Displacement of the Tape–angle Transducer Contact Point Statistical Parameters, µm				
Mean	Standard Deviation	Minimum	Maximum	Range (Maximum–Minimum)
−0.00244	0.0106	−0.0476	0.039	0.086
Longitudinal tape velocity Statistical parameters, mm/s				
Mean	Standard Deviation	Minimum	Maximum	Range (Maximum–Minimum)
0.0122	0.0244	−0.0777	0.0859	0.163

The assessment of the results of the displacement of the tape–angle transducer contact point (Figure 8 and Table 2) revealed that the value of the displacement of tape–angle transducer contact point varies by about 0.086 µm. The analysis of the received spectral density results revealed that the the following frequencies appeared in the tape displacement graphs: 1.67; 4.04; 19.9; 32.7; 40; 80.1; 120.9 and 160 Hz.

The assessment of the tape velocity results (Figure 9 and Table 2) revealed that the tape velocity value varies by about 0.163 mm/s. The analysis of the obtained spectral density results showed that the following frequencies appeared in the tape displacement graphs: 6.17; 10.04; 17.1 19.9; 32.7; 40; 80.1; 120.9 and 160 Hz.

The analysis of the received results showed that frequencies appearing in tape displacement and velocity frequency graphs in Figures 8 and 9 were very close, and the assessment of the tape slip in respect of the angle transducer allows stated that there was no slip in the frequency range from 0 to 200 Hz.

The analysis of graphs in Figures 8c and 9c revealed that the displacement and velocity amplitudes at the same frequencies dominated (examining the frequency range up to 50 Hz), which indicated that the band motion patterns were transmitted to the tape–gauge roll assembly and there was no slip between the tape and gauge rolls.

4. Numerical Investigation of Tape Displacement

A model for calculating mechanical properties (Figure 10) was developed for the tape displacement measurement system, which was one of the most important components of a raster generation device. It should be noted that the numerical simulation model presents one very important node in the experimental setup illustrated in Figure 4c.

Five Hypalon plates were affixed to the bottom and top of the steel tape. Both top and bottom contact pairs were monitored during simulation. Figure 11 illustrates the contact pairs. The iterative solution method was selected. Two selections of boundaries that could not penetrate each other under deformation were chosen for the analysis of the contact pair boundary condition.

The contact pairs defined boundaries for parts that could come into contact as shown in Figure 11a,b, for upside and downside contacts, respectively. The augmented Lagrangian was selected as the algorithm for computing contact in the normal direction. Characteristic stiffness in current simulation was equivalent to Young's modulus. Penalty factor controls the stability and stiffness of the interface surface during iterations. Moreover, an additional control over solver cutbacks, such as trigger cutback criterion in numerical simulation, was not used because there was no destination boundary movement history. The exponential dynamic Coulomb friction model using both static and dynamic friction coefficients were included in the time dependent study for both contact pairs (Figure 11).

In addition, the steel tape had a spring foundation condition from the left boundary and uploaded velocity from the opposite boundary as shown in Figure 10. Experimental measurement tape velocity data were used us a set velocity boundary condition in simulation as shown in Figure 12. The analytic velocity function of steel tape's end from the experiment (Figure 12) was used for the set velocity as one of boundary conditions for simulation as shown in Figure 10.

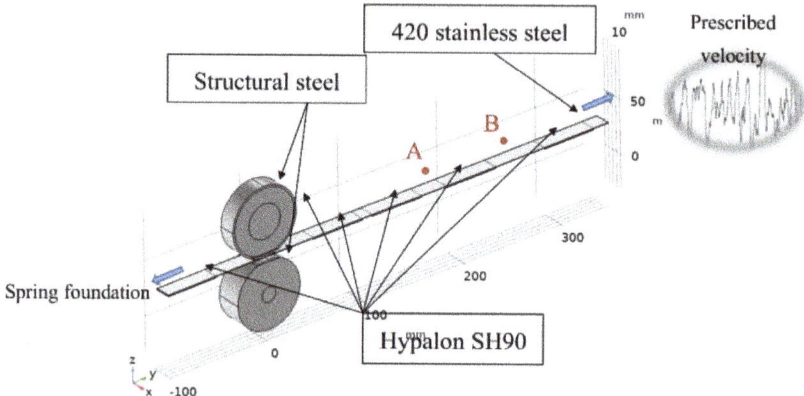

Figure 10. Roller–steel tape unit modeled using Comsol Multiphysics.

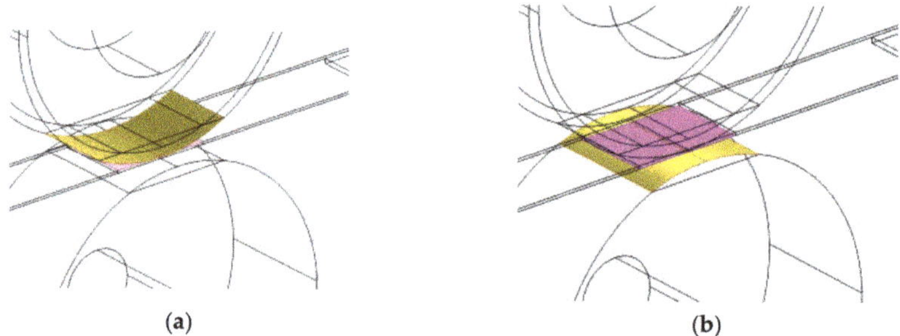

(a)   (b)

Figure 11. Upside contact pair (**a**) and downside contact pair (**b**).

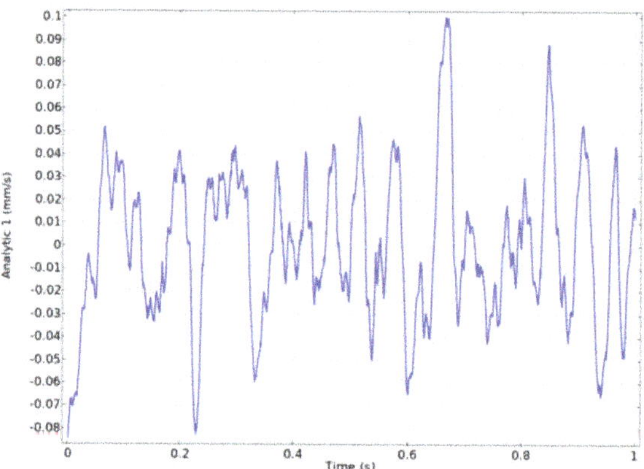

Figure 12. Set velocity curve for simulation of tape movements (experiment data).

Two-point probes similar to the experimental setup sensor position (Figure 4c) were chosen for postprocessor analysis after simulation. They are illustrated in Figure 10 as points A and B.

Figure 13 illustrates fixed constraints and the body load of the current model. Five contact zones between the steel tape and support from Hypalon material are presented in Figure 13a. Dynamic friction and gravity of these contact pairs only were taken into consideration during simulation. DOF (Degree of Freedom) by z and x axes in contact zones was restricted. Two coordinated restrict the set velocity (Figure 10), and the steel tape had a freedom of movement along the length only. Gravity was also included in the simulation. The simulation was based on the parametric sweep of the top roller preload by–z axes (Figure 13b) and a change from 10 to 40 N with 10-N steps. This range was chosen based on the experimental 30-N preload setup.

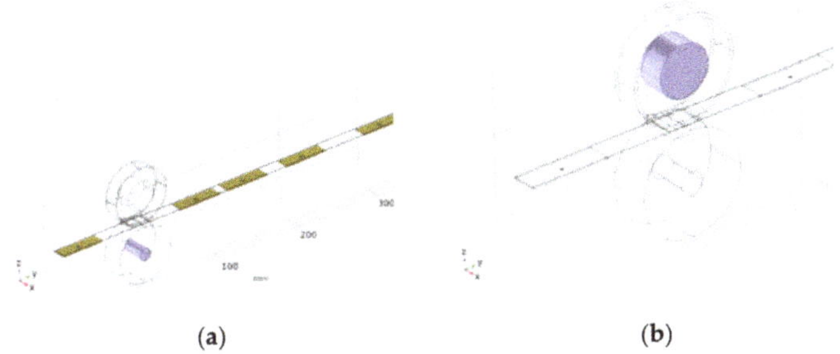

(a)   (b)

Figure 13. Fixed constraints (**a**) and body load (**b**).

Mechanical properties of materials used in numerical simulation are presented in Table 3.

Table 3. Mechanical properties of used materials.

	Structural Steel	Hypalon SH90	Aisi 420
Density, kg/m^3	7850	1200	7700
Young's modulus, Pa	200e9	20e6	215e9
Poisson's ratio	0.30	0.40	0.28

Figure 14 below presents a compiled mesh of finite elements. The steel tape unit was modeled using COMSOL Multiphysics software. The quality of mesh elements maximizes in the contact zone. In the simulation, 0.48 and 0.93, respectively, were the minimum and the average element qualities.

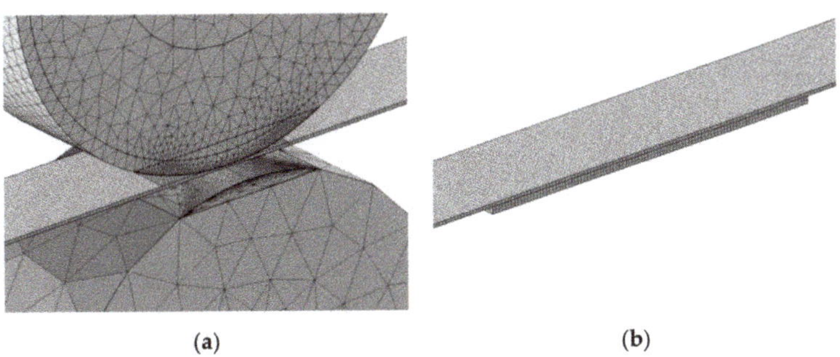

(a)   (b)

Figure 14. Finite element mesh of the steel tape unit in the contact zone between the rollers and the steel tape (**a**) and Hypalon plates and steel tape contact pairs (**b**).

Having generated the finite element mesh for geometric model, the next step was modeling the problem solution. In order to model the problem solution, equation system type and solution method were selected for the steel tape unit. A time-dependent solution type was chosen, then solving the contact and dynamic movement problems.

The time-dependent study was limited to 1 second because of the calculation time and data storage limitation. The longitudinal displacement and velocity in point B (Figure 10) under different preloads between the top roller and the steel tape are presented in Figure 15, respectively. Dependences confirmed the feasibility of the simulation model and the experimental results of the measurement point B displacement and velocity (Figures 8a, 9a and 15).

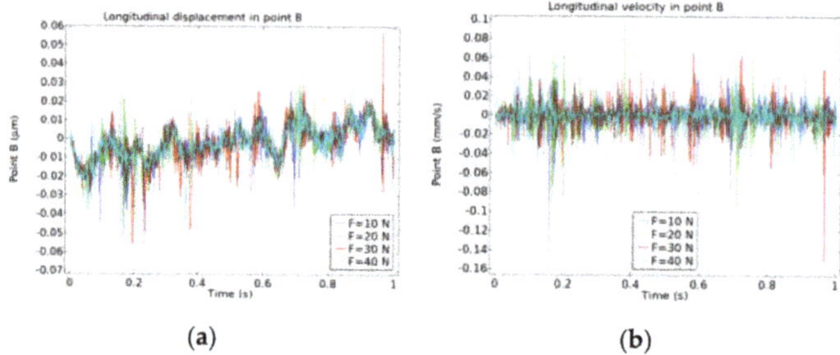

Figure 15. (**a**) Longitudinal displacement and (**b**) longitudinal velocity in point B, under variable preloads from 10 to 40 N.

Of course, the longitudinal parasitic vibration of the steel tape had a greater impact on the pitch formation precision, however, we believe that the taping vibration of the steel tape should also be taken into account. Figure 16 illustrates the taping displacement of points A and B (left and right). Displacement values were much higher than those in the longitudinal direction, and the effect on pitch precision was the sum of the three coordinates of the total parasitic vibration of the steel tape.

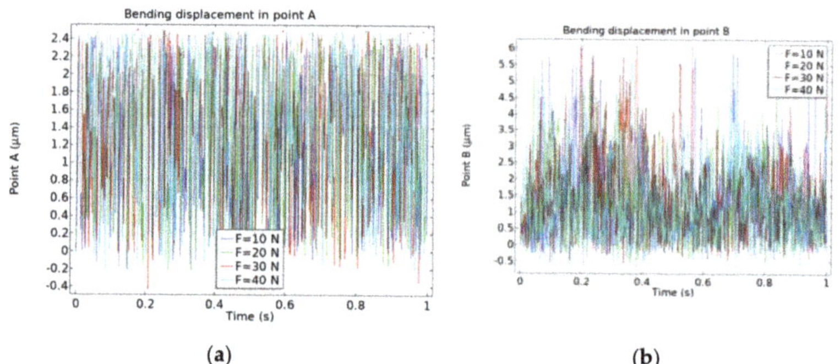

Figure 16. Bending displacement in point A (**a**) and B (**b**) under different top roller preloads.

The received results show that the roller-tape contact pressure responded to the dynamic reaction of the steel tape during movement and must be choosen carefully.

5. Conclusions

The article examines analytical modeling of one of the main components of the raster generation system (tape displacement measurement unit) and presents the experimental research carried out.

A mechanical model of the steel tape and rollers subjected to loads was developed in the article. The article presents the modeling, which defines deformation of a steel tape model under preloads with the variable boundary conditions.

The measurements of the moving tape of the precision raster generation system in the vertical and longitudinal directions revealed that the value of the displacement of tape point when the tape was supported from the bottom varied by about 37.5 μm during operation, and ranged to about 53.5 μm when the tape was not supported from the bottom. The frequency of tape oscillations was very low (0.04 Hz) and came as a result of deviations in the shape of the tape.

The main parameters that affected parasitic longitudinal and vertical displacements of the mesh were the number and the position of mesh supports and the tensile and clamping forces of the mesh.

Author Contributions: Conceptualization, A.K. (Artūras Kilikevičius) and A.K. (Albinas Kasparaitis); methodology, K.K.; software, A.F.; validation, S.B., J.M. and A.F.; formal analysis, K.K.; investigation, A.K. (Artūras Kilikevičius); resources, A.K. (Albinas Kasparaitis); data curation, K.K.; writing—original draft preparation, A.K. (Artūras Kilikevičius); writing—review and editing, J.M.; visualization, S.B.; supervision, A.K. (Artūras Kilikevičius); project administration, J.M. All authors have read and agreed to the published version of the manuscript.

Funding: This research received no external funding.

Conflicts of Interest: The authors declare no conflicts of interest.

References

1. Brake, M.R.; Wickert, J.A. Frictional vibration transmission from a laterally moving surface to a traveling beam. *J. Sound Vib.* **2008**, *310*, 663–675.
2. Ghayesh, M.H. Stability characteristics of an axially accelerating string supported by an elastic foundation. *Mech. Mach. Theory* **2009**, *44*, 1964–1979.
3. Shin, K.-H.; Kwon, S.-O. The Effect of tension on the lateral dynamics and control of a moving web. *IEEE Trans. Ind. Appl.* **2007**, *43*, 403–411.
4. Yang, H.; Engelen, J.B.C.; Pantazi, A.; Häberle, W.; Lantz, M.A.; Müftü, S. Mechanics of lateral positioning of a translating tape due to tilted rollers: Theory and experiments. *Int. J. Solids Struct.* **2015**, *66*, 88–97.
5. Tsutsumida, N.; Rodríguez-Veiga, P.; Harris, P.; Balzter, H.; Comber, A. Investigating spatial error structures in continuous raster data. *Int. J. Appl. Earth Obs. Geoinform.* **2019**, *74*, 259–268.
6. Sato, A.K.; Martins, T.C.; Gomes, A.M.; Tsuzuki, M.S.G. Raster penetration map applied to the irregular packing problem. *Eur. J. Oper. Res.* **2019**, *279*, 657–671.
7. Samourgkanidis, G.; Kouzoudis, D. Characterization of magnetoelastic ribbons as vibration sensors based on the measured natural frequencies of a cantilever beam. *Sens. Actuators Phys.* **2020**, *301*, 111711. [CrossRef]
8. Ma, L.; Cheng, L. Numerical and experimental benchmark solutions on vibration and sound radiation of an Acoustic Black Hole plate. *Appl. Acoust.* **2020**, *163*, 107223. [CrossRef]
9. Nandakumar, H.; Mallick, S.P.; Srivastava, S. Sensing high frequency sub-nanometer vibrations using optical coherence tomography with real-time profilometry of multiple inner layers. *Opt. Lasers Eng.* **2020**, *127*, 105992. [CrossRef]
10. Blanchet, T.A.; Grzyboski, S.M. Conduction modeling of multi-track raster scanning of heat over a rectangular surface region. *J. Manuf. Process.* **2020**, *52*, 165–180.
11. Chen, S.; Qian, X.; Zhou, M. Research of Automation Steel Tape Measurement System Based on PLC and Computer Vision Technology. In Proceedings of the 2009 WRI World Congress on Software Engineering, Xiamen, China, 19–21 May 2009; pp. 251–255.
12. Jin, X.; Shui, H.; Shpitalni, M. Virtual sensing and virtual metrology for spatial error monitoring of roll-to-roll manufacturing systems. *CIRP Ann.* **2019**, *68*, 491–494.
13. Li, Y.; Wang, Z.; Chen, X.; Lu, R. Novel Zero Calibration Method of Steel Tape. In Proceedings of the 2017 First International Conference on Electronics Instrumentation & Information Systems (EIIS), Harbin, China, 3–5 June 2017; pp. 1–5.
14. Di Bartolomeo, M.; Lacerra, G.; Baillet, L.; Chatelet, E.; Massi, F. Parametrical experimental and numerical analysis on friction-induced vibrations by a simple frictional system. *Tribol. Int.* **2017**, *112*, 47–57.
15. Han, F.; He, R.; Yan, H.; Xiong, F. Lateral motion of the endless flat belt in a two-pulley belt system. *Adv. Mech. Eng.* **2017**, *9*. [CrossRef]
16. Brake, M.R.; Wickert, J.A. Tilted guides with friction in web conveyance systems. *Int. J. Solids Struct.* **2010**, *47*, 2952–2957.

Appl. Sci. **2020**, *10*, 4041

17. Raeymaekers, B.; Talke, F.E. Measurement and sources of lateral tape motion: A review. *J. Tribol.* **2009**, *131*, 011903. [CrossRef]
18. Tarnopolskaya, T.; Gates, D.J. Analysis of the effect of strip buckling on stability of strip lateral motion with application to cold rolling of steel. *J. Dyn. Syst. Meas. Control* **2008**, *130*, 011001. [CrossRef]
19. Zen, G.; Müftü, S. Stability of an axially accelerating string subjected to frictional guiding forces. *J. Sound Vib.* **2006**, *289*, 551–576.
20. Pijl, A.; Bailly, J.-S.; Feurer, D.; El Maaoui, M.A.; Boussema, M.R.; Tarolli, P. TERRA: Terrain extraction from elevation rasters through repetitive anisotropic filtering. *Int. J. Appl. Earth Obs. Geoinf.* **2020**, *84*, 101977. [CrossRef]
21. Hao, J.; Zong, G.; Liu, Y.; Song, Y.; Wang, W.; Cheng, H.; Zhang, W. Study on jute twine/polypropylene pre-preg tapes: The effects of surface modification and impregnation. *Constr. Build. Mater.* **2020**, *247*, 118089. [CrossRef]
22. Zheng, T.; Zhang, Y.; Ke, Q.; Wu, H.; Heng, L.W.; Xiao, D.; Zhu, J.; Pennycook, S.J.; Yao, K.; Wu, J. High-performance potassium sodium niobate piezoceramics for ultrasonic transducer. *Nano Energy* **2020**, *70*, 104559. [CrossRef]
23. Qin, J.-Q.; Yin, J.-H.; Zhu, Z.-H.; Tan, D.-Y. Development and application of new FBG mini tension link transducers for monitoring dynamic response of a flexible barrier under impact loads. *Measurement* **2020**, *153*, 107409. [CrossRef]
24. Sun, H.; Urayama, R.; Uchimoto, T.; Takagi, T.; Hashimoto, M. Small electromagnetic acoustic transducer with an enhanced unique magnet configuration. *NDT E Int.* **2020**, *110*, 102205. [CrossRef]
25. Crowder, R. 4—Velocity and position transducers. In *Electric Drives and Electromechanical Systems*, 2nd ed.; Crowder, R., Ed.; Butterworth-Heinemann: Cambridge, MA, USA, 2020; pp. 107–134.

 *applied sciences*

Article

Analysis of the Effect of Shape Factor on Cork–Rubber Composites under Small Strain Compression

Helena Lopes [1], Susana Silva [2] and José Machado [1,*]

[1] University of Minho, MEtRICs Research Center, Campus of Azurém, 4800-058 Guimarães, Portugal; id7466@alunos.uminho.pt
[2] Innovation Department, Amorim Cork Composites, Rua de Meladas, 260, 4535-186 Mozelos VFR, Portugal; susana.silva@amorim.com
* Correspondence: jmachado@dem.uminho.pt

Received: 16 September 2020; Accepted: 9 October 2020; Published: 15 October 2020

Abstract: Like other types of elastomers, different geometries of the same cork–rubber material present different mechanical behaviour when subject to compression between bonded plates. To validate the application of Hooke's Law on cork–rubber materials, under compression at small strains, a set of experimental and numerical analyses has been conducted. Using finite element analysis, a methodology is described to relate frictionless and frictional compression between a cork–rubber sample and loading plates. Based on that, the performance of square cross-section blocks with other dimensions can be evaluated. The results obtained by this approach showed a good agreement with experimental compression tests and with outputs from other models available in the literature relating Young and apparent compression moduli.

Keywords: cork–rubber composites; compression; apparent compression modulus; Young's modulus; bonded condition

1. Introduction

One of the application areas of cork–rubber-composites is vibration isolation. This type of elastomers is composed of a rubber matrix filled with granules of cork, which can be utilised as isolation pads for systems subjected to the presence of dynamic loads, such as buildings, industrial machines and floating floors [1,2]. Cork is a natural origin material that is mostly used in wine stoppers. The insertion of cork granules into a rubber compound contributes to its recovery improvement when submitted to compression loads [3].

The use of numerical methods, such as finite element analysis (FEA), has proven to be an advantageous tool to predict the mechanical behaviour of many materials. The application of FEA related to elastomers and other cork composites has been utilised to access static, dynamic loading and impact behaviour [4–8]. One of the first requirements for the application of isolation pads is to evaluate their capacity to support static loadings. Regarding elastomers, one of the crucial steps during FEA is the definition of material properties [9]. Typically for large strains, elastomer's properties are defined through the application of non-linear models. However, if a linear stress–strain relationship, at small strains, is observed, Hooke's Law can be adopted for that strain range [10].

Concerning rubber materials, the compression behaviour of a specimen between bonded surfaces can be quantified using apparent compression modulus (E_c) [9,11]. This parameter is associated with the rubber sample's geometry. In the case of compression without bonded surfaces, the rubber block presents axial and uniform lateral deformation, as depicted in Figure 1a. With a bonded condition, the lateral surfaces deform, assuming a barrel shape due to incompressibility, as presented in Figure 1b. The compression of rubber materials, considering friction between specimen and loading instrument surfaces, will also cause a similar shape effect. However, the rubber's stiffness will be smaller when compared to a bonded block due to some slippage of the edges [9].

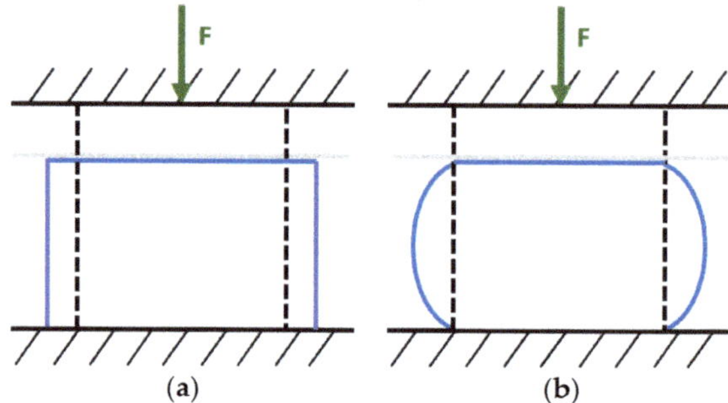

Figure 1. Compression behaviour of a rubber block: (**a**) no friction between surfaces; (**b**) between bonded surfaces.

For example, considering frictional contact compression on two blocks with the same area and different thicknesses, the sample with lower thickness will present higher stiffness. This effect is related to the ratio between loaded area and total free area, also known as shape factor [10,12]. For disks and rectangular blocks, it can be determined by Equations (1) and (2), respectively:

$$f_{disk} = \frac{D}{4T} \tag{1}$$

$$f_{block} = \frac{LW}{2T(L+W)} \tag{2}$$

with L and W are the length and width of a block, T corresponds to the thickness and D to the diameter of a cylinder.

Another aspect that has been considered by several authors is the effect of the rubber's hardness. A softer material presents higher deformation at a certain load level compared with a material with higher hardness. To describe this effect, mathematical relationships between hardness and Young's modulus (E_0) have been developed by several authors with applications on different types of rubbers [13–15].

Mathematical models relating the compression behaviour with bonded contact and Young's modulus can also be found in the literature. Several works present different types of expressions for the approximated solution of the relation between apparent compression and Young's moduli [12,16–21]. Some of these studies are described in the literature review section. However, most of the analytical models developed only focus on disk geometry and/or assume a condition of material incompressibility, restricting its application to cork–rubber materials.

The existence of a relation between moduli can be useful, for example, to estimate Young's modulus value of a cork–rubber material and use it to evaluate the performance of an isolation pad with other dimensions, reducing the need of several experimental testing. Thus, the goal of the present study focuses on finding a relationship between compression with and without frictional surfaces. A set of different square cross-section blocks composed of cork–rubber materials, characterised by a linear region below 10% strain, was chosen to determine a relation between Young and apparent compression moduli, based on finite element analysis results. A comparison of the proposed methodology with other analytical models, relating Young and apparent compression moduli, was also conducted for validation purposes.

2. Literature Review on the Relation between Young and Apparent Compression Moduli

One of the first models was proposed by Gent and Lindley [12]. Considering a disk sample and assuming a condition of total incompressibility, the resultant apparent compression modulus can be determined by Equation (3), considering the geometry of the specimen [12]:

$$E_c = E_0\left(1 + 2f^2\right) \tag{3}$$

where f is the shape factor. Other cross-section geometries and loading modes are studied in [22]. Another aspect of the relation between moduli noticed, and accounted for in the work of Gent and Lindley [12], was the influence of bulk modulus for blocks with large shape factors.

The study presented by Horton et al. [19] did not follow the same assumption of a parabolic deformation of lateral surfaces considered in [12]. Thus, another approach to the problem of compression loading was performed based on a superposition method. The expression determined by the authors also included the effect of bulk modulus (K), as presented in Equation (4), for the case of circular cross-section blocks:

$$\frac{1}{E_c} = \frac{1}{E_0}\left[1 - f\sqrt{\frac{2}{3}}tanh\left(\frac{1}{f}\sqrt{\frac{3}{2}}\right)\right] + \frac{1}{K}. \tag{4}$$

Lindley [16] developed theoretical relations for the compression moduli of blocks with circular and cross-section for soft elastic materials, following the same assumptions used for incompressible materials in the Gent and Lindley's study [12]. The approach was validated by the results' agreement with finite element analysis. For circular cross-section blocks, the resultant analytical expressions for the determination of apparent compression modulus are presented:

$$E_c = 2G + \frac{\lambda G}{E_3}\left[1 + \frac{3}{8}\frac{\lambda}{E_3}\left(\frac{w}{h}\right)^2\left(1 - \frac{G\left(\frac{w}{h}\right)^2}{2E_3 + \frac{33}{32}G\left(\frac{w}{h}\right)^2}\right)\right], \quad w < w_2 \tag{5}$$

$$E_c = \lambda + 2G - \frac{\lambda^2}{15E_3}\frac{w_2}{w}\left(8 - \frac{w_2}{w}\right), \quad w \geq w_2 \tag{6}$$

where G is shear modulus, λ is second Lamé constant, w is the width of the cross-section and h is the thickness of the block. The parameters E_3 and w_2 are calculated using Equations (7) and (8):

$$E_3 = \lambda + G \tag{7}$$

$$w_2 = \sqrt{\frac{64E_3h^2}{15G}} \tag{8}$$

A simpler model to determine the relationship between apparent compression and Young's moduli was proposed by Williams and Gamonpilas [20] for disks. The resulting equation is only dependent on Poisson's ratio (v) and aspect ratio (S)—ratio between the radius and thickness—and is presented in Equation (9):

$$\frac{E_c}{E_0} = \frac{1 + 3v\left(\frac{1-v}{1+v}\right)S^2}{1 + 3v(1 - 2v)S^2} \tag{9}$$

Discussion about the compression behaviour of elastomer materials continues as other conditions are considered to the problem solving, and mathematical models' performance is compared against experimental testing data [23,24]. Other conditions studied include the influence of the boundary condition type, such as the existence of frictional contact between sample and plates [21,25–28] and the application of non-linear models for large strains [29], for example.

3. Materials and Methods

The methodology followed in this study is presented in Figure 2. Given a certain geometry and Young's modulus and Poisson's ratio values obtained through an experimental compression test—without friction between sample and loading plates—a simulation of the compression behaviour considering a frictional contact can be performed through Finite Element Analysis (FEA). The results obtained from the numerical analysis can then be used for determining a relationship between the two testing configurations. A detailed description of the methodology steps is given below.

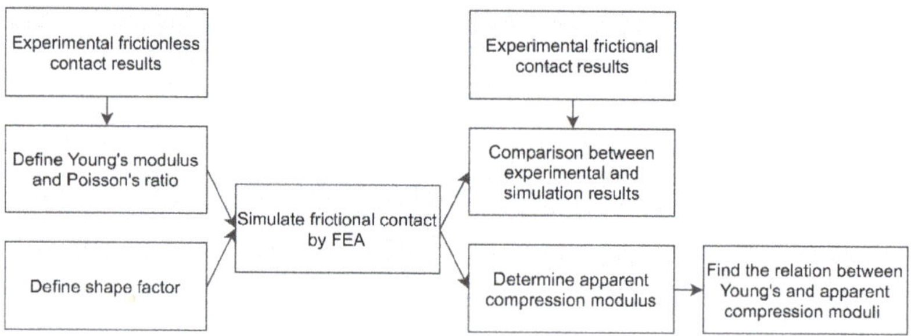

Figure 2. Methodology scheme.

Regarding experimental methods, two compression tests were performed, considering frictionless and frictional contact between sample and loading plates. To simulate frictionless contact, lubricated plates were used. The value of Young's modulus was calculated based on load-displacement data from that experiment. Poisson's ratio of cork–rubber materials was also determined based on an internal experimental procedure. This value was obtained through the measurement of lateral deformation, using a probe indicator until a maximum of 20% axial strain. In this study, assuming an isotropic material behaviour, the Poisson's ratio obtained for the cork–rubber composites had a value of 0.31. About the experimental testing with frictional contact, no lubricant or rough surface was applied between the sample and compression plates. All the samples tested have the same square cross-section, differing only in thickness.

Based on the values of Young's modulus obtained by experimental tests, the compression behaviour of different geometries of a cork–rubber compound was simulated, employing finite element analysis. Simulations were performed using ANSYS Mechanical. The experimental setup and 3-D finite element model are presented in Figure 3. The finite element model was composed of a solid block, representing the cork–rubber specimen, placed between two surfaces. The numerical analysis consisted of simulating the compression machine displacement in contact with one of the surfaces of a cork–rubber block, recording as output the reaction force on the opposite surface. For that, rigid joints were employed on each surface: one with a fixed condition on all degrees of freedom, and another with only free translation on the load direction. Two contact options—rough and frictionless—were also employed to simulate the experimental method, according to the respective testing condition.

In this study, to model the compression behaviour of cork rubber composites, Hooke's Law was considered, since these materials present a linear region at small strains [22]. These values of deformation usually correspond to the application range of the cork–rubber vibration isolation pads, which simplifies the problem under study. However, at higher strains, the use of non-linear models is recommended [4,7,20]. The value of Poisson's ratio used as an input for numerical analysis was the same for all simulations performed since all tested materials presented similar results for that parameter.

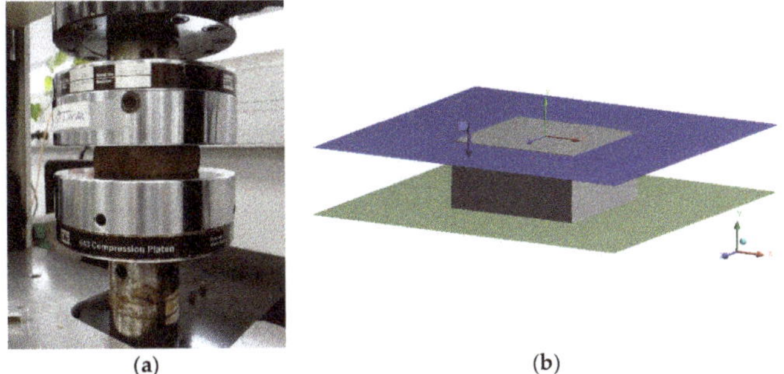

(a)   (b)

Figure 3. Compression of a cork–rubber sample: (**a**) Experimental setup; (**b**) Finite element model.

With the results from the numerical analysis of frictional contact between surfaces-stress and strain data, the apparent compression modulus was calculated for different geometries, using Equation (10):

$$E_c = \frac{\sigma}{\varepsilon} \tag{10}$$

where σ is stress and ε is strain. Then, the apparent compression modulus was related to Young's modulus for each shape factor.

Finally, for validation purposes, the results obtained by FEA for frictional contact were compared to samples tested experimentally. In addition, the results concerning the application of the methodology proposed in this article were compared against results obtained by the application of Lindley [16] and Williams' [20] models.

4. Results

The results from experimental and simulated compression tests are described first, according to the contact type for comparison and validation of the approach developed. Then, using simulation results, the relation between Young and apparent compression moduli for each shape factor defined is presented. Figure 4 demonstrates a typical stress–strain plot for a cork–rubber composite, considering a frictional contact. These materials exhibit a linear behaviour of up to approximately 10% strain. Considering this, all results presented in this study correspond to a maximum strain of 10%, since all samples demonstrated a linear behaviour until this point.

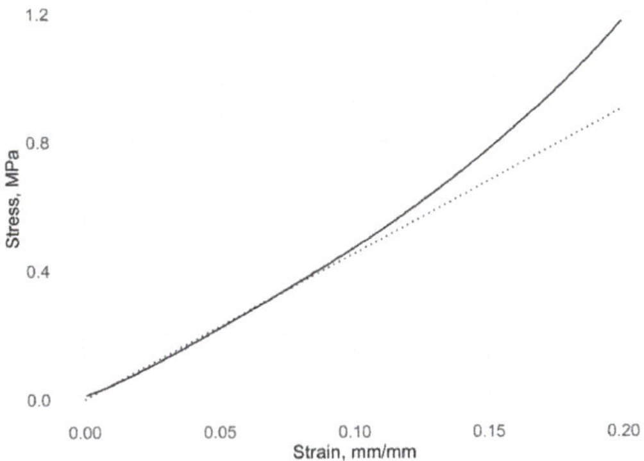

Figure 4. Stress–strain plot of a cork–rubber composite with frictional contact between loading surfaces (solid line).

4.1. Comparison between Experimental and Simulation Results: Frictionless Contact

An experimental compression test using lubricated plates was conducted to validate the application of the finite element analysis for this problem. From the same cork–rubber compound, three blocks, with an equal square cross-section and different thicknesses, were tested. Based on experimental force-deflection data, Young's modulus was estimated for small deformations, under 10% strain. The results obtained from the experiments, in terms of stress–strain curves, are presented in Figure 5. As it was possible to verify, with a frictionless condition, the results between different samples were independent of the geometry or shape factor. Furthermore, comparing experimental with numerical results, for each sample, it was possible to observe that it had very similar results (Figure 6).

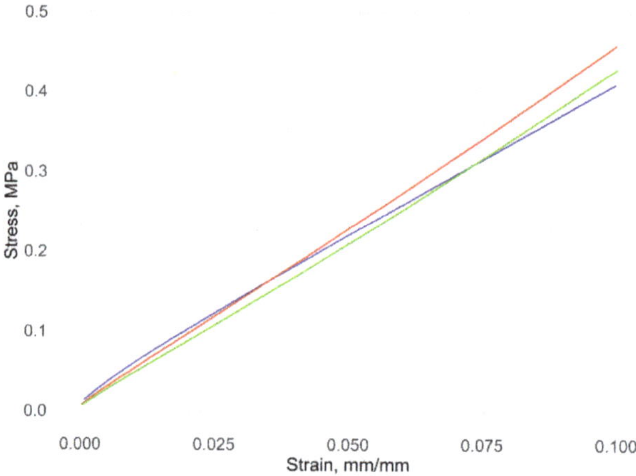

Figure 5. Experimental results for different shape factors: 0.75 (blue), 0.5 (red) and 0.3 (green).

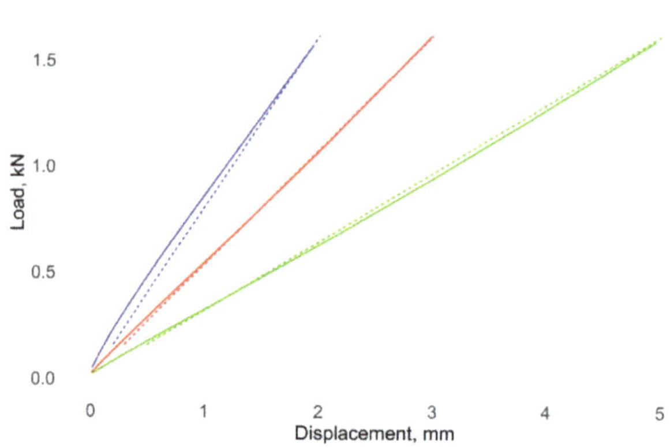

Figure 6. Comparison between experimental and simulation results for different shape factors: 0.75 (blue), 0.5 (red) and 0.3 (green).

4.2. Relation between Young and Apparent Compression Moduli

In the matter of frictional contact between sample and plates, different values of Young's modulus and shape factors were introduced as inputs for simulating compression behaviour. The values of Young's modulus were based on the results obtained for four different cork–rubber materials with distinct hardness. Regarding geometry, all models considered had the same square cross-section differing only in thickness. With each analysis results, the apparent compression modulus was calculated by the ratio between stress obtained for a certain strain—in this study, a 10% strain was considered.

A plot between Young and apparent compression moduli, for each shape factor, is presented in Figure 7a. A linear relationship between the two parameters is noticeable. The ratio between Young and apparent compression moduli, for each shape factor, is presented in Figure 7b.

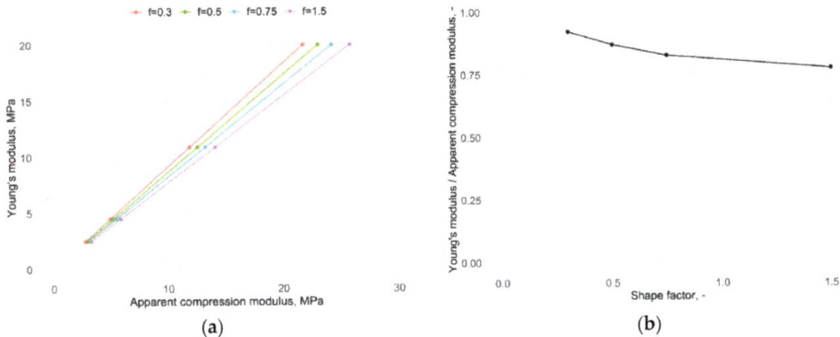

Figure 7. (**a**) The relation between Young and apparent compression moduli; (**b**) Ratio between moduli according to shape factor.

Figure 8 presents stress–strain plots for two values of Young's modulus, for several shape factors, to highlight differences between distinct stiffness material compounds. With higher stiffness materials, there seemed to exist more significant differences between shape factors.

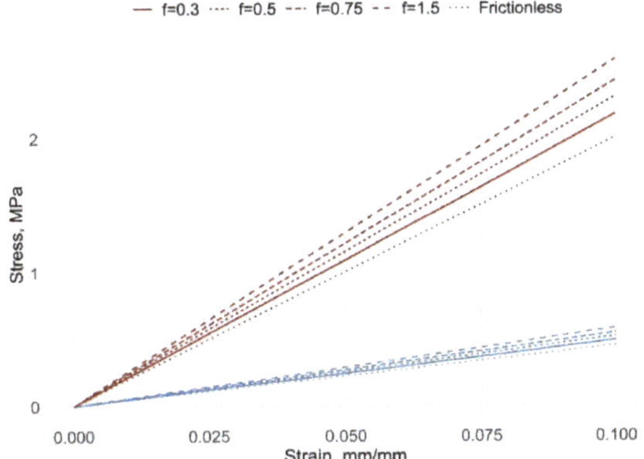

Figure 8. Stress–strain plots of samples with different Young's modulus and shape factors.

4.3. Comparison between Experimental and Simulation Results: Frictional Contact

Regarding the compression without lubricant between rubber block samples and plates, a comparison between experimental and simulation results was also performed to evaluate the performance of the proposed simulation approach.

Based on the experimentally determined apparent compression modulus, Young's modulus used as input for the FEA experiment was determined using the relation obtained according to the specimen's shape factor, as presented in Figure 7b. The resulting reaction forces to approximately 10% strain were compared with the results from experimental compression tests. Figure 9 presents the results for samples with 1.5 and 0.75 shape factors values. As is it possible to observe, the results of both approaches were relatively close, although there was a better agreement regarding the lower shape factor example.

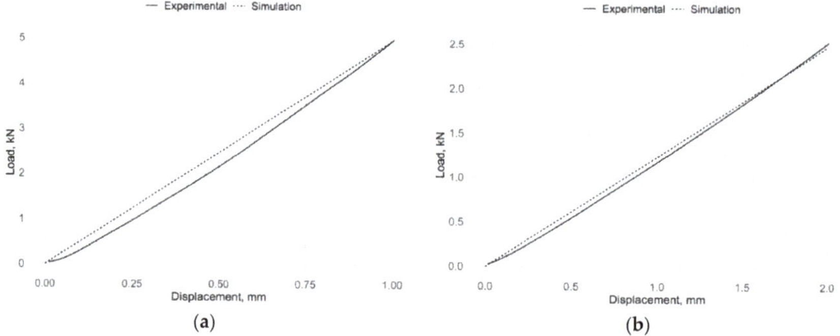

Figure 9. Load-deflection results for square cross-section blocks: (**a**) shape factor of 1.5; (**b**) shape factor of 0.75.

4.4. Comparison between Experimental and Simulation Results: Frictional Contact

The ratio of apparent compression modulus and Young's modulus, obtained from the presented methodology, was compared with the results obtained by the application of models developed by Lindley [16] and Williams' [20] models. Table 1 presents the FEA results for two types of cross-section-circular and -square, and the results obtained from the application of the two theoretical models regarding a cylinder shape. There were few differences in the relation between Young (E_0) and apparent compression moduli (E_c) between FEA data concerning the shape of the cross-section. The parameters' values and maximum differences between analytical and numerical approaches are presented in Table 1.

Table 1. Comparison of models to estimate the ratio of apparent compression modulus and Young's modulus.

Thickness	Shape Factor	FEA		Lindley [16]		Williams [20]	
		Square	Circular				
		E_c/E_0	E_c/E_0	E_c/E_0	Error	E_c/E_0	Error
10	1.5	1.29	1.28	1.28	0.9%	1.29	−0.3%
20	0.75	1.21	1.19	1.19	1.9%	1.17	3.5%
30	0.5	1.15	1.13	1.13	2.2%	1.10	4.5%
50	0.3	1.09	1.07	1.06	2.3%	1.04	4.1%

For the range of shape factors considered, it was possible to notice that differences between the application of analytical models and the finite element approach did not exceed 5%. Although the maximum error occurs when the geometry considered for the theoretical models and finite-element simulations are not identical-based on the assumption that, in this case, the same shape factor geometries had similar compression behaviour, there seemed to be a good correspondence between results. Overall, the FEA approach seemed to show a closer agreement with results from Lindley's model, rather than Williams', probably due to the increased complexity of the first theoretical model.

5. Conclusions

To determine the axial deformation of cork–rubber square cross-section blocks when subject to compression between two surfaces with frictional contact, a methodology for estimating the relation between Young and apparent compression moduli was presented. Although there are many analytical models available in the literature for rubber materials, the use of finite element analysis introduces the advantage of providing results according to specific cork–rubber composite properties and geometries.

Based on experimental data of frictionless compression tests, the relation between Young's modulus and apparent compression modulus was determined for some shape factors, recurring to FEA.

The results from numerical analysis indicated a linear relationship between the two moduli for the same shape factor. In addition, regarding frictional contact, the results showed an increase in stiffness for higher shape factors, as is verified and reported in other theoretical models. For the range of shape factors studied, Young's modulus of cork–rubber composites corresponded to more than 75% of the value of apparent compression modulus. The outputs retrieved from the analysis were compared against experimental compression results with frictional contact. Although relatively close, a better agreement was achieved for the sample with the lowest shape factor. A comparison between the FEA approach and two theoretical models used to evaluate cylinder shapes, was performed. The maximum error between FEA and analytical models was below 5%. Furthermore, for the case study presented, the differences between the use of a disk or a square cross-section block were small, considering the same shape factor and thickness.

The results obtained in this study are limited to a specific type of cork–rubber materials, where the assumption of Hooke's Law under small strain compression applies. Furthermore, only four different shape factors were analysed, and it was assumed that the behaviour of frictional contact was very similar to a bonded contact type. Future research should address the effect of higher shape factors, other cross-section shapes (rectangular and other polygons), and the friction coefficient between sample and loading surfaces. Moreover, applying this knowledge and relating it to the dynamic compression behaviour of isolation pads could be a topic of interest.

Author Contributions: Conceptualization, H.L., S.S. and J.M.; Data curation, H.L.; Funding acquisition, S.S. and J.M.; Investigation, H.L.; Methodology, H.L., S.S. and J.M.; Project administration, S.S. and J.M.; Supervision, J.M.; Validation, S.S. and J.M.; Writing—original draft, H.L.; Writing—review & editing, S.S. and J.M. All authors have read and agreed to the published version of the manuscript.

Funding: The authors are grateful to FCT–Fundação para a Ciência e Tecnologia who financially supported this work through scholarship SFRH/BD/136700/2018 and to Amorim Cork Composites for providing all materials and some physical and human resources. This work has been supported by the FCT–Fundação para a Ciência e Tecnologia within the RD Units Project Scope: UIDP/04077/2020 and UIDB/04077/2020.

Conflicts of Interest: The authors declare no conflict of interest.

References

1. Gil, L. Cork Composites: A Review. *Materials (Basel)* **2009**, *2*, 776–789. [CrossRef]
2. Silva, S.P.; Sabino, M.A.; Fernandes, E.M.; Correlo, V.M.; Boesel, L.F.; Reis, R.L. Cork: Properties, capabilities and applications. *Int. Mater. Rev.* **2005**, *50*, 345–365. [CrossRef]
3. Mestre, A.; Vogtlander, J. Eco-efficient value creation of cork products: An LCA-based method for design intervention. *J. Clean. Prod.* **2013**, *57*, 101–114. [CrossRef]
4. Antunes, P.J.; Dias, G.R.; Coelho, A.T.; Rebelo, F.; Pereira, T. Hyperelastic modelling of cork-polyurethane gel composites: Non-linear FEA implementation in 3D foot model. *Mater. Sci. Forum* **2008**, *587–588*, 700–705. [CrossRef]
5. Soleimanloo, H.S.; Barkhordari, M.A. Effect of shape factor and rubber stiffness of fiber-reinforced elastomeric bearings on the vertical stiffness of isolators. *Trends Appl. Sci. Res.* **2012**, *8*, 14–25. [CrossRef]
6. Koblar, D.; Boltežar, M. Evaluation of the frequency-dependent Young's modulus and damping factor of rubber from experiment and their implementation in a finite-element analysis. *Exp. Tech.* **2016**, *40*, 235–244. [CrossRef]
7. Fernandes, F.A.O.; Pascoal, R.J.S.; Alves de Sousa, R.J. Modelling impact response of agglomerated cork. *Mater. Des.* **2014**, *58*, 499–507. [CrossRef]
8. Cole, D.; Forrester, S.; Fleming, P. Mechanical characterisation and modelling of elastomeric shockpads. *Appl. Sci.* **2018**, *8*, 501. [CrossRef]
9. Gent, A.N. *Engineering with Rubber: How to Design Rubber Components*; Carl Hanser Verlag GmbH & Co. KG: Munich, Germany, 2012; ISBN 978-3-446-42764-8.

10. Schaefer, R.J. Mechanical Properties of Rubber. In *Harris' Shock and Vibration Handbook*; Harris, C.M., Piersol, A.G., Eds.; McGraw-Hill.: New York, NY, USA, 2002.

11. Fediuc, D.O.; Budescu, M.; Fediuc, V.; Venghiac, V.-M. Compression Modulus of Elastomers. In *Buletinul Institutului Politehnic din Iasi. Sectia Constructii, Arhitectura*; 2013; Volume 59, pp. 157–166. Available online: http://www.bipcons.ce.tuiasi.ro/Content/ArticleInformation.php?ArticleID=369 (accessed on 1 July 2020).

12. Gent, A.N.; Lindley, P.B. The compression of bonded rubber blocks. *Proc. Inst. Mech. Eng.* **1959**, *173*, 111–122. [CrossRef]

13. Gent, A.N. On the relation between indentation hardness and Young's modulus. *Rubber Chem. Technol.* **1958**, *31*, 896–906. [CrossRef]

14. Qi, H.J.; Joyce, K.; Boyce, M.C. Durometer hardness and the stress-strain behavior of elastomeric materials. *Rubber Chem. Technol.* **2003**, *76*, 419–435. [CrossRef]

15. Kunz, J.; Studer, M. Determining the Modulus of Elasticity in Compression via the Shore a Hardness. *Kunstst. Int.* **2006**, *96*, 92–94.

16. Lindley, P.B. Compression moduli for blocks of soft elastic material bonded to rigid end plates. *Anal. Eng. Des.* **1979**, *14*, 11–16. [CrossRef]

17. Kakavas, P.A.; Blatz, P.J. Effects of voids on the response of a rubber poker chip sample. III. *J. Appl. Polym. Sci.* **1991**, *43*, 1081–1086. [CrossRef]

18. Tsai, H.-C.; Lee, C.-C. Compressive stiffness of elastic layers bonded between rigid plates. *Int. J. Solids Struct.* **1998**, *35*, 3053–3069. [CrossRef]

19. Horton, J.M.; Tupholme, G.E.; Gover, M.J.C. Axial loading of bonded rubber blocks. *J. Appl. Mech.* **2002**, *69*, 836–843. [CrossRef]

20. Williams, J.G.; Gamonpilas, C. Using the simple compression test to determine Young's modulus, Poisson's ratio and the Coulomb friction coefficient. *Int. J. Solids Struct.* **2008**, *45*, 4448–4459. [CrossRef]

21. Suh, J.B.; Kelly, S.G. Stress Response of a Rubber Block under Vertical Loading. *J. Eng. Mech.* **2012**, *138*, 770–783. [CrossRef]

22. Gent, A.N.; Meinecke, E.A. Compression, Bending, and Shear of Bonded Rubber Blocks. *Polym. Eng. Sci.* **1970**, *10*, 48–53. [CrossRef]

23. Anderson, M.L.; Mott, P.H.; Roland, C.M. The compression of bonded rubber disks. *Rubber Chem. Technol.* **2004**, *77*, 293–302. [CrossRef]

24. Haji, Z.N.; Oyadiji, S.O.; Samami, H.; Farrell, O. Assessment of Analytical Equations for the Derivation of Young's Modulus of Bonded Rubber Materials. *Int. J. Mech. Mater. Eng.* **2019**, *13*, 462–466.

25. Gent, A.N.; Discenzo, F.M.; Suh, J.B. Compression of Rubber Disks Between Frictional Surfaces. *Rubber Chem. Technol.* **2009**, *82*, 1–17. [CrossRef]

26. Polukoshko, S.; Gonca, V.; Martinovs, A.; Sokolova, S. Boundary Conditions Influence on Compressive Stiffness of Elastomeric Isolators. In Proceedings of the 15th International Scientific Conference Engineering for Rural Development, Jelgava, Latvia, 25–27 May 2016; pp. 737–744.

27. Dong, J.; Huan, Y.; Liu, W.; Feng, Y.H.; Dai, Y.J.; Hao, S.W. The Influence of Boundary Conditions on Modulus Measurement in Uniaxial Compression Tests. *Exp. Tech.* **2017**, *41*, 327–330. [CrossRef]

28. Bechir, H.; Djema, A.; Bouzidi, S. On Poisson's functions of compressible elastomeric materials under compression tests in the framework of linear elasticity. *Acta Mech.* **2019**, *230*, 2491–2504. [CrossRef]

29. Lalo, D.F.; Greco, M.; Meroniuc, M. Numerical Modeling and Experimental Characterization of Elastomeric Pads Bonded in a Conical Spring under Multiaxial Loads and Pre-Compression. *Math. Probl. Eng.* **2019**. [CrossRef]

MDPI

St. Alban-Anlage 66

4052 Basel

Switzerland

Tel. +41 61 683 77 34

Fax +41 61 302 89 18

www.mdpi.com

Applied Sciences Editorial Office

E-mail: applsci@mdpi.com

www.mdpi.com/journal/applsci